International Series in Operations Research & Management Science

Volume 204

This book was recommended by Dr. Price

For further volumes:
http://www.springer.com/series/6161

Elaine Chew

Mathematical and Computational Modeling of Tonality

Theory and Applications

Springer

Elaine Chew
Centre for Digital Music
Queen Mary University of London
London
UK

ISSN 0884-8289 ISSN 2214-7934 (electronic)
ISBN 978-1-4614-9474-4 ISBN 978-1-4614-9475-1 (eBook)
DOI 10.1007/978-1-4614-9475-1
Springer New York Heidelberg Dordrecht London

Library of Congress Control Number: 2013953210

Printed on acid-free paper

Springer is part of Springer Science+Business Media (www.springer.com)

This book is dedicated to

my father, Chew Kim Lin, who instilled in me a love for mathematics,

George Dantzig, in memoriam, who gave me my first glimpse into research, and

Jeanne Bamberger, who showed me I could combine it all with music

Preface

Blending ideas from operations research, music psychology, music theory, and cognitive science, this book aims to tell a coherent story of how tonality pervades our experience, and hence our models, of music.

The story is told through the developmental stages of the Spiral Array model for tonality, a geometric model designed to incorporate and represent principles of tonal cognition, thereby lending itself to practical applications of tonal recognition, segmentation, and visualization. Mathematically speaking, the coils that make up the Spiral Array model are in effect helices, a spiral referring to a curve emanating from a central point. The use of "spiral" here is inspired by spiral staircases, intertwined spiral staircases: nested double helices within an outer spiral.

The book serves as a compilation of knowledge about the Spiral Array model and its applications, and is written for a broad audience ranging from the layperson interested in music, mathematics, and computing to the music scientist–engineer interested in computational approaches to music representation and analysis, from the music–mathematical and computational sciences student interested in learning about tonality from a formal modeling standpoint to the computer musician interested in applying these technologies in interactive composition and performance. Some chapters assume no musical or technical knowledge, and some are more musically or computationally involved.

I am extremely pleased that this book is to appear fifteen years after the eureka moment that gave rise to the Spiral Array model, and five years—and five house moves, including one cross-country and one cross-Atlantic—after the book proposal, formulated and accepted while I held the Edward, Frances, and Shirley B. Daniels Fellowship at the Radcliffe Institute for Advanced Study. The collaborators who have contributed to this volume include Alexandre R. J. François, Ching-Hua Chuan, and Yun-Ching Chen; our joint work forms the basis of the chapters on visualization, audio key finding, and pitch spelling. Alex is additionally author of the MuSA_RT Mac App, an interactive visualization software based on the Spiral Array that is a part of the supplemental material for this book.

This compendium would not have been possible without the support of Jeanne Bamberger, who has been a mentor well beyond my doctoral research, Kim Lin Chew, my father and the only person I know willing to proofread equations, and other long-suffering members of my family. I thank Jordan Smith for his last-minute voluntary proofreading, Doug Keislar for his (in)voluntary edits to the book draft, the late Lindy Hess for her generous advice on the book proposal, Matthew Amboy for speedy feedback on the book drafts, Camille Price, incoming series editor, for her steadfast encouragement over the years, and Fred Hillier, a former teacher, whose optimism and impending departure as Series Editor provided the catalyst to finally complete the book.

Last but not least, I thank the anonymous student who asked the seemingly innocuous question, "What do you mean by key?," that started this whole undertaking.

London, Singapore, Los Angeles, Boston, August 2013 Elaine Chew

Contents

Part I
Introduction

Chapter 1
Tonality

Abstract Tonality, the underlying principles of tonal music, is both an elusive and pervasive property of the music with which we are familiar. Elusive because its effects can be felt without the mind being consciously aware that it is actively construing the pitch relations that define tonality; and pervasive because it underpins most of the music that we hear. Its effects can be quickly demonstrated by the ability of the listener to sense when a piece has ended: try humming only the first three phrases of "Happy Birthday." The chapter begins by motivating the study of tonality from the practical standpoint of the listener and of the music practitioner, namely the performer and the composer. It then proceeds to describe the genesis of the project in the pianolab of MIT, when a student asked, "What do you mean by key?," reflecting on what it is that allows a listener to ascertain the most stable pitch in a sequence. In the spirit of Bugliarello's new trivium and quadrivium, in which "no domain can any longer be considered and learned in isolation," the chapter describes how the book bridges the disparate disciplines of music theory and operations research. Spanning C.P. Snow's two cultures, the book mixes mathematical formalisms with qualitative descriptions, mingling intimate and subjective case studies with impartial large-scale and quantitative testing of algorithms. The chapter proceeds to trace the development of the Spiral Array model from its inception through the applications that have followed, thereby providing a narrative of the remaining chapters of the book and the way in which they are interlinked.

Martha Argerich crashes into the rumbling arpeggios that augur the beginning of the quasi cadenza of Strauss' *Burleske*. The orchestra led by Claudio Abbado fades to silence. A pair of doubled octaves strike with a resounding crash. The audience awaits the soloist's virtuosic display in concentrated stillness. It is the Berlin Philharmonic New Year's Eve Concert in 1992.

This chapter incorporates material from the Introduction (Chapter 1) of "Towards a Mathematical Modeling of Tonality" by Elaine Chew, an MIT PhD dissertation, Cambridge, Massachusetts (2000) https://dspace.mit.edu/handle/1721.1/9139

E. Chew, *Mathematical and Computational Modeling of Tonality*, International Series in Operations Research & Management Science 204, DOI: 10.1007/978-1-4614-9475-1_1,

Thunderous trills announce blistering streams of arpeggios, each outdoing its precursor by further elaboration or by reaching new heights. The excitement mounts, pushing the cadenza to its apex. The peak is sustained, but not for long. The music falls from the climax; only to be hit by another surge of energy, and another drop, this time to melt away.

A lone melodic line enters, stepping inexorably downward. The soloist pauses. The audience waits, knowing that the cadenza is not done. The orchestra musicians are poised in readiness, awaiting their cue. The soloist holds on to the note, drawing out the suspense, pushing to the limit the audience's focused participation. Then, she turns to the conductor, and smiles to signal her acquiescence. Together, the orchestra and soloist gracefully fall in step to usher in the lyrical theme, and the listener lets out a sigh of contentment ... aah.

Underneath the obvious technical displays of the cadenza lies a complex network of pitch relations that gives inner logic and coherence to the music. Composers—by choosing which note to write and where (i.e. when the note is played in relation to others)—and performers—by choosing which notes or silences to emphasize and how—alike work this system to choreograph and manipulate the listener's expectations. Re-consider the previous anecdote, now embellished with a running subtext describing the pitch relations and the expectations they engender.

In the preceding passages, D minor has been established as the main key, meaning that the pitch D is the most significant pitch, and the note material of the piece, all drawn from the D minor scale, are heard in relation to this reference pitch. Martha Argerich crashes into the rumbling arpeggios that *embellish the G minor triad, a chord (iv) which is composed of the simultaneous combination of three notes based on the fourth degree ($\hat{4}$) of the D minor scale, and* augur the beginning of the quasi cadenza of Strauss' *Burleske*. The orchestra led by Claudio Abbado fades to silence. A pair of doubled octaves strike *the unison A's, the fifth degree ($\hat{5}$) of the D minor scale,* with a resounding crash, *the G and A outlining the first two chords of one of the most prototypical cadential sequence, iv-V-i, setting up the expectation that the chord based on the first degree of the scale, the tonic, also the pitch of greatest stability ($\hat{1}$), is to return.* The audience awaits the soloist's virtuosic display in concentrated stillness, *expecting the technical pyrotechnics that typically accompanies the prolongation of the V chord in the cadenza of a concerto.* It is the Berlin Philharmonic New Year's Eve Concert in 1992.

Thunderous trills announce blistering streams of arpeggioss, *anchored by re-iterations of the A octaves, re-enforcing the root of the V chord in the iv-V-i sequence*, each outdoing its precursor by further elaboration or by reaching new heights *and depths, the A octave eventually hitting the lowest note on the keyboard.* The excitement mounts, pushing the cadenza to its apex, *an emphatic A octave followed by an enigmatic half-diminished chord.* The peak is sustained, but not for long, *the half-diminished chord resolves to an equally unstable fully diminished chord.* The music falls from the climax *following the natural voice leading to the B♭*; only to be hit by another surge of energy, and another drop *returning to B♭*, this time to melt away *into a simple A major triad, A being the natural consequence of B♭.*

A lone melodic line enters, stepping inexorably downward, *starting with the now familiar pitch B♭ followed by A*. The soloist pauses *on the A, emphasizing its connection to the earlier octaves and the fact that we again hear the fifth degree of the scale, ($\hat{5}$)* The audience waits *for the resolution back to the tonic D, not knowing when the suspense will be over, only* knowing that the cadenza is not done. The orchestra musicians are poised in readiness, awaiting their cue. The soloist holds on to the note *A, lingering on it, deliberately prolonging the wait and* drawing out the suspense, *the bar is stretched to over three times the average length of a bar in the first half of the cadenza* pushing to the limit the audience's focused participation. Then, she turns to the conductor, and smiles to signal her acquiescence. Together, the orchestra and soloist gracefully fall in step to usher in the lyrical theme, *which not only begins with the chord of D minor (i) in the key of D minor, but also with the melody note D ($\hat{1}$) increasing the satisfaction of and pleasure at the expectation fulfilled*, and the listener lets out a sigh of contentment ... aah.

The system of pitch relations that underlie this ebb and flow of expectations is called *tonality*. In the following chapters, I shall describe the theoretical underpinnings of the Spiral Array model, a spatial representation of the relations embodied in tonality, and present applications of the model to a variety of problems in automatic music analysis.

1.1 What is Tonality?

Tonality refers to the underlying principles of tonal music, and is one of the principal ways by which listeners intuit form and structure in music; it is also one of the primary means by which music evokes psychological feelings in listeners [24]. According to Bamberger [3], "tonality and its internal logic frame the coherence among pitch relations in the music with which [we] are most familiar."

The study of tonality has a long and illustrious history dating back to Rameau's 1722 "Treatise on Harmony" [39]. Music theorists, composers, mathematicians, philosophers, and psychologists have sought to uncover the nature of, and formalize the concept of, tonality from a variety of disciplinary perspectives. Monographs written on the subject that are most directly related to this book, and that have influenced its development, include: Krumhansl's "Cognitive Foundations of Musical Pitch" [26], Lewin's "Generalized Musical Intervals and Transformations" [31], and Temperley's "The Cognition of Basic Musical Structures" [47]. Another important contribution that is closely related to this book is Lerdahl's "Tonal Pitch Space" [30]. Scholars who have specifically proposed mathematical formulations for aspects of tonality include Mazzola [34] and Tymoczko [49]. Well-known composers who have added their theories and thoughts include Hindemith [37] and Schoenberg [41].

Dahlhaus [17], in his "Studies in the Origin of Harmonic Tonality," wrote, "In common usage the term [tonality] denotes, in the broadest sense, relationships between pitches, and more specifically a system of relationships between pitches having a "tonic" or central pitch as its most important element." He further quotes Fétis' 1844 definition of tonality as including the "necessary successive or simultaneous relationships between the notes of a scale." It is worth explicitly noting that tonality impacts relationships between not only simultaneous and successive pitches, but also between pitches or sets of pitches across small and large expanses of time. Indeed, time is a critical element in the experience of tonality, as music itself unfolds and is heard in time.

The term tonality is sometimes synonymized with key, which in turn is often further said to refer to adherence to the pitch set of a major or minor scale. Dahlaus [17] offers a broader definition of tonality:

> ... tonality reaches further than the note content of a major or minor scale, through chromaticism, passing reference to other key areas, or wholesale modulation: the decisive factor in the tonal effect is the functional association with the tonic chord (emphasized by functional theory), not the link with a scale (which is regarded as the basic determinant of key in the theory of fundamental progressions). A tonality is thus an expanded key.

Extending beyond the definition of tonality as an expanded key, Dahlhaus states, "Tonality [is] the underlying element of a tonal structure, the effective principle at its heart."

1.2 Elusive Yet Pervasive

Tonality is both an elusive and pervasive property of music.

Elusive because its effects can be felt without the mind being consciously aware that it is construing the relationships amongst the pitches, without the listener deliberately deciphering what is the tonic now, and how are the other tones related to the tonic. Less visceral than the sensing of time structures such as pulse, rhythm, and meter, the understanding of scale degrees (a basic part of tonality) has been noted by Huron [24] to be a cognitive rather than a perceptual phenomenon, as it is an act of the mind in "interpret[ing] physically sounding tones, rather than how the tones are in the world." Equating this unconscious cognitive activity with arithmetic, Leibniz [40] said in 1712, "The pleasure we obtain from music comes from counting, but counting unconsciously. Music is nothing but unconscious arithmetic."

Elusive because while it is a principal music feature that evokes senses both intellectual and hedonistic, the feelings it engenders and how they arise are hard to pin down and to describe in words. In fact, one may argue that time structures such as rhythm and meter give rise to much more visceral sensations that lend themselves more readily to description and communication.

Pervasive because it underpins most of the music that we hear. Listeners all know it, a fact that can be verified with a simple experiment: hum "Happy Birthday" and stop 3/4 of the way through the song—i.e., hum the notes corresponding to the words "happy birthday to you, happy birthday to you, happy birthday to Lisa." The sense of incompleteness is palpable. Stronger than the unfinished prose of the unsung text, or the asymmetry of the phrases, is the unfulfilled longing to hear the tonic on a strong beat at the end of the song, of any song. The unresolved expectation leaves a void in the gut and a strong urge to hum the next phrase. But not just the next phrase in the well known song; any phrase that ends with the tonic on a strong beat can finish the song, some better than others. In fact, any phrase that ends with the tonic can also finish it on the third phrase, the satisfaction of a stable ending overriding the asymmetry of the phrases.

Nearly all music is tonal. One exception to tonality's pervasiveness would be music that is unpitched, for example, music that is based entirely on noise or purely on rhythmic patterns, in which case other organizing principles replace that of tonality. It is difficult to contrive music that is entirely devoid of pitch, for even percussive sounds are often inherently pitched, such as the high and low pitches of palms coming together in Steve Reich's *Clapping Music*.

Because tonality pervades our experience of music, some composers have viewed it as oppressive, and espoused what is sometimes call atonal music, music that deliberately avoids creating the sense of a tonal context. Schoenberg [42], the inventor of twelve-tone music—compositions based on an impartial ordering of the twelve pitches of the chromatic scale in order to avoid the trappings of traditional tonal music—argues eloquently against the existence of atonal music:

> Permit me to point out that I regard the expression atonal as meaningless, and shall quote from what I have already expounded in detail in my *Harmonielehre*. 'Atonal can only signify something that does not correspond to the nature of tone.' And further: 'A piece of music will necessarily always be tonal in so far as a relation exists from tone to tone, whereby tones, placed next to or above one another, result in a perceptible succession. The tonality might then be neither felt nor possible of proof, these relations might be obscure and difficult to comprehend, yes, even incomprehensible. But to call any relation of tones atonal is as little justified as to designate a relation of colors spectral or complementary. Such an antithesis does not exist.'

A major obstacle to creating atonal music is that listeners cannot help but generate mental associations amongst pitches, and in particular, "relationships between pitches having a "tonic" or central pitch as its most important element." Even when confronted with a random sequence of pitches, the mind will construct a tonal context. For example, based on the pitches that have come before, a listener might imagine two consecutive pitches to be the subdominant ($\hat{4}$) and leading-tone ($\hat{7}$) implying a tonic ($\hat{1}$).

Just as tonality can refer to other kinds of tonality in western music, like Hindemith's non-diatonic tonal system, tonality can also refer to similar systems in music of non-western cultures, where the specifics of the inter-pitch associations may differ from that of western classical music, but nevertheless generate interpretations of varying stability amongst the pitches. The thaat system in North Indian classical music is a case in point, where a tonal hierarchy emerges over the course of a raag, with the *vadi* as the most stressed tone, and *samvadi* (typically the fourth or fifth degree of the scale) the second most stressed tone, see [12].

1.3 Inception

This project began as an attempt to formally describe the generating of a tonal center. The quest led to a mathematical model, on which this book is based, that incorporates music knowledge and that can be used to mimic the human decision-making process in apprehending tonality. Nevertheless, the underlying questions that provided the impulse for the extended project remain as important motivating issues.

In my first semester as a pianolab instructor—pianolab was a keyboard skills class for students enrolled in Music Fundamentals and Composition courses—at MIT in 1998, I encountered a few students who had no prior musical training. I asked one such student, after he carefully traced out the melody of "Yankee Doodle," "What is the key of this piece?" He responded with a reasonable question: *"What do you mean by key?"*

The obvious, but not altogether accurate, answer was to look at the key signature at the beginning of the piece and accidentals in the passage. There are several problems with this approach. *Looking* at the number of sharps and flats in the key signature ignores the fact that the human cognition of key is an aural experience, and for musicians, a physical one as well. Accidentals, although helpful, are not the primary cues for the key of the passage. One could easily find an example with no accidentals that does not generate the feel of C major or A minor, which are the two keys with no sharps or flats.

For example, in Fig. 1.1, a segment from the spiritual "Nobody Knows the Trouble I've Seen," the notes have no sharps or flats or other accidentals, but this melody sounds distinctly in the key of F major, and not C major (or A minor) as suggested by the key signature and the absence of accidentals. The mind does not need many notes to judge the key of a melody. In the mind's ear, so to speak, the first four notes of "Nobody Knows" already suggest strongly F major. I shall shortly attempt to outline some of the decision process I undergo as a listener when encountering this melody. For now, I continue with the narration of the story of the student who asked, "What do you mean by key?"

To any musician who plays an instrument, each key has its own distinctive terrain on their instrument. The key of a piece of music confers a physical shape, a unique topography, to the moving hands of the musician. An approach sometimes used by instrumentalists to get a sense of the tonality is to play a short excerpt of the piece, and based on the way the piece feels in the hands, determine the key; not only the key, the tonal center, the tonic, but also how the various notes they are playing or are about to play relate to that tonic. This approach relies on the physical experience of

Fig. 1.1 Excerpt from "Nobody Knows the Trouble I've Seen." A musical example with no accidentals, that is actually in the key of F

making music, drawing upon embodied knowledge through the physicality of playing an instrument. But what of the aural experience? A listener who is not playing an instrument, or who does not have the experience of playing an instrument, too, can sense tonality.

I jumped at the next idea that came to mind. I hummed the piece, and stopped mid-stream. I asked the student if he could sing me the note on which the piece should end. Without a second thought, he sang the correct pitch, the tonic. The success of this method raised more questions than it answered. These questions are aptly described by Bamberger in *Developing Musical Intuitions* (p.155):

> How can we explain this tonic function which seems so immediately intuitive? While theorists have argued about answers to this question, most agree that for listeners who have grown up in western musical culture, the stable function of the tonic derives primarily from its relation to the other pitches which surround it. Thus, the tonic function that a pitch acquires is entirely an internal affair: a pitch acquires a tonic function through its contact with a specific collection of pitches, the particular ordering and rhythmic orientation of this collection as each melody unfolds through time.

What is it we know that causes us to hear one pitch as being more stable than another? How does the function of the tonic evolve over the unfolding of a piece? A closer examination of "Nobody Knows" might shed some light on this matter.

The melody "Nobody Knows the Trouble I've Seen" serves to demonstrates that many factors contribute to the listener's perception of tonality. These factors include interval relations, pitch durations, and meter. The first four notes of 'Nobody Knows' set up a most stable pitch, and already give a strong indication of the key. I will outline my own experience in determining this most stable pitch through the first four notes of "Nobody Knows."

The descending major sixth interval between the first and second note, A and C respectively, strongly hints that the most stable pitch is F. This knowledge is the result of experience in listening to western tonal music. Many other tonal melodies begin with two pitches that are a major sixth interval apart. For example, the traditional Scottish folksong "My Bonnie Lies Over The Ocean" and Chopin's Nocturne Op. 9 No. 2 in E♭, see Fig. 1.2. In both cases, the two pitches separated by an interval of a major sixth surround and point to the pitch that is a major third interval below the upper pitch as the stable pitch, the likely tonic.

Thus, the mind immediately determines that the first two notes in "Nobody Knows" very likely can be assigned the (movable *do*) solfège syllables *mi* and *sol*, or scale degrees $\hat{3}$ and $\hat{5}$. The distance between the pitches of the second and fourth note in the melody forms an interval of a perfect fourth. The rising fourth, from C to F, suggests the scale degree assignments ($\hat{5} - \hat{1}$), further reinforcing the F as tonic. Further examples of this rising fourth interval in other melodies are given in Fig. 1.3. Together, the first, second and fourth notes outline the F major triad, implying an affinity to F major.

In "Nobody Knows," the rhythm in the melody also reinforces the tonic implied by the interval relations. The note of longest duration in the first half of Fig. 1.1 is the fourth note in the melody, and its pitch is F. In addition, the indicated meter places a

Fig. 1.2 Excerpts from "My Bonnie Lies Over the Ocean" and Chopin's Nocturne Op. 9 No. 2, both beginning with a rising major sixth interval

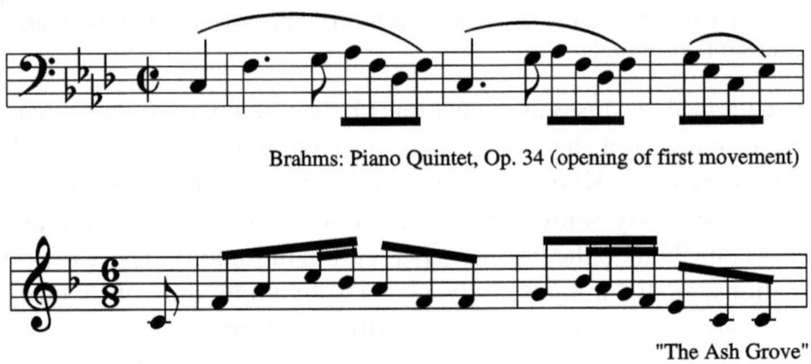

Fig. 1.3 Excerpts from Brahms' Piano Quintet and "The Ash Grove," both beginning with a rising perfect fourth interval

downbeat on this F. Although the last note in Fig. 1.1 has a longer duration than the F, its onset begins on a weak beat (the fourth beat of the third bar).

What is it we know that causes us to hear one pitch as being more stable than another? How does the function of the tonic evolve over the unfolding of a piece? Is there a way to describe formally the framework of pitch inter-relations that determines the key? Thus one student's seemingly innocent question of how does one find the key led to my quest for a concise and effective model for the generating of tonal centers.

1.4 Design Principles

In the following chapters, I shall describe a model for tonality called the Spiral Array. It behoves me to mention here some of the characteristics and design principles of the model.

The model is generative, and designed to provide a concise description of the characteristics of tonality. The hierarchical structure of tonality is embedded in the model: key representations are generated from chords, and chord representations from component pitches. The model is related to Riemann's theory of tonality [17] in that significant tonal relationships are established by means of the chord functions of the tonic, the dominant and the subdominant. The model can also be used to generate note and chord sequences, although that aspect is not a focus of this book.

Distinct from other geometric and network models for tonal relations, the Spiral Array represents pitches, intervals, chords and keys in the same spatial framework. Derived from the *tonnetz* (tone network), first attributed to Euler [16] and used extensively in neo-Riemannian theory (for example [6, 11, 14–16, 20, 22, 31]), the model fuses both the original network model with properties of continuous space. In this space, any collection of pitches can generate a center of effect, that is essentially a mathematical sum of its parts. Thus, while hierarchical structure is incorporated into the model, at the same time, the model flattens this hierarchy by representing elements from all levels in the same space.

Distance in the model corresponds to perceived closeness in tonal music. Two pitches may be physically close on the piano keyboard, for example F and G♭, but perceptually quite distant. Whereas, two pitches that are farther apart on the piano, for example F and C, are perceived to be closely related. This has something to do with the simple fractional relationships between the frequencies of the pitches, but suffice it to say that the model aims to encapsulate the auditory sense of closeness rather than the configuration of distance on an instrument.

Because of the egalitarian representation of elements of all hierarchical levels in the same space, literal (and conceptual) distance can be measured between any two elements, from any hierarchical level, in the Spiral Array. The model thus offers a way to re-conceptualize tonal relationships. Because keys are represented in the model, the model also offers a way to envision the generating of tonal centers.

A computational model, the Spiral Array raises pertinent questions regarding, and produces insights that illuminate, some basic issues in traditional music theory. I shall demonstrate the versatility of the model by applying it to a number of fundamental problems in the cognition and analysis of western tonal music: that of finding keys (from MIDI, event-based, and digital audio information), pitch spelling (ascribing tonally consistent letter names and accidentals to pitches), and determining key boundaries (searching for modulations, shifts in tonality). The majority of these algorithms are designed for real-time processing. Furthermore, I present the model as a visualization tool, aided by the real-time aspect of the methods as well as the three-dimensional geometry of the Spiral Array space.

Being able to study the nature of these fundamental problems and their solutions is critical to the understanding of human cognition and analysis of tonal music, and also to pedagogical issues pertaining to these problems. Solving these basic tonal analysis problems computationally is a precursor to any computer analysis of western tonal music, and automated systems that interface computer-generated music with real-time performance. Since the model was invented, the field of music information retrieval has burgeoned, fueled by the proliferation of digital music and music streamed over the Internet. Thus, mathematical representations that lend themselves readily to efficient computational implementation have become ever more important for retrieval tasks such as music similarity assessment, segmentation and summarization.

Being able to characterize the relationships that generate a tonal center is crucial to the understanding and the making of composition as well as performance decisions. In composition and improvisation, it affects the choice of notes and note sequences; in performance it impacts the choice of prosody in expressive musical communication.

Knowledge of tonality provides information as to which notes or chords are structurally more stable, and will give the sense of solidity when further emphasized, or are not as stable, and in need of greater stress (or de-emphasis of competing notes) in order to afford the same presence. Knowing the tonality also provides a map of the listener's tonal expectations, revealing which notes can be prolonged (or elaborated upon) to delay the onset of a known and inevitable outcome (typically of closure); which notes can be emphasized to underscore the thwarting of the listener's expectations; and, which notes can be glossed over because they concur with the listener's expectations, or because the primary functions which they support are undertaken by other structurally more significant notes.

1.5 Straddling Cultures

One of the main contributions of this book is the bridging of two disparate disciplines, namely that of music theory and operations research. Operations research is the science of decision-making using mathematical models which integrate the operating criteria of the system in question. According to George Dantzig[1], the inventor of the Simplex Method and father of Linear Programming, *"OR is the wide, wide world of mathematics applied to anything!"* Music theory describes the underlying principles that govern the system of relations organizing the cognition and analysis of tonal music compositions. As quoted earlier, Leibniz (1646–1716), the German philosopher, physicist, and mathematician, has gone so far as to say: *"Music is nothing but unconscious arithmetic."* It would seem natural, then, to utilize the techniques in operations research to model effectively the perceptual problem solving inherent in the comprehension of western tonal music.

The two disciplines are bridged by a computational geometric model that is inspired by operations research techniques, and is built upon the framework of tonal relations based in music theory. The model, in turn, will provide insights into the symmetries and other relationships of the tonal system. While a subset of music theorists have invoked in their scholarly work mathematical techniques, typically from the more abstract branch of pure mathematics, the methods laid out in this book have a practical engineering flavor characteristic of operations research approaches to problem solving, see [36]. An early precursor of this kind of engineering

[1] Personal communication, 7 November 1999

approach to fundamental musical practice is Xenakis' thesis [51] on stochastic and other mathematical applications of mathematics to music composition. While core signal processing techniques applied to music audio employ selected optimization algorithms, they rarely accommodate more than a cursory nod to music theoretic knowledge. The model at the heart of this book owes its genesis to music theoretic models, and its development to operations research techniques. This interdisciplinary effort forms a core contribution of this book.

This book project itself has been an exercise in bridging C.P. Snow's two cultures [44]—on the one side the arts and humanities, and the other science—in open acknowledgement of the deep connections that cut across disciplines. A major synthesis of over 10 years' work, this book blends carefully curated selections from earlier writings with newer insights. The styles have been tailored to a mixed audience. Mathematical formalisms commingle with qualitative descriptions. Writing in the first vs. third person varies across some chapters. Up close and subjective case studies exist alongside impartial large-scale and quantitative testing of algorithms. Sometimes, a compromise is struck between humanities and scientific practices. The divide that continues to exist between the two was an important reason for the extended interval between the publication of my dissertation in 2000 and this current book; while book publishing is considered de rigueur in the humanities, a non-peer reviewed publication is of considerably lower value in the sciences. Nevertheless, it is with great pleasure that I find myself in the position to work on this book at this stage.

As Bugliarello [5] wrote in his treatise on a new trivium and quadrivium, "no domain can any longer be considered and learned in isolation." Computational music analysis is an excellent example of one such blended domain of study; it is, by definition, an interdisciplinary study linking human perception and cognition, mathematical and computational modeling, and music theory. A confluence of the three could ideally result in fruitful research leading to an enrichment of our understanding of all three disciplines. Desain et al. [18] documents several successful attempts to bridge pairs of these disciplines in the past decades. I have summarized in Fig. 1.4 a small sample of some interdisciplinary research in computational music analysis as an example. The development and construction of my model draws upon all three disciplines.

Music is not an easy domain within which to design effective computational models that describe changing tonalities and harmonies. The human cognition of tonality utilizes both top-down and bottom-up analyses [18]. When assessing tonal (and rhythmic) structure, the human mind contemporaneously considers several different structural levels in the music. The mental ability to simultaneously scale up and down allows the listener to gather information at all levels.

Furthermore, music can inherently be described structurally in multiple and equally valid ways [2, 3, 29]. For example, Bamberger [2] showed that even young children are capable of focussing on different but legitimate hearings of the same rhythm which she terms figural and formal. This is directly related to Lerdahl and Jackendoff's [29] demonstration of the fundamental difference between the musical elements of grouping and meter.

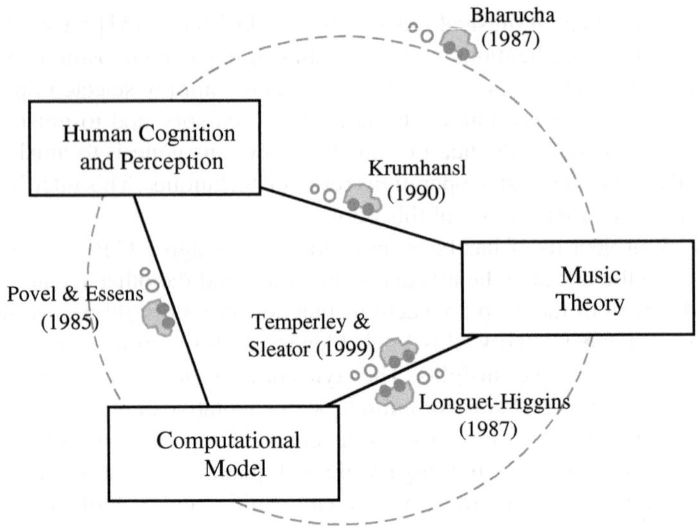

Fig. 1.4 Bridging the disciplines: a sampling of some interdisciplinary research in computational music analysis

At the time of the Spiral Array's invention, several theses had been devoted to the modeling of tonal perception, including Krumhansl's treatise on "The Psychological Representation of Musical Pitch in a Tonal Context" [25] which presents a behavioral approach, Laden's "A Model of Tonality Cognition which Incorporates Pitch and Rhythm" [27] using connectionist methods, and Temperley's thesis on "The Perception of Harmony and Tonality: An Algorithmic Perspective" [45] grounded in cognitive science. Soon after the publication of my dissertation, a special issue on tonality induction edited by Leman and Vos appeared in Music Perception [28]. Dissertations that have followed since include Aarden's "Dynamic Melodic Expectancy" [1], Honingh's "The Origin and Well-Formedness of Tonal Pitch Structures" [21], and Milne's "A Computational Model of the Cognition of Tonality" [35]. Where relevant, newer publications have been incorporated into the literature review of various chapters.

1.6 Overview

The contents of this book represent the culmination of research amassed over fifteen years. The first six chapters, including parts of the present one, are based on material from my dissertation [7]. The next five chapters are based on extensions of the model to post-tonal music, to non-score based digital music information, and data visualization, and derive from a handful of articles selected to illustrate the model's core applications.

Chapter 2 provides some relevant background and survey of models and methodologies that have influenced the design of the Spiral Array and its computational approach. The chapter begins with an intuitive and illustrated overview of the Spiral Array model and the Center of Effect Generator (CEG) key-finding algorithm as a basis for comparison. I review some spatial models of musical pitch that have impacted the model's configuration of pitch representations; next, I describe von Neumann's Center of Gravity algorithm and Dantzig, G. B.Dantzig's bracketing technique that inspired the CEG method. No attempt is made to provide a comprehensive summary of the literature.

Chapter 3 introduces the Spiral Array model, explaining how pitch, chord and key representations are generated in this structure. In addition, some symmetries in the model are highlighted. Later, in Appendix A, I present an example of how the model parameters can be calibrated so that the model represents the cognition of inter-pitch, inter-chord, pitch-chord, and pitch-key distances. The model is sufficiently well defined in Chap. 3 that the calibration details can be skipped without detriment to understanding of the model's applications in later chapters.

Chapter 4 introduces the first computational application that uses the Spiral Array: the problem of key-finding in melodies. I formally propose the CEG key-finding algorithm, explaining how it works by applying it to an example, "Simple Gifts." Step-by-step scientific visualizations accompany the ranked key outputs and distances. I compare the CEG algorithm to those by two other researchers in Chap. 5, giving detailed analyses of the comparison results when applied to the 24 fugue subjects in Book 1 of Bach's *Well-Tempered Clavier*. I show that the CEG algorithm surpasses previous ones in its average performance, and is close to optimal. The MATLAB code for the CEG algorithm, including programs to generate the Spiral Array model, and detailed key and distance output for each fugue subject can be found in Appendix B.

In Chap. 6, I propose an algorithm for determining modulations, the Boundary Search Algorithm (BSA). The algorithm is applied to two examples, Bach's Minuet in G and March in D, both from his *"A Little Notebook for Anna Magdalena"*. The conclusion suggests more sophisticated variations on this basic algorithm. Chap. 7 presents another segmentation algorithm based on concepts borrowed from Control Theory; the method is named Argus, for the all-seeing giant of Greek mythology, because it scans backward and forward in time to determine if the degree of change has exceeded a threshold, giving rise to a boundary.

In order to apply the tonal analysis algorithms to music data such as MIDI or audio, one must first turn numeric pitch information to tonally consistent letter names, the subject of the next chapter. The key determines pitch spelling, and the spelling of the pitches reveals the key; a chicken and egg problem. Chapter 8 describes a bootstrapping algorithm based on the Spiral Array, work done with Yun-Ching Chen, that presents a solution to this conundrum.

Chapter 9 presents MuSA.RT, "Music on the Spiral Array. Real-Time," an interactive tonal analysis and visualization system based on the Spiral Array model, and incorporating the pitch-spelling and CEG key-finding algorithms, as well as an algorithm for chord recognition similar to that for key-finding. This work was conducted in collaboration with Alexandre R. J. François, whose software architecture style enabled the concurrent processing of multiple data streams—music, video, camera control, and computational algorithms. François has since created a MuSA_RT App [19]—freely available from the Mac App Store—that is the focus of supplemental material to this volume at http://musa-rt.blogspot.com. Inspired by David Huron's study on music-engendered laughter [23], Chap. 10 uses MuSA.RT to conduct an analysis of tonally grounded devices employed by "P. D. Q. Bach" (Peter Schickele) to generate musical humor.

While all of the above techniques have been presented and illustrated using score-based and MIDI information, each can be extended and applied to music audio signals. A score-based tonal analysis algorithm can be extended to MIDI by applying pitch spelling to the note numbers; and a MIDI-based algorithm can be expanded to music audio signals by converting the signals to spectral or pitch-class data. Chapter 11 applies the CEG algorithm to audio key-finding, describing aspects of the system design—including signal processing techniques for transforming audio signals to pitch-class numbers, adaptations of the Spiral Array weights to audio data, and different key determination policies—and presenting an analysis of the system's sensitivity to these model choices.

There is some degree of repetition, especially with regard to summary descriptions of the Spiral Array model in Chap. 6 through 11. I have deliberately left these sections intact so as to improve readability of these later chapters as standalone sources. The general reader, who may not be familiar with some of the musical terms, may find the glossary at the end of the book helpful.

For focus and unity of the volume, this book does not cover a number of applications of the Spiral Array algorithms. The Spiral Array is compared to Lerdahl's Tonal Pitch Space and Krumhansl's and Krumhansl and Kessler's spatial representations of pitch class and key relations in [8]. The model is used to reveal symmetries in Webern's "Sehr Schnell" [10], and pedaling strategies that reinforce tonality [9]. Key and chord distributions and sequences determined with center of effect algorithms can be used to identify musical variations [33, 50] in music information retrieval. Key information forms a first step to automatic learning and generating of melody harmonizations [13]. Tonal novelty quantified using the Spiral Array is one parameter explored in the explaining of the cognition of boundaries in music [43].

Finally, while this volume synthesizes research on the Spiral Array and its applications to date, I view this volume not as the culmination of a decade and a half of work on the Spiral Array, but as an introduction to the model and the beginning of the breadth of computational possibilities offered by the model. The model itself can be further optimized for specific applications, the parameters themselves can be learned from data, they can also be time varying to track our changing cognition of musical structures. Chapter 11 hints at this: the policies for deciding key are subject to debate. As in Temperley's Bayesian reinterpretation of Krumhansl and Schmuckler's probe tone profile method for key-finding [48], the Spiral Array lends itself to extension via spatial probability models.

References

1. Aarden, B.: Dynamic melodic expectancy. Ph.D. Dissertation, Ohio State University, Columbus (2003)
2. Bamberger, J.S.: The Mind Behind the Musical Ear. Harvard University Press, Cambridge, MA (1991)
3. Bamberger, J.S.: Developing Musical Intuition. Oxford University Press, New York (2000)
4. Bharucha, J.: Music cognition and perceptual facilitation: a connectionist framework. Music Percept. **5**(1), 1–30 (1987)
5. Bugliarello, G.: A new trivium and quadrivium. Bull. Sci. Tech. Soc. **23**(2), 106–113 (2003)
6. Callender, C.: Voice-leading parsimony in the music of Alexander Scriabin. J. Music Theory **42**, 219–233 (1998)
7. Chew, E.: Towards a Mathematical Model of Tonality. Ph.D. Dissertation, Massachusetts Institute of Technology, Cambridge, MA (2000)
8. Chew, E.: Out of the grid and into the spiral: geometric interpretations of and comparisons with the Spiral-Array model. Comput. Musicol. **15**, 51–72 (2007)
9. Chew, E., François, A. R. J.: MuSA.RT and the pedal: the role of the sustain pedal in clarifying tonal structure. In: Proceedings of the 10h International Conference on Music Perception and Cognition, Sapporo, Japan (2008)
10. Chew, E.: Pitch symmetry and invariants in Webern's Sehr Schnell from variations Op.27. In: Klouche, T., Noll, T. (eds.) Mathematics and Computation in Music, CCIS, vol. 37, pp. 240–246. (2009)
11. Childs, A.: Moving beyond Neo-Riemannian triads: exploring a transformational model for seventh chords. J. Music Theory **42**, 181–193 (1998)
12. Chordia, P., Rae, A.: Raag recognition using pitch-class and pitch-class dyad distributions. In: Proceedings of the International Conference on Music Information Retrieval, Vienna (2007)
13. Chuan, C.-H., Chew, E.: Generating and evaluating musical harmonizations that emulate style. Comp. Music J. **35**(4), 64–82 (2011)
14. Cohn, R.: Maximally smooth cycles, hexatonic systems, and the analysis of late-romantic triadic progressions. Mus. Anal. **15**, 9–40 (1996)
15. Cohn, R.: Neo-Riemannian operations, parsimonious trichords, and their tonnetz representations. J. Mus. Theory **41**, 1–66 (1997)
16. Cohn, R.: Introduction to Neo-Riemannian theory: a survey and historical perspective. J. Mus. Theory **42**(2), 167–180 (1998)
17. Dahlhaus, C.: Studies in the Origin of Harmonic Tonality. Robert O. Gjerdingen (Trans.) Princeton University Press, Princeton, New Jersey (1990)
18. Desain, P., Honing, H., Vanthienen, H., Windsor, L.: Computational modeling of music cognition: problem or solution? Music Percept. **16**(1), 151–166 (1998)
19. François, A.R.J.: MuSA_RT (2012) https://itunes.apple.com/ca/app/musa-rt/id506866959?mt=12. Accessed 30 Aug 2013
20. Gollin, E.: Some aspects of three-dimensional tonnetze. J. Mus. Theory **42**, 195–206 (1998)
21. Honingh, A.: The origin and well-formedness of tonal pitch structures. Ph.D. Dissertation, University of Amsterdam, Amsterdam (2006)
22. Hook, J.: Uniform triadic transformations. Ph.D. Dissertation, Indiana University, Bloomington (2002)
23. Huron, D.: Music-engendered laughter: an analysis of humor devices in PDQ bach. In: Lipscomb, S., Ashley, R., Gjerdingen, R.O., Webster, P. (eds.) Proceedings of the 8th International Conference on Music Perception and Cognition, pp. 700–704 (2004)
24. Huron, D.: Sweet Anticipation: Music and the Psychology of Expectation. MIT Press, Cambridge (2006)
25. Krumhansl, C.L.: The psychological representation of musical pitch in a tonal context. Ph.D. Dissertation, Stanford University, Stanford, CA (1978)
26. Krumhansl, C.L.: Cognitive Foundations of Musical Pitch. Oxford University Press, New York (1990)

27. Laden, B.: A model of tonality cognition which incorporates pitch and rhythm. Ph.D. Dissertation, University of Washington, Seattle, WA (1989)
28. Leman, M., Vos, P. (eds.): Special issue on tonality induction. Mus. Perc. **17**(4) (2000)
29. Lerdahl, F., Jackendoff, R.: A Generative Theory of Tonal Music. MIT Press, Cambridge, MA (1983)
30. Lerdahl, F.: Tonal Pitch Space. Oxford University Press, New York (2001)
31. Lewin, D.: Generalized Musical Intervals and Transformations. Yale University Press, New York (1987)
32. Longuet-Higgins, H.C.: Mental Processes. MIT Press, Cambridge, MA (1987)
33. Mardirossian, A., Chew, E.: Music summarization via key distributions: analyses of similarity assessment across variations. In: Proceedings of the International Conference on Music Information Retrieval (2006)
34. Mazzola, G.: The Topos of Music. Birkhäuser, Basel (2002)
35. Milne, A.: A computational model of the cognition of tonality. Ph.D. Dissertation, Open University, Milton Keynes (2013)
36. Morse, P.M.: Mathematical problems in operations research. Bull. Amer. Math. Soc. **54**(7), 602–621 (1948)
37. Ortmann, O.: An analysis of Paul Hindemith's "Unterweisung im Tonsatz". Bull. Am. Mus. Soc. **4**, 26–28 (1940)
38. Povel, D.J., Essens, P.: Perception of temporal patterns. Mus. Perc. **2**(4), 411–440 (1985)
39. Rameau, J.-P.: Treatise on Harmony. Philip Gossett (Trans.) Dover Publications, New York (1722)
40. Sacks, O.: The Man who Mistook his Wife for a Hat. Touchstone, New York (1985)
41. Schoenberg, A.: Structural Functions of Harmony. Norton & Co. Inc., New York (1969)
42. Schoenberg, A.: Style and Idea: Selected Writings of Arnold Schoenberg. Leonard Stein (ed.), Leo Black (Trans.) St. Martins Press, New York (1975)
43. Smith, J.B.L., Chuan, C.-H., Chew, E.: Audio Properties of Perceived Boundaries in Music. In: Li, T., Ogihara, M., Tzanetakis, G. (eds.) Special issue on music data mining, IEEE Transactions on Multimedia (2014)
44. Snow, C.P.: The Two Cultures—C. P. Snow's epochal essay published online for the first time. New Statesman (2013) Available via http://www.newstatesman.com/cultural-capital/2013/01/c-p-snow-two-cultures. Accessed 12 Aug 2013
45. Temperley, D.: The perception of harmony and tonality: an algorithmic perspective. Ph.D. Dissertation, Columbia University, New York (1996)
46. Temperley, D., Sleator, D.: Modeling meter and harmony: a preference-rule approach. Comp. Mus. J. **23**(1), 10–27 (1999)
47. Temperley, D.: The Cognition of Basic Musical Structures. MIT Press, Cambridge (2001)
48. Temperley, D.: Music and Probability. MIT Press, Cambridge (2010)
49. Tymoczko, D.: A Geometry of Music-Harmony and Counterpoint in the Extended Common Practice. Oxford University Press, New York (2011)
50. Unal, E., Chew, E., Georgiou, P., Narayanan, S.: A preplexity based cover song matching system for short length queries. In: Proceedings of the International Conference on Music Information Retrieval (2011)
51. Xenakis, I.: Formalized Music: Thought and Mathematics in Composition. Indiana University Press, Bloomington (1971)

Chapter 2
An Abbreviated Survey

Abstract This chapter weaves together a backdrop of related work in music theory, cognitive science, and operations research that has inspired and influenced the design of the Spiral Array and its associated algorithms. The chapter begins with an overview of the Spiral Array model and its core ideas. This is followed by a review of some spatial models for musical pitch that have informed the model's design, and an overview of the Harmonic Network (a.k.a. the *tonnetz*) and some of its applications. The idea of the *center of effect* (CE) is central to the Spiral Array and its associated algorithms. The idea of the CE draws inspiration from interior point methods in linear optimization. The second part of the chapter describes the von Neumann Center of Gravity algorithm and Dantzig's bracketing technique to speed convergence, and then draws analogies between the algorithm and the CEG method.

This book is centered on the Spiral Array model, a spatial construct that represents the interrelations among musical pitches. This chapter gives a brief survey of research that has inspired and influenced the design of the Spiral Array model and its associated tonal analysis algorithms. As motivation, I first give an overview of the intuition behind the Spiral Array model and the Center of Effect Generator (CEG) key-finding algorithm.

This chapter is based, in part, on "Dantzig's Indirect Contribution to Music Research: How the von Neumann Center of Gravity Algorithm Influenced the Center of Effect Generator Key Finding Algorithm" by Elaine Chew, an article in the INFORMS Computing Society Newsletter (Spring 2008), and on the Background (Chapter 2) of "Towards a Mathematical Modeling of Tonality" by Elaine Chew, an MIT PhD dissertation, Cambridge, Massachusetts (2000) https://dspace.mit.edu/handle/1721.1/9139

E. Chew, *Mathematical and Computational Modeling of Tonality*,
International Series in Operations Research & Management Science 204,
DOI: 10.1007/978-1-4614-9475-1_2,

2.1 Spiral Array Overview

2.1.1 Tonality in a Nutshell

The Spiral Array is a mathematical model for tonality. More specifically, it is a geometric model that represents elements of the tonal system underlying the music with which we are familiar. These elements include (1) pitches, sounds of a given fundamental frequency; (2) chords, simultaneous sounding of multiple pitches (a chord with three pitches is called a triad); and (3) keys, a collections of pitches, the sequences of which generate particular patterns of perceived stability. The pitch set of a key can be unambiguously defined by three triads. The name of the key is simultaneously the pitch name of the most stable tone. Tonality refers to this set of interrelations among pitches and sets of pitches.

The existence of the tonal system is one of the main reasons why we are able to form expectations, and have them resolved, when listening to music—for example, we can hear when a melody is ended or not (try singing only the first three phrases of "Happy Birthday.") There are many reasons to model tonality—for example, to create a representation on which to base algorithms for computational music analysis, and to understand human perception and cognition. Examples of investigations into mental representations of tonality include [34, 36]. The Spiral Array and associated algorithms are motivated primarily by computational analysis. Algorithms for automated analysis drive systems for automatic accompaniment and computer-assisted composition, and for analysis and synthesis of expressive music performances.

2.1.2 The Spiral Array

The Spiral Array aims to model aspects of tonality. It consists of an array of nested helices that each represents a kind of tonal entity—pitches, major/minor chords, and major/minor keys—in music. Higher level constructs are generated, successively, as the convex sum of their components.

Figure 2.1 shows some of the components of the Spiral Array model. Figure 2.1a depicts the outermost pitch class helix. Pitch representations are spaced evenly at each quarter turn of the helix. Each node represents a pitch class, i.e., a C is not just the middle C on a keyboard, but all Cs at different octaves above and below it. Frequencies of pitches in the same class are related by powers of two. Neighboring pitch classes along the helix contain pitches with frequency ratios of approximately 2:3, and vertical neighbors have ratios of approximately 4:5.

The pitch class spiral is a helical configuration of Longuet-Higgins' Harmonic Network [29, 30], which is also known in music theoretic circles as the *tonnetz* (tone network) [8, 27]. Section 2.2 describes some spatial representations of musical pitches that preceded the Spiral Array, followed by a more detailed treatment of the Harmonic Network.

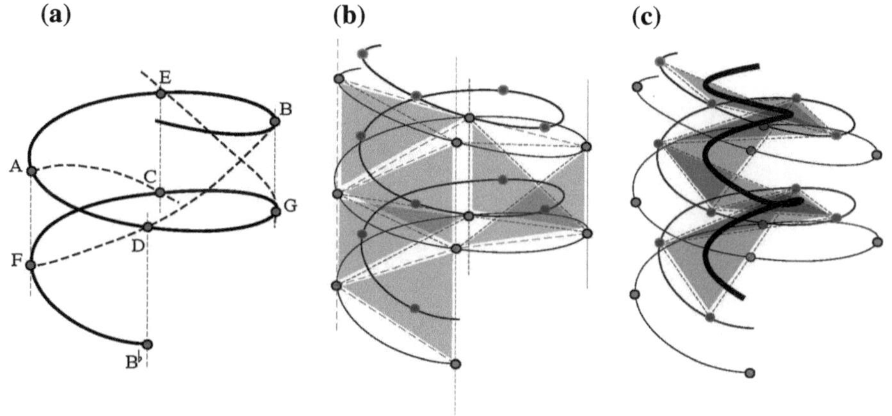

Fig. 2.1 The Spiral Array model: successive generation of major chords from pitches, and major keys from major chords. **a** Pitch class representations. **b** Major chord representations. **c** Major key representations

Distinct from previous models that employ network (or their dual) representations of pitch classes, for example [27], the Spiral Array uses the interior space to define spatial representations of chords and keys as weighted sums of their components. Thus, the Spiral Array captures, spatially and mathematically, the idea that higher level representations are a composite of their lower level elements. Another important trait is that tonal elements from different hierarchical levels reside within the same framework, in the same space.

Starting from the pitch class spiral, the Spiral Array goes on to define points in the interior of the helix that represent higher level tonal constructs. Chords are represented as points that are the convex combination of their component pitches, a centroid of sorts. For example, each triad is represented as a point on the triangle outlined by its component pitches; Fig. 2.1b shows the major triad representations, which themselves lie on an inner helix. Minor triads and their corresponding helix are defined in a similar fashion.

The major key representations are generated as weighted combinations of their defining triads, which would be three adjacent major triads for a major key, as shown in Fig. 2.1c. The minor key helix is produced in a similar way. Figure 2.1c also depicts the nested helices for pitch classes, major triads, and major keys, in decreasing order of radii. Figure 2.1c is repeated in Fig. 2.2 alongside the corresponding nested helices for pitch classes, major and minor triads, and minor keys.

The weights in the Spiral Array can be calibrated so that model distances concur with particular principles of tonal cognition, as shown in Appendix A. Like the Harmonic Network, the Spiral Array places in the closest proximity pitches related by intervals of a perfect fifth (P5), a major third (M3), and a minor third (m3). As in the work of Shepard [34] and Krumhansl [22], to be reviewed in Sect. 2.2, close tonal relations are mirrored by spatial proximity among their corresponding representations.

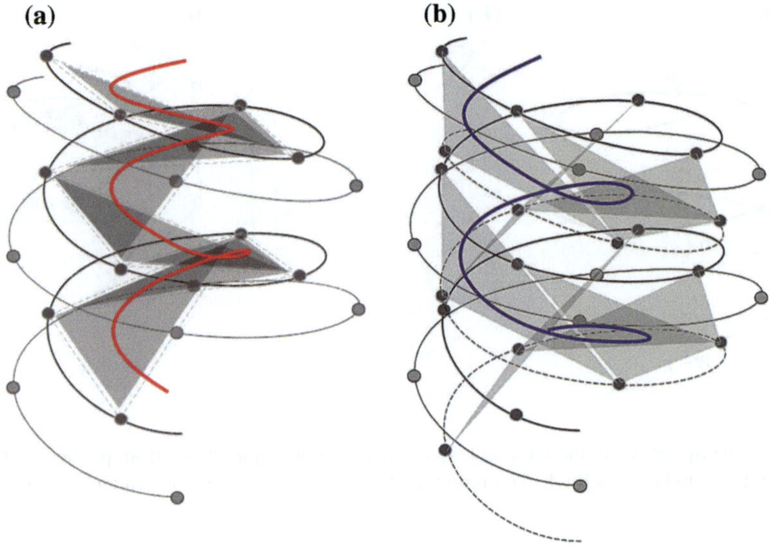

Fig. 2.2 The Spiral Array model: nested helices representing pitch classes, major/minor chords, and major/minor keys (shown separately on two diagrams for clarity). **a** Pitch class, major triad, and major key helices. **b** Pitch class, major and minor triad, and minor key helices

2.1.3 The Center of Effect

This idea of moving off the grid of the Harmonic Network to define points in the interior of the helix is central to the Spiral Array and its algorithms—the key-finding algorithm in Chap. 4, the segmentation algorithms in Chaps. 6 and 7, and the pitch-spelling algorithm in Chap. 8, to name a few.

Any collection of pitch representations can be appropriately weighted to generate a barycenter in the interior of the helix that represents the composite effect of the pitches, called the *center of effect* (CE). For example, in the Center of Effect Generator (CEG) key-finding algorithm (described in Chap. 4), an input music stream is mapped to its pitch representations, with each note weighted by its duration to get a CE; the key is then identified by searching for the key representation nearest to the CE.

The CEG algorithm can be illustrated simply with a melody. A melody consists of a sequence of note events, each note having the properties of pitch and duration. The algorithm generalizes to more complex music with simultaneous tones at any given time. In the Spiral Array, the tonal context of a segment of music is represented by a summary point, the center of effect, CE, of the pitches. A CE of a collection of notes can be generated as the sum of the pitch positions, weighted by their respective durations.

Given a melody, the CEG algorithm successively generates CEs as each note event occurs, thus updating itself as it gravitates toward the key representation. The distance between CE and key need not decrease monotonically; the CE trace can move toward, or away from, a key. The key at any given time is determined by a nearest-neighbor search for the closest key representation.

Figure 2.3 provides a pictorial guide to the CEG algorithm. Figure 2.3a shows the initial CE at the first note of the melody, which is the pitch representation of the first note. At the second note, which is of the same duration as the first, and a perfect fourth up as shown in Fig. 2.3b, the CE moves to the midpoint between the first and second pitches. Suppose the third note is the same as the second, then the CE simply moves closer to the pitch of the second/third note, as shown in Fig. 2.3c. As the iterations continue, a trajectory is traced in the interior of the model. Suppose that, at the state shown in Fig. 2.3i, one wishes to determine the key. The key is found by searching for the nearest key representation. The solution key and the convex hull of its pitch set are shown in Fig. 2.3j.

When tested on a small test set of the fugue subjects of all 24 fugues in Bach's *Well-Tempered Clavier* Book I, ignoring correct answers on the first note, the algorithm

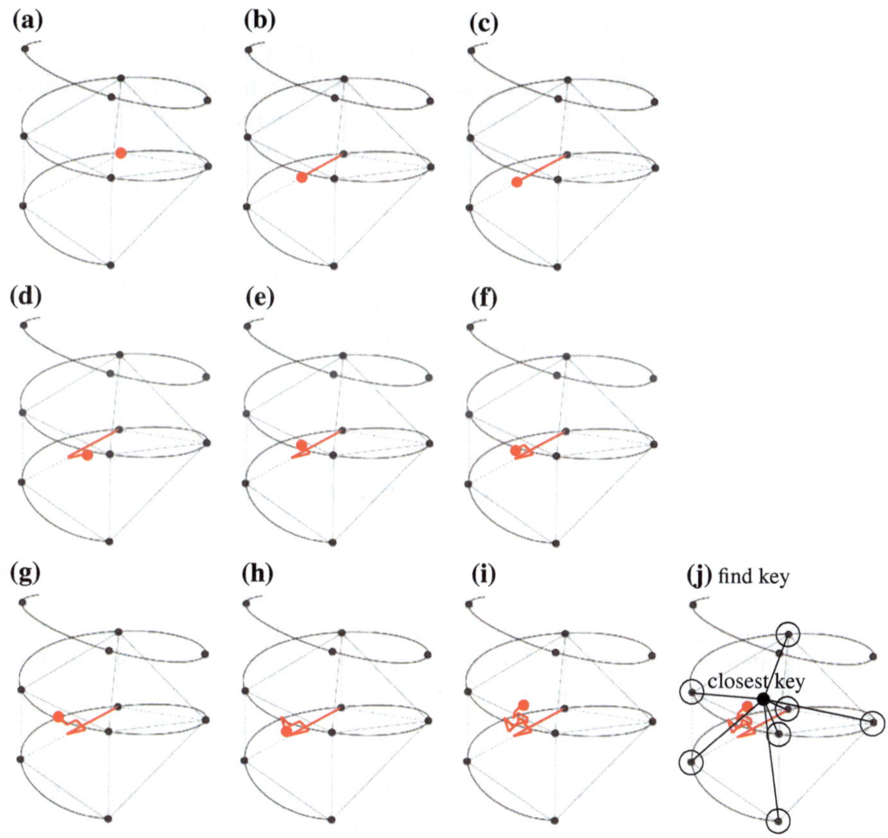

Fig. 2.3 Illustration of the Center of Effect Generator algorithm

found the key of the fugue in 3.75 note events on average, which is on par with key-finding performance by humans, and faster than previous methods.

The inspiration for the idea of the CE stems from interior point algorithms in Operations Research. Interior point algorithms deviate from well-known methods for solving linear optimization problems, namely Dantzig's Simplex Method (see [18] for an introduction), by moving away from the vertices of a polytope to search through the interior of the space for the solution that optimizes an objective function. The CEG algorithm was motivated by my experiences working with George Dantzig on one of the earliest interior point algorithms, the Center of Gravity algorithm [14] proposed by von Neumann. The Center of Gravity algorithm will be described in Sect. 2.3.

2.2 Spatial Models for Pitch Relations

Spatial analogs of physical and psychological phenomena are known to be powerful tools for solving abstract intellectual problems [34]. Biologists have long used geometric models for DNA structure and protein folding. Chemists study structural models for chemical bonding. In mathematics, it was George Dantzig's adroitness at geometry that inspired him to invent the Simplex Method for solving linear optimization models.

Some have argued that problems in music perception and cognition can be reduced to that of finding an optimal data representation [35]. *Computing in Musicology* [17], volume 11, covers an assortment of pitch encodings for computer comparisons of melodies. Tonality describes a highly structured system of pitch relations, but few representations incorporate the functional relations among pitches that generate a tonal center.

2.2.1 Cognitive Representations of Musical Pitch

In 1982, Shepard [34] stated that "the cognitive representation of musical pitch must have properties of great regularity, symmetry, and transformational invariance." Recognizing that perfect fifth (P5) and major third (M3) intervals were perceptually important pitch relations, Shepard sought to design a spatial model in which these relations have direct counterparts in the geometric structure.

In the tradition of spiral models for pitch dating as far back as 1855, Shepard proposed a model that spaced all 12 chromatic pitches equally over one full turn of a spiral (see Fig. 2.4). The equal spacing, distinct from previous models, emphasized the close relationship of pitches related by octave intervals. Further extensions to incorporate perfect fifth interval relations resulted in double-helix structures that still did not provide for the major third.

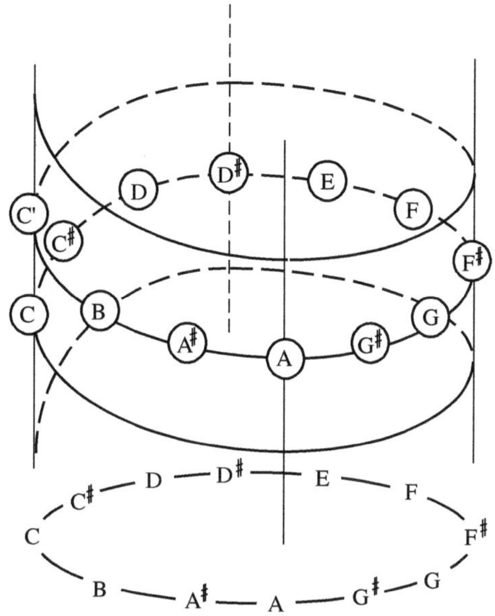

Fig. 2.4 Shepard's spiral model of chromatic pitches

Krumhansl's [20] doctoral thesis attempted to uncover the structure of pitch relations in tonality using experimental data. In this and later publications [21–23], Krumhansl and colleagues used two main ideas in analyzing data garnered from listener-based experiments. The first is the use of Probe Tone Profiles, judgements with respect to how well each of the 12 pitch classes fit into a given key context. The second is the use of Shepard's Multidimensional Scaling technique [32, 33], which maps each profile onto a point in Euclidean space such that high correlation is mirrored by spatial proximity.

Based on statistical analysis of such experimental data, Krumhansl proposed a conical structure for pitch relations [20]. The pitches of the major triad were located on a plane closest to the vertex, other diatonic pitches on a second plane, and the remaining pitches in the most distant plane. This conical structure does not contradict the Spiral Array arrangement. In fact, the planar version of the key spirals falls neatly out of the Multidimensional Scaling of the listeners' probe tone profile ratings. Lerdahl's Tonal Pitch Space [25] would later have direct parallels with Krumhansl's pitch class cone and key space—see [4] for an exposition and Fig. 2.5 for an illustration comparing Krumhansl's and Lerdahl's pitch cones. In addition, Krumhansl deduced that an established tonal context changes the pitch relationships, thus suggesting that a geometric representation may have to be altered depending on the key context.

In [21, 23], Krumhansl used the Probe Tone Profile method and Multidimensional Scaling to perform studies of listener judgements of triadic proximity, and used empirical data to corroborate the psychological reality of neo-Riemannian transformations, which brings us to the next section on the neo-Reimannian *tonnetz* (tone network).

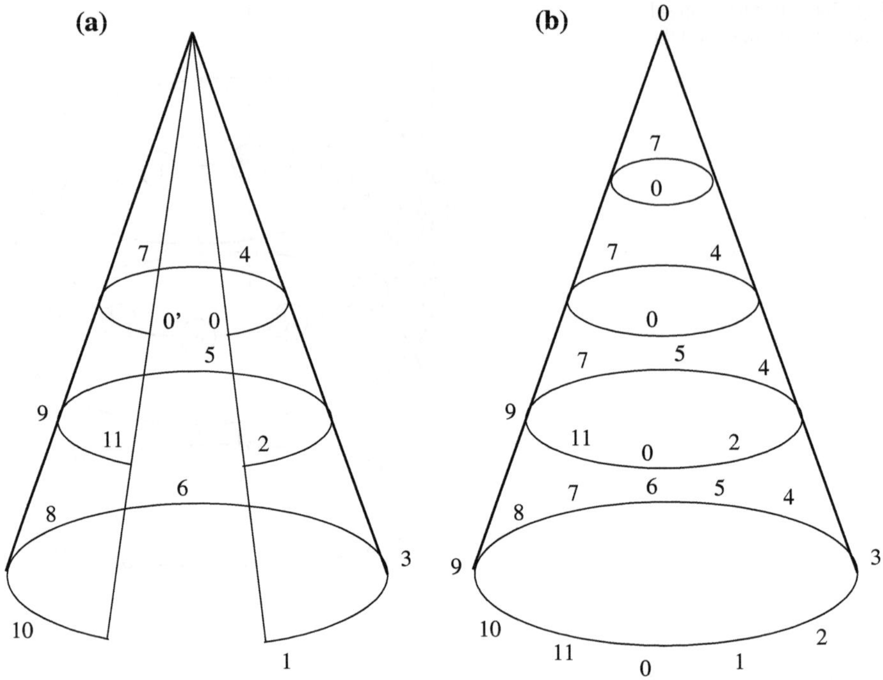

Fig. 2.5 Pitch class cones (figures from [4]): **a** Krumhansl's pitch cone (inverted); **b** Lerdahl's pitch class cone

2.2.2 The Tonnetz

Tonality describes a highly structured system of pitch relations, a system that has been studied by theorists over the past centuries. Many theories have been proposed to explain the relations implied in tonal music, and many theorists have chosen to represent these tonal structures geometrically (see review in [24]). One such theory, proposed by the nineteenth-century music theorist Riemann, posits that tonality derives from the establishing of significant tonal relationships through chord functions [10]. This idea has influenced a wide range of research in Music Theory.

Riemann's theory agrees with Shepard's intuitions that the most significant interval relations are the perfect fifth (P5) and the major/minor third (M3/m3). Riemann represented these relations in a tone network called the *tonnetz*, shown in Fig. 2.6. Cohn [8] has traced the earliest version of this network of pitches related by perfect fifth and major/minor third intervals to the eighteenth-century mathematician Euler. In the nineteenth century, this representation was appropriated by music theorists such as Oettingen and Riemann.

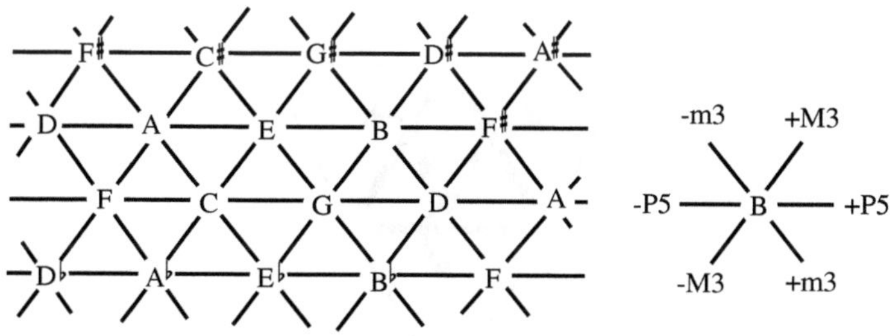

Fig. 2.6 The *Tonnetz* (figure reproduced from [4])

2.2.3 Isomorphic Representations

Longuet-Higgins [29, 30] showed that frequency ratios of musical intervals take the form $2^x.3^y.5^z$, where $x, y, z \in \mathbb{Z}$. He used this relationship to generate a three-dimensional grid of pitches. Since pitches related by powers of two are considered to be equivalent, the projection of the space onto the y–z plane gives the Harmonic Network, shown in Fig. 2.7. The Harmonic Network is effectively the same as the tonnetz, and I shall use the terms Harmonic Network and tonnetz interchangeably.

Noting that pitches in a key are positioned in a compact neighborhood on the Harmonic Network—see for example, the C major pitches highlighted in bold in Fig. 2.7—Longuet-Higgins and Steedman [31] proposed a Shape-Matching Algorithm (SMA) for key-finding using this table of harmonic relations. The SMA is closely related to the Spiral Array Center of Effect Generator (CEG) key-finding algorithm, and will be described in greater detail in Chap. 5, where it will be compared to the CEG.

Longuet-Higgins' work has inspired further mathematical analyses of the Harmonic Network, such as that of Balzano [1], which favors the major third-minor third (M3/m3) axes as opposed to the perfect fifth-major third (P5/M3) arrangement. However, one can demonstrate that the perfect fifth-major third (P5/M3), major

E	B	F♯	C♯	G♯	D♯	A♯	E♯	B♯
C	G	D	**A**	**E**	**B**	F♯	C♯	G♯
A♭	E♭	B♭	**F**	**C**	**G**	**D**	A	E
F♭	C♭	G♭	D♭	A♭	E♭	B♭	F	C
D♭♭	A♭♭	E♭♭	B♭♭	F♭	C♭	G♭	D♭	A♭

Fig. 2.7 The Harmonic Network

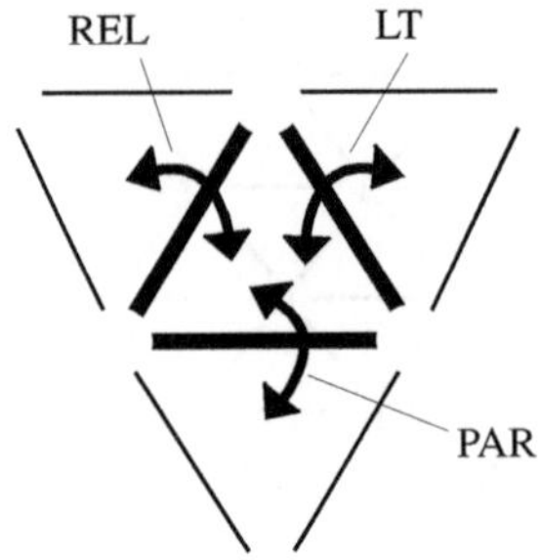

Fig. 2.8 Transformations on the Harmonic Network

third-minor third (M3/m3), and the perfect fifth-minor third (P5/m3) representations are equivalent.

Observe that the pitches displayed on the Harmonic Network repeat periodically. Rolling up the planar network so that the repeated pitch names are aligned one with another, one gets the pitch class helix of the Spiral Array. Like the Harmonic Network, neighboring pitches (on the Spiral Array's pitch class helix) are a perfect fifth (P5) apart, and pitches vertically above each other are a major third (M3) apart. If one considers equal temperament, e.g. A$\sharp$ = B$\flat$, the spiral would close to form a torus.

The added dimension (above the planar configuration) provided by the spiral arrangement allows computationally efficient and cognitively accurate algorithms to be designed that use the interior of the helix to succinctly represent different pitch collections. These will be described in subsequent chapters. Prior to the Spiral Array, researchers have alluded to the inherent toroid structure of the Harmonic Network but only used the planar versions in applications to music analysis.

2.2.4 Transformational Theory

Transformational Theory, also known as Neo-Riemannian Theory, is a branch of music theory that views transitions between chords as group theoretic transformations. In the 1980s, Lewin [26, 27] was instrumental in reviving the use of the tonnetz in music analysis, thus planting the seed for the emerging field of Transformational Theory for the analysis of triadic post-tonal music [9].

In the Harmonic Network, similar triads (where similarity is measured in terms of common pitches and parsimonious voice leading) form triangles that have common sides and are consequently near each other. Lewin [26] proposed a set of transformations on the Harmonic Network that mapped similar triads one to another.

The three main types of transformations are parallel (PAR), relative (REL), and leading tone exchange (LT), as shown in Fig. 2.8. PAR maps a major (minor) triad to its parallel minor (major) triad. REL maps a major (minor) triad to its relative minor (major) triad. And, LT lowers the root of the triad to its leading tone to form a new triad; in this case the minor triad is rooted on the mediant, the third scale degree ($\hat{3}$), of the major triad, and the major triad is based on the submediant, the sixth scale degree ($\hat{6}$), of the minor triad.

Each transformation links two triads that differ from each other only by one pitch. The PAR transformation links two triads that share the root and the fifth. In the REL transformation, the root and third of the major triad is the third and fifth of the minor triad. In the LT transformation, the third and fifth of the major triad is the root and third of the minor triad. This minimal change property of the transformations results in parsimonious voice leading; thus, smooth harmonic movements in a piece of music trace continuous paths in this space. This same principle of distance-minimizing motion is exploited in the Spiral Array space.

Paths and cycles in the Harmonic Network that are a result of a string of transformations defined by Lewin correspond to triadic movements in tonal music. They can also be viewed as trajectories in the dual tonnetz space, see Fig. 2.9. An entire body of literature has emerged based on such transformations in the dual space.

In 1996, Cohn focused on the parsimonious voice-leading property [7, 8] of the transformations to abstract hexatonic and octatonic transformation cycles on the tonnetz. In his essays, he showed how these patterns corresponded to triadic movements in chromatic music of the nineteenth century. Such models have been generalized [19, 28], and extended to meta-cycles (cycles of cycles) [15] and to tetrachords [5, 15, 16]. Computationally, the finding of transformation cycles was recently automated using finite state transducers [2]. Neo-Riemannian theory has also been applied to the analysis [3] of, and to the generating of tonally idiomatic chord progressions [6] in, pop-rock music.

These mathematical approaches are primarily graph-theoretic in nature. Several allude to the inherent toroid structure of the tonnetz; none has explored the use of the helical configuration of the Harmonic Network.

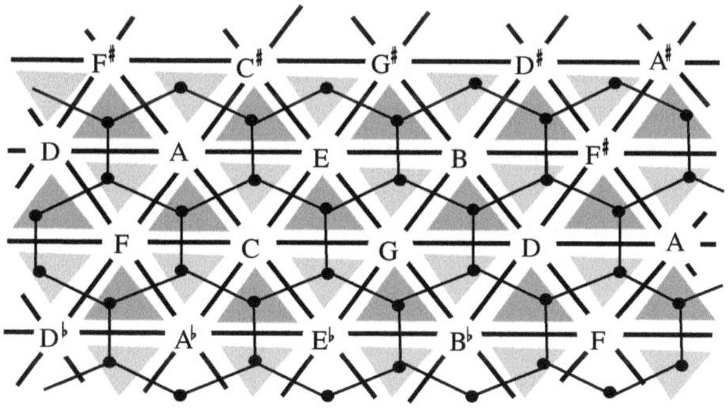

Fig. 2.9 Network showing the dual graph of the tonnetz (figure from [4])

Back to the Spiral Array

The Spiral Array preserves the Harmonic Network's preference for perfect fifth (P5) and major/minor third (M3/m3) intervals. Similar triads are still physically close in the Spiral Array. At the same time, the new configuration offers the benefits of modeling pitch relations in a higher dimension.

The Spiral Array allows one to calculate barycenters for any pitch collection that are distinct from the pitches themselves, in the interior of the helical space. As a result, one can define representations for higher level elements such as intervals, triads, tetrachords, and keys. Similar elements—where similarity is measured in terms of common pitches and perfect fifth or major/minor third interval relations—are still physically close in space.

Note that elements that are spatially close in the Spiral Array are not necessarily close in space on a keyboard or fingerboard. Pitches that are near one to another on a keyboard, such as two pitches a half step apart, are represented by points in space that are relatively far apart. The Spiral Array (and the Harmonic Network) stresses harmonic relations (perfect fifth and major/minor third intervals), and not linear relations such as pitches separated by a half step.

2.3 An Early Interior Point Algorithm

Leaving behind the music theory and psychology music domains, we now transition to the field of operations research. In the summer of 1991, I was fortunate to work with George Dantzig on an undergraduate research project focusing on von Neumann's Center of Gravity algorithm, an early interior point method for solving linear optimization problems.

The Simplex Method

Linear programming problems are typically solved using the Simplex Method invented by Dantzig (see [18] for an introduction). A typical linear programming problem is expressed as a system of equations or inequalities consisting of a linear objective function that must be maximized (or minimized) subject to a number of linear constraints.

Suppose there are n variables and m constraints. If the problem is well-behaved, and a feasible solution exists, the problem translates to one of finding the variables that maximize (or minimize) the objective function value within a feasible region in n-dimensional space, typically a polytope with m sides that is bounded by hyperplanes defined by the constraints. The optimal solution resides in a corner of the polytope, i.e., is one of the corner point feasible solutions (CPFSs).

The Simplex Method can be illustrated using a simple three-dimensional example. Starting with a known CPFS—perhaps the origin, as depicted in Fig. 2.10a—as the

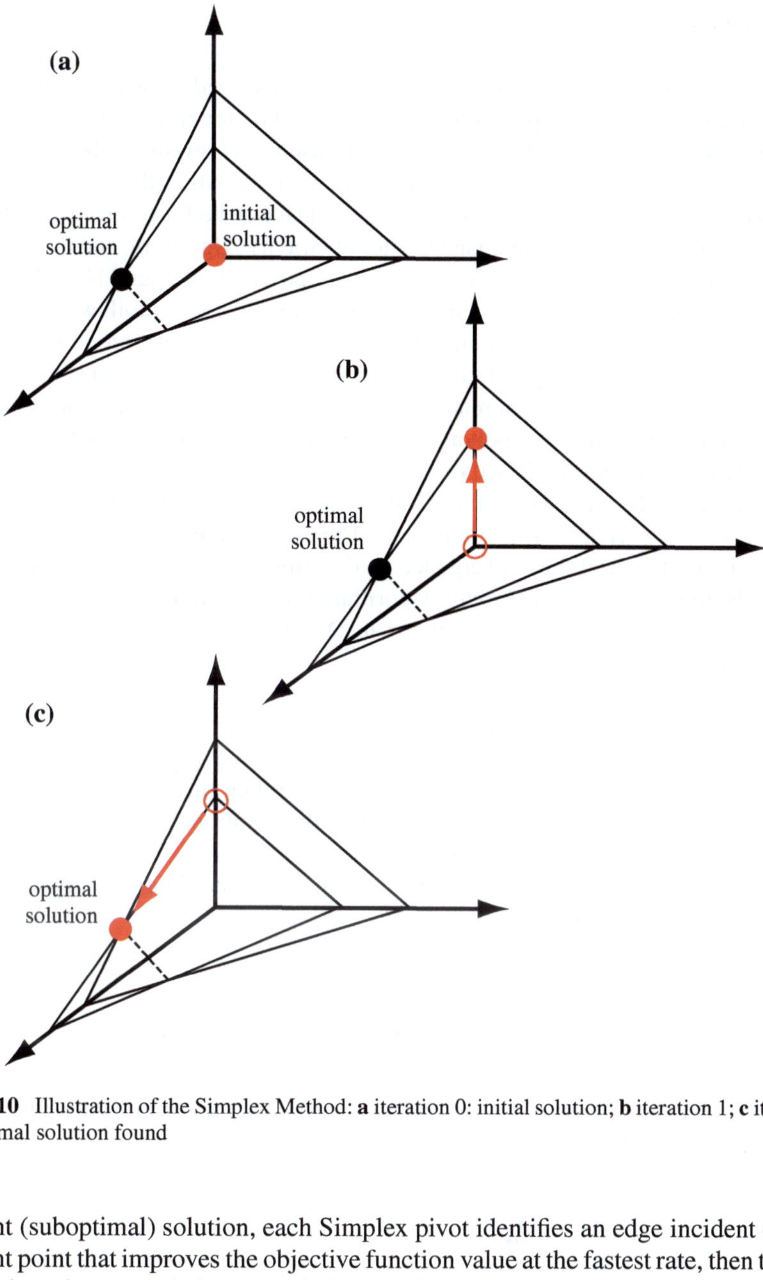

Fig. 2.10 Illustration of the Simplex Method: **a** iteration 0: initial solution; **b** iteration 1; **c** iteration 2: optimal solution found

current (suboptimal) solution, each Simplex pivot identifies an edge incident on the current point that improves the objective function value at the fastest rate, then travels along that edge to reach the next solution, as shown in Fig. 2.10b. The algorithm stops when traveling down any of the incident edges cannot improve the objective value function. The optimal solution is thus found, as in Fig. 2.10c.

Although the Simplex Method is theoretically exponential in computational time, the algorithm is hardly ever this inefficient.

Interior Point Methods

In the late 1980s, interior point methods were popularized by Karmarkar's much-publicized discovery of a polynomial time algorithm for linear programming. In the wake of this discovery, George Dantzig himself was revisiting the first interior point algorithm, the von Neumann Center of Gravity Algorithm, communicated verbally to him in 1948 by John von Neumann. Dantzig had just documented proof of convergence for the von Neumann Center of Gravity Algorithm, and his proposed extension of the algorithm that had a guaranteed polynomial bound [11, 12].

The von Neumann Center of Gravity Algorithm solves the following problem: given n points on a unit sphere centered on the origin in m dimensions, find nonnegative weights so that the weighted sum of the n points is the origin. Each von Neumann iteration could be calculated in relatively few computational steps. Dantzig showed that the algorithm had lower polynomial complexity (degree 2) than Karmarkar's, but a much higher constant based on the required precision.

While the von Neumann Center of Gravity Algorithm was guaranteed to converge, it did so very slowly after the initial steps. Dantzig's modification—to bracket the target—aimed to speed the convergence. A description of, and theorems related to, the von Neumann Center of Gravity Algorithm would later be incorporated into the chapter on Early Interior Point Methods in Dantzig and Mukund Thapa's *Linear Programming 2* [14].

In tests I conducted [13], the bracketing method was shown to work well for most problems, yielding a solution with very few von Neumann iterations. The bracketing method performed best when solving linear programs having many more variables than constraints, and a large radius of feasibility. When the radius of feasibility is very small and the number of constraints large, the iterations converged slowly, like the original Center of Gravity algorithm.

The following sections describe von Neumann's Center of Gravity algorithm and Dantzig's bracketing technique.

2.3.1 von Neumann's Center of Gravity Algorithm

The von Neumann Center of Gravity Algorithm solves linear optimization problems that have been recast in the form:

$$\begin{aligned} \sum_{j=1}^{n} P_j x_j &= 0, \\ \sum_{j=1}^{n} x_j &= 1, \\ \|P_j\|_2 &= 1 \quad \forall\, j = 1 \ldots n, \\ x_j &\geq 0 \quad \forall\, j = 1 \ldots n. \end{aligned}$$

The P_j's are points on a unit ball centered on the origin, and the goal is to find the combination of weights, x_j's, on these points so that they sum to the origin.

A two-dimensional example demonstrates the von Neumann Center of Gravity Algorithm. Assume that $n = 5$ and that the five P_j's are situated on the unit circle as shown in Fig. 2.11a. Any of these five points on the circumference can serve as the initial solution, A^0, such as the P_j that is colored red in Fig. 2.11a.

At iteration t, draw a line from the current solution, A^t, through the center of the circle, as shown in Fig. 2.11b. Find the P_j that makes the smallest acute angle, θ, with this line, P_{acute}, as shown in Fig. 2.11c. If there is no P_j on the opposite side of the circle, i.e., $\theta > \pi/2$, then the problem is infeasible. Drop a perpendicular from the origin to the line through A^t and P_{acute} to get the new solution point, A^{t+1}. Update the weights on the P_j's accordingly. This step is shown in Fig. 2.11d.

Dantzig [12] proved that, independent of the number of rows, m, and columns, n, in the problem, a precision of ε can be guaranteed with less than $1/\varepsilon^2$ iterations, where ε is the distance between the solution and the origin. Thus, convergence can be slow if the solution must be very close to the origin.

2.3.2 Dantzig's Bracketing Technique

The bracketing technique proposed in [11] aims to speed the convergence of the von Neumann Center of Gravity Algorithm by providing larger targets that are centered on the vertices of a simplex inside the convex hull of the P_j's.

This bracketing technique is shown in two dimensions in Fig. 2.12. Instead of targeting the origin, the bracketing technique applies the von Neumann Center of Gravity Algorithm $m + 1$ times, with each vertex of a simplex inside the unit ball as target. Such a simplex is shown in Fig. 2.12. For each vertex, the algorithm iterates until the approximate solution converges to a point within a given radius of the vertex.

The vertices, and the corresponding radius defining the required precision for these targets, are chosen so that the balls circumscribing each vertex fall within the circle (or sphere) that lies inside the convex hull of the P_j's. One such circle that resides inside the convex hull is indicated by the dotted line in Fig. 2.12. In practice, one does not know the radius of the dotted circle, and different radii were tested empirically in the experiments that I ran.

Once approximate solutions inside the distributed targets are found, the origin is bracketed by these solutions, and one can then solve for the origin with straightforward linear operations.

In an experiment involving 47 data files, with matrices varying in size from 3×4 to 28×40, it was found that the von Neumann Center of Gravity Algorithm with bracketing works well for most problems, yielding a solution with relatively few von Neumann steps. For problems that have a small radius of feasibility, the iterations converged slowly, and the bracketing extension does little to improve on the original von Neumann Center of Gravity Algorithm.

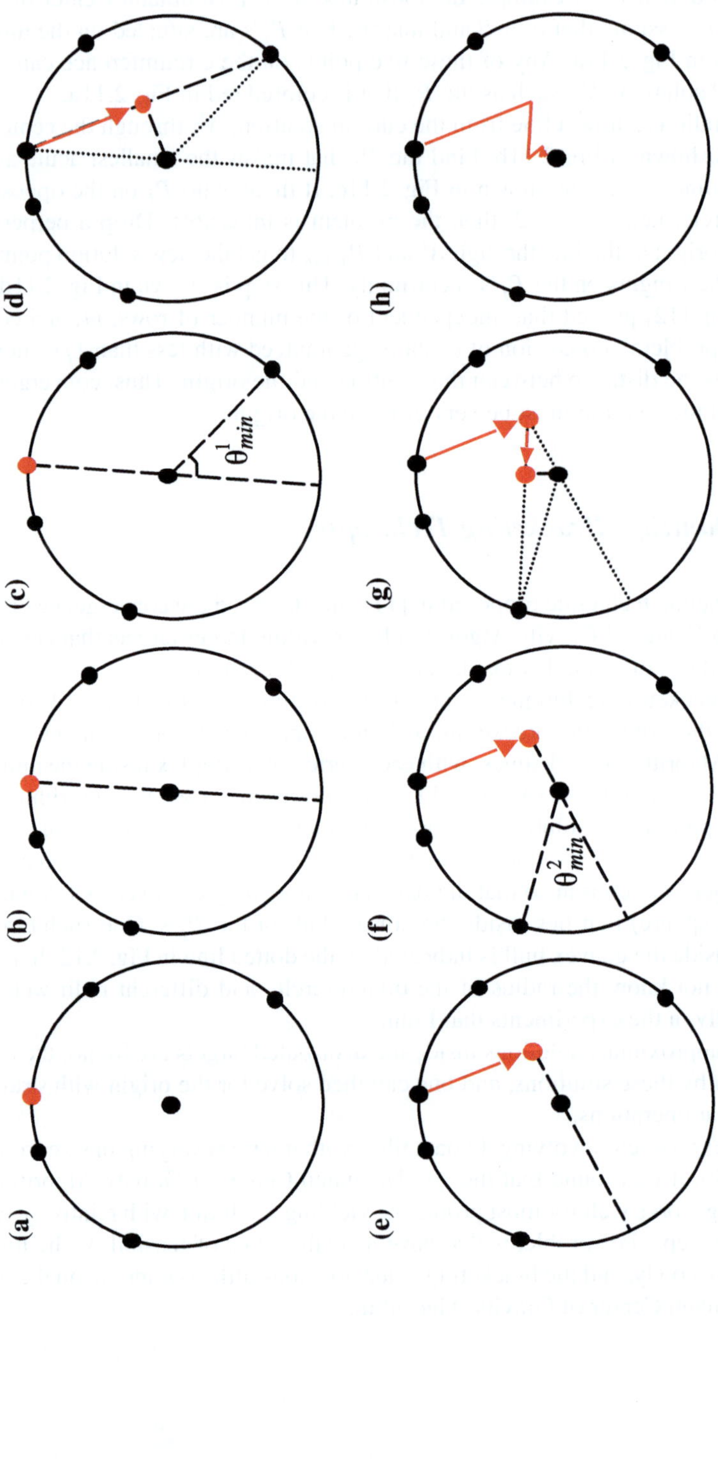

Fig. 2.11 An illustration of the von Neumann Center of Gravity Algorithm: **a** initial solution; **b** shoot line through center; **c** find point making smallest angle; **d** drop perpendicular; **e** shoot line through center; **f** find point making smallest angle; **g** drop perpendicular; **h** iterate

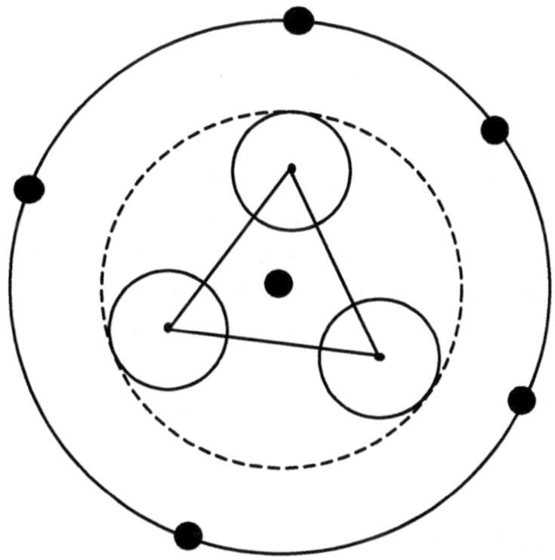

Fig. 2.12 Dantzig's bracketing technique

Back to the Center of Effect

The Spiral Array model represents tonal entities in music—pitches, chords, and keys—as points on nested helices in the same three-dimensional space. Higher level constructs are generated by successive aggregation, as weighted sums of their components. Central to the model and its algorithms is the idea of the center of effect, CE. The tonal context of a segment of music is represented by a summary point, the CE, a weighted sum of the pitch set. As more music is heard, the position of the CE generally gravitates toward the key, and the key is computed through a nearest-neighbor search. The CEG method was shown to converge faster, than other existing methods to the key (in 3.75 steps, which is on par with human hearing) when applied to a classical test set.

While I did not set out to deliberately employ ideas from the von Neumann algorithm and its extension, in retrospect, the Spiral Array model and CEG algorithm was very much influenced by the kind of geometric and interior point approach embodied in the von Neumann Center of Gravity Algorithm.

The most obvious similarity between the von Neumann Center of Gravity Algorithm and the Spiral Array CEG algorithm is the geometric and interior point approach that underlies both methods. The von Neumann Center of Gravity Algorithm is an early interior point algorithm, while the Spiral Array CEG algorithm can be loosely considered an interior point approach to key-finding. A difference is that, in the CEG algorithm, convergence is not guaranteed; the goal of music is not to converge monotonically to a key, but to create interest through the varying of distances to different keys.

The von Neumann algorithm works best in problems where the number of variables is extremely large, compared to the number of constraints. The Spiral Array is presently defined only in three dimensions, thus there exists a fixed limit to the problem size. The CEG algorithm is thus highly amenable to real-time applications, as in the MuSA.RT analysis and visualization system, described in Chap. 9.

Direct implementation of the idea of bracketing in key-finding still requires some thought. The idea exists in music cognition: notes that sound imply certain chords, and these chords in turn point to the key context. The challenge in employing parallel optimization to multiple targets in music analysis is the management of time. Music unfolds in a single stream that is experienced over time. It is unclear whether the iterations toward multiple targets should be implemented simultaneously or in series in a real-time system that mimics human key-finding abilities. A possible CEG algorithm with bracketing could be to determine the recent chords (the distributed targets) and use them to determine the key (the bracketed ultimate target).

References

1. Balzano, G.J.: The group-theoretic description of 12-fold and microtonal pitch systems. Comp. Music J. **4**(4), 66–84 (1980)
2. Bragg, J., Chew, E., Shieber, S.: Neo-Riemannian cycle detection with weighted finite-state transducers. In: Proceedings of the International Conference on Music Information Retrieval (2011)
3. Cappuzo, G.: Neo-Riemannian theory and the analysis of pop-rock music. Music Theor. Spectr. **26**(2), 177–199 (2004)
4. Chew, E.: Out of the grid and Into the spiral: geometric interpretations of and comparisons with the Spiral Array model. In: Hewlett, W.B., Selfridge-Field, E., Correia Jr, E. (eds.) Tonal Theory for the Digital Age, Computing in Musicology, vol. 15. CCARH, Stanford (1999)
5. Childs, A.: Moving beyond Neo-Riemannian triads: exploring a transformational model for seventh chords. J. Music Theor. **42**, 181–193 (1998)
6. Chuan, C.-H., Chew, E.: Generating and evaluating musical harmonizations that emulate style. Comput. Music J. **35**(4), 64–82 (2011)
7. Cohn, R.: Maximally smooth cycles, hexatonic systems, and the analysis of late-romantic triadic progressions. Music Anal. **15**, 9–40 (1996)
8. Cohn, R.: Neo-Riemannian operations, parsimonious trichords, and their tonnetz representations. J. Music Theor. **41**, 1–66 (1997)
9. Cohn, R.: Introduction to Neo-Riemannian theory: a survey and a historical perspective. J. Music Theor. **42**(2), 167–180 (1998)
10. Dahlhaus, C.: Studies in the Origin of Harmonic Tonality. Trans. Robert, O. Gjerdingen. Princeton University Press, Princeton (1990)
11. Dantzig, G.B.: Converting a converging algorithm into a polynomially bounded algorithm. Systems Optimization Laboratory Technical Report, Report SOL 915, Stanford University, Stanford (1991)
12. Dantzig, G.B.: An ε-precise feasible solution to a constraint in $1/\varepsilon^2$ iterations independent of problem size. Systems Optimization Laboratory Technical Report, Report SOL 925, Stanford University, Stanford (1992)
13. Dantzig, G.B.: Bracketing to speed convergence illustrated on the von Neumann algorithm for finding a feasible solution to a linear program with a convexity constraint. Systems Optimization Laboratory Technical Report, Report SOL 926, Stanford University, Stanford (1992)
14. Dantzig, G.B., Thappa, M.N.: Von Neumann's method. In: Linear Programming 2: Theory and Extensions, pp. 7084. Springer, New York (2003)
15. Douthett, J., Steinbach, P.: Parsimonious graphs: a study in parsimony, contextual transformations, and modes of limited transposition. J. Music Theor. **42**(2), 241–264 (1998)

16. Gollin, E.: Some aspects of three-dimensional tonnetze. J. Music Theor. **42**, 195–206 (1998)
17. Hewlett, W.B., Selfridge-Field, E. (eds.) Special Issue on Similarity in Music. Computing in Musicology, vol. 11, CCARH, Stanford (1999)
18. Hillier, F.S., Lieberman, G.J.: Introduction to Operations Research, 9/e. McGraw-Hill, New York (2010)
19. Hook, J.: Uniform triadic transformations. Ph.D. Dissertation, Indiana University, Bloomington (2002)
20. Krumhansl, C.L.: The psychological representation of musical pitch in a tonal context. Ph.D. Dissertation, Stanford University, Stanford (1978)
21. Krumhansl, C.L., Kessler, E.J.: Tracing the dynamic changes in perceived tonal organisation in a spatial representation of musical keys. Psychol. Rev. **89**(4), 334–368 (1982)
22. Krumhansl, C.L.: Cognitive Foundations of Musical Pitch. Oxford University Press, New York (1990)
23. Krumhansl, C.L.: Perceived triad distance: evidence supporting the psychological reality of Neo-Riemannian transformations. J. Music Theor. **42**(2), 265–281 (1998)
24. Krumhansl, C.L.: The geometry of musical structure. Theoretical ACM computers in entertainment **3**(4) (2005) doi: 10.1145/1095534.1095542
25. Lerdahl, F.: Tonal Pitch Space. Oxford University Press, New York (2001)
26. Lewin, D.: A formal theory of generalized tonal functions. J. Music Theor. **26**(1), 32–60 (1982)
27. Lewin, D.: Generalized Musical Intervals and Transformations. Yale University Press, New York (1987)
28. Lewin, D.: Cohn functions. J. Music Theor. **40**(2), 181–216 (1996)
29. Longuet-Higgins, H.C.: Letter to a musical friend. Music Rev **23**, 244–248 (1962)
30. Longuet-Higgins, H.C.: Second letter to a musical friend. Music Rev **23**, 271–280 (1962)
31. Longuet-Higgins, H.C., Steedman, M.J.: On interpreting Bach. In: Meltzer, B., Michie, D. (eds.) Machine Intelligence, vol. 6, pp. 221–241. Edinburgh University Press, Edinburgh (1971)
32. Shepard, R.N.: The analysis of proximities: multidimensional scaling with an unknown distance function (Part I). Psychometrika **27**(2), 125–140 (1962)
33. Shepard, R.N.: The analysis of proximities: multidimensional scaling with an unknown distance function (Part II). Psychometrika **27**(3), 219–246 (1962)
34. Shepard, R.N.: Structural representations of musical pitch. In: Deutsch, D. (ed.) The Psychology of Music, pp. 335–353. Academic Press, New York (1982)
35. Tanguiane, A.S.: Artificial Perception and Music Recognition. Lecture Notes in Artificial Intelligence, vol. 746. Springer, New York (1993)
36. Zatorre, R.J., Krumhansl, C.L.: Mental models and musical minds. Science **298**(5601), 2138–2139 (2002)

Part II
The Model

Chapter 3
The Spiral Array

Abstract This chapter presents the musical and geometric reasoning behind, and mathematical formulation of, the Spiral Array model, showing how the model successively generates representations for higher level tonal elements as a composite of each entity's lower level components. The concept of the *center of effect* is defined, wherein any element in the space can generate a higher level construct, modeled, in the same space, as the centroid, a mathematical sum, of its lower level members. For example, chord representations are generated from their component pitches as the center of effect of their member pitches, and key representations are similarly derived from their defining chords. Intuitive images illustrate the construction of the model, beginning from the "rolling up" of the Harmonic Network, and through the stages of defining higher level entities in the interior of the model. Theorems describe mathematical properties of the different levels of representations. The chapter concludes with a summary of the definitions and a visual depiction of the resulting array of spirals: nested helices comprising the outermost pitch class spiral, the major/minor chord double helix inside the pitch class spiral, and the major/minor key double helix inside the major/minor chord double helix.

This chapter presents the mathematical formulation of the components of the Spiral Array model, giving the musical and geometric reasoning behind them. The Spiral Array models the relations embodied in tonality. Each pitch (in this case, pitch class, since I assume octave equivalence[1]), chord, and key have a spatial counterpart in this geometric framework.

The Spiral Array is a *generative* model. In the model, key representations are generated from their component chords, and chord representations from their component pitches. These representations are distinct from the root of the chord and the

---

This chapter is a revision of The Spiral Array Model (Chapter 3) of "Towards a Mathematical Modeling of Tonality" by Elaine Chew, an MIT PhD dissertation, Cambridge, Massachusetts (2000) https://dspace.mit.edu/handle/1721.1/9139

[1] *Octave equivalence* means that pitches of the same letter name are considered to be the same, to belong to the same pitch class.

E. Chew, *Mathematical and Computational Modeling of Tonality*,
International Series in Operations Research & Management Science 204,
DOI: 10.1007/978-1-4614-9475-1_3, © Springer Science+Business Media New York 2014

tonal center (tonic) of the key, entities that give chords and keys their names. These chord and key representations attempt to capture spatially the idea that the higher level representations generated are a composite of their lower level elements. Distinct from other models, these different hierarchical elements can reside within the same framework in the same space.

Here, in this chapter, I introduce the Spiral Array model and show how the model successively generates representations for higher level tonal elements as a composite of the entities' lower level components. For example, chord representations are the result of their component pitch positions, and key representations are derived from their defining chords. Later, in Appendix A, I show how the model can be adjusted so that distances between the tonal representations match the cognition of their closeness. Thereafter, I show how the Spiral Array can be used as a framework on which to base the design of algorithms for tonal analysis, such as finding keys, naming chords, and determining modulations.

3.1 The Pitch Class Spiral

The starting point of the Spiral Array is a three-dimensional realization of the Harmonic Network.The Harmonic Network is an array of pitches separated by perfect fifth intervals along the x-axis, and by major third intervals along the y-axis, as shown in Fig. 3.1. This table of pitch names is the same as that shown in Fig. 2.7, and equivalent to Fig. 2.6 under an affine transformation, in this case, a simple shear. The pitch names repeat, resulting in periodicity in the network, and redundant representations. Inherent in the Harmonic Network is a spiral model without redundancies.

3.1.1 Description 1: A Derivation of the Harmonic Network

Take the planar Harmonic Network and roll it up as shown in Fig. 3.1 so that the A lines up with the A four columns away, and B♭ coincides with the B♭ four columns away. The result of this transformation is the pitch class spiral of the Spiral Array,

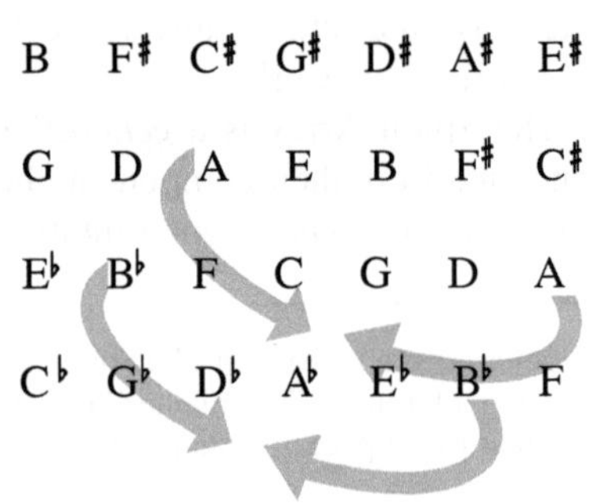

Fig. 3.1 "Rolling up" the harmonic network to form the Spiral Array

in which pitch name redundancies due to the periodicity in the Harmonic Network have been eliminated.

Alternatively, one could also generate this structure by wrapping a line of pitches ascending by perfect fifth intervals on a vertical cylinder so that pitches a major third apart line up above one another four steps later.

3.1.2 Description 2: A Closer Look

In the Spiral Array's pitch class spiral, pitch classes are represented by points on a spiral. Adjacent pitch class representations are positioned at each quarter turn of the spiral as shown in Fig. 3.2. Following along the ascending spiral, neighboring pitches are a perfect fifth apart. Along the vertical axis, neighboring pitches are related by major third intervals. Pitches separated by octaves are assumed to be equivalent and share the same position.

In Fig. 3.2, follow the winding spiral from the lowest pitch shown, B♭, up a quarter turn to the next pitch. This pitch is a perfect fifth above the previous one, that is, it is the pitch F. Continuing along the spiral, one comes to C (a perfect fifth above F) and G (a perfect fifth above C) then D (a perfect fifth above G). By the pitch D, one has advanced four quarter turns up the spiral after B♭, making one full turn. As a result, D lines up vertically above the starting point B♭. D and B♭ are related by an interval of a major third. Similarly, starting from F, four steps up along the spiral brings one to A, which is a major third above F. Continuing along, one can see that

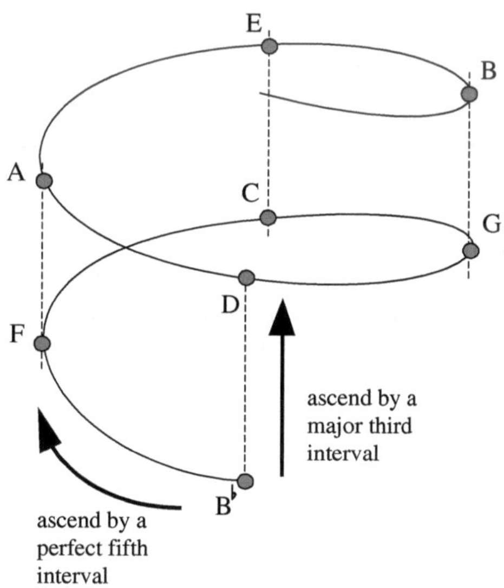

Fig. 3.2 Pitch class spiral of the Spiral Array

pitches vertically above each other are always a major third apart, shown in Fig. 3.2 by dotted lines.

3.1.3 Discussion 1: How do the Intervals Add Up?

Two pitches related by an interval of a perfect fifth are four steps apart along the major (or minor) scale starting on the lower pitch. However, if one considers all the pitches that lie between the pitches defining a perfect fifth interval, the perfect fifth spans seven half steps (or semitones). Note that this definition of a perfect fifth interval simply gives a method of identifying the interval's span, and is not limited to only the first and fifth degrees of a scale. In fact, in a major scale, the scale degree pairs ($\hat{1}$, $\hat{5}$), ($\hat{2}$, $\hat{6}$) and ($\hat{3}$, $\hat{7}$) are but some of the perfect fifth intervals among its member pitches.

Two pitches related by an interval of a major third are separated by two scale steps along the major scale starting on the lower pitch, which is the same as four half steps. Four perfect fifths stacked on top of one another make a total of twenty-eight half steps. Since the model assumes octave equivalence, one can subtract octaves (octave intervals are defined by pitches twelve steps apart) from the twenty-eight half steps. This is equivalent to performing modulo twelve arithmetic.

Hence, the twenty-eight half steps are equivalent to four half steps, which is the distance between pitches defining a major third interval.

$$\begin{aligned} \text{4 perfect fifths} &= 4 \times 7 \text{ half steps} \\ &= 28 \bmod 12 \text{ half steps} \\ &= 4 \text{ half steps} \;=\; 1 \text{ major third} \end{aligned}$$

Similarly, a minor third can be represented by three perfect fifths in the opposite direction, that is to say, three pitch-class representations down along the spiral. Two pitches related by an interval of a minor third are separated by two steps along the minor scale starting on the lower pitch, which is the same as three half steps. Three perfect fifths piled up beneath one another, which is equivalent to traveling down the spiral in the opposite direction, make a total of twenty-one half steps in decreasing frequencies. Invoking octave equivalence (modulo twelve arithmetic), twenty-one half steps equate to three half steps, which is the distance between pitches defining a minor third interval.

$$\begin{aligned} -3 \text{ perfect fifths} &= -3 \times 7 \text{ half steps} \\ &= -21 \bmod 12 \text{ half steps} \\ &= 3 \text{ half steps} \;=\; 1 \text{ minor third} \end{aligned}$$

3.1.4 Discussion 2: The Inherent Toroid Structure

If one considers equal temperament, the spiral would close to wind itself around a toroid. I choose to keep the Spiral Array in its cylindrical form for both pragmatic and music conceptual reasons.

In equal temperament, enharmonic pitches sound the same, that is to say, A♯ sounds exactly the same as B♭ as is the case on a piano keyboard, where both A♯ and B♭ are sounded by hammers hitting the same strings. If the spatial representation of the A♯ pitch class were identical to that of the B♭ pitch class, then the spiral would fold back on itself to form a torus. I argue that the context in which a pitch occurs determines its spelling, with the corresponding tonal (such as voice leading) implications; similarly, the spelling of a pitch, regardless of whether A♯ is acoustically equivalent to B♭ or not, shapes the tonal context by implying certain directional tendencies, for example B♭ leans towards A, and A♯ is more likely to resolve to B. Thus, it is important to keep the spiral in the cylindrical configuration, hence distinguishing between A♯ and B♭, to preserve this tonally critical information.

The cylindrical orientation of the Harmonic Network has the added advantage of simplicity over the torus. Distance in a cylinder, which is more akin to our experience in the physical world, is more readily understood than that inside a torus, especially where distances between adjacent pitches must be invariant and independent of position on the torus.

3.1.5 Mathematical Representation

Pitches are indexed by their number of perfect fifths from C. (Here, I shall use the terms pitches and pitch classes interchangeably.) For example, D has index two because C to G is a perfect fifth, and G to D is another perfect fifth. **P**(k) denotes the point on the spiral representing a pitch of index k. Since the spiral makes one full turn every four pitches, each pitch can be defined in terms of transformations from its previous neighbor—a rotation and a translation as shown in Fig. 3.3.

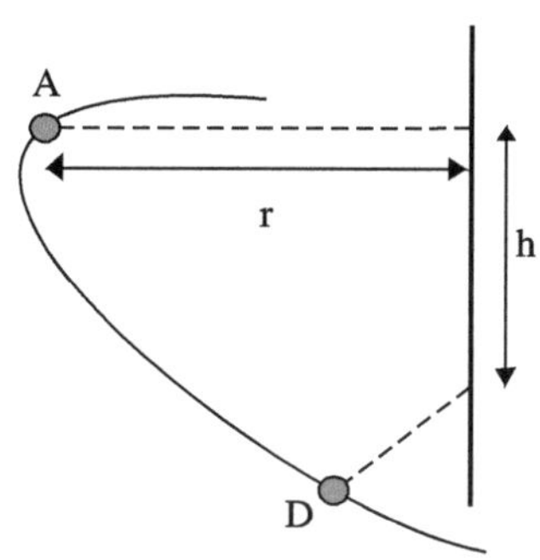

Fig. 3.3 The two parameters that uniquely identify pitch position: the radius (r), and vertical step (h)

Note that C has been arbitrarily chosen as the reference pitch. Any other pitch could have served as the reference, **P**(0). Furthermore, the following mathematical definition has arbitrarily fixed the pitch C at the point [0,1,0]. This, too, is an arbitrary choice. It is the relation between the pitch representations that is of utmost importance; where the spiral begins is of little consequence.

Definition 1 *Each pitch position can be described as a rotation of the previous one by 90° clockwise on the horizontal plane, and an elevation by h units in the vertical direction:*

$$\mathbf{P}(\mathrm{k}+1) \stackrel{\text{def}}{=} \mathbf{R}\cdot\mathbf{P}(\mathrm{k})+\mathbf{h}, \text{ where } \mathbf{R} = \begin{bmatrix} 0 & 1 & 0 \\ -1 & 0 & 0 \\ 0 & 0 & 1 \end{bmatrix} \text{ and } \mathbf{h} = \begin{bmatrix} 0 \\ 0 \\ \mathrm{h} \end{bmatrix}. \quad (3.1)$$

Each pitch position **P**(k) can also be parametrically defined as a function of the radius of the cylinder encasing the pitch class spiral, and the elevation at each step (see Fig. 3.3). Appendix A discusses how these parameters can be selected to reflect the cognitive interval relations.

Definition 2 *Two parameters, the radius of the cylinder, r, and the height gain per quarter rotation, h, uniquely define the position of a pitch representation, which can be described as*

$$\mathbf{P}(\mathrm{k}) \stackrel{\text{def}}{=} \begin{bmatrix} x_\mathrm{k} \\ y_\mathrm{k} \\ z_\mathrm{k} \end{bmatrix} = \begin{bmatrix} \mathrm{r}\sin\frac{\mathrm{k}\pi}{2} \\ \mathrm{r}\cos\frac{\mathrm{k}\pi}{2} \\ \mathrm{kh} \end{bmatrix}. \quad (3.2)$$

Since the spiral makes one full turn every four pitches to line up vertically above the starting pitch position, positions representing pitches four indices, or a major third, apart are related by a simple vertical translation. For example, C and E are a major third apart, and E is vertically above C.

Property 1 *Positions representing pitches four indices, or a major third, apart are related by a vertical translation of four elevation steps:*

$$\mathbf{P}(\mathrm{k}+4) = \mathbf{P}(\mathrm{k}) + 4\cdot\mathbf{h}. \quad (3.3)$$

3.2 Intervals in the Spiral Array

The Spiral Array is, in essence, generated from interval relations. Many of these interval relations are inherited from the Harmonic Network, and a few are a result of the new configuration. In this section, I will discuss some of the consequences of the Spiral Array's helical pitch class arrangement.

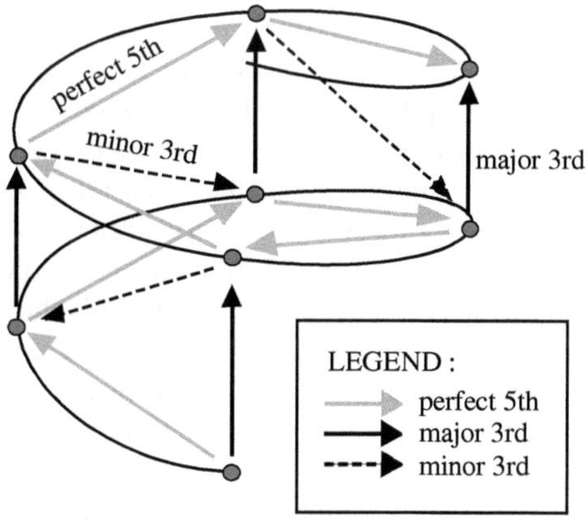

Fig. 3.4 Perfect fifth, major third, and minor third interval representations in the Spiral Array

3.2.1 Relations Inherited from the Harmonic Network

The model inherits the major/minor third and perfect fifth interval relationships of the Harmonic Network. In the Spiral Array, the interval representations undergo a new twist as shown in Fig. 3.4.

For any given pitch representation, its left neighbor (looking on from the outside) is the pitch a perfect fifth above and its top neighbor a major third above. Because the major third interval is equivalent to four perfect fifth intervals, starting from a given pitch, the pitch a major third above is four steps up along the pitch class spiral. In addition, diagonals represent minor third relations. This spiral lattice of adjacent perfect fifths and vertical major thirds affects other interval relations as well. For example, because of the placement of pitches, a major second interval is now represented by pitch positions diametrically opposite each other on the Spiral Array.

3.2.2 Interval Distances

In the Spiral Array, interval relations correspond to Euclidean distances between pitch representations. Each pair of pitch positions is separated by some spatial distance. And since pitch pairs define intervals, these spatial distances represent, spatially, different interval distances. One of the design strategies employed in the Spiral Array is to select parameters so that interval distances in the model correspond to interval distances interpreted by the mind.

Note that distance in the model does not correspond to distance on a keyboard, or any other musical instrument for that matter. The closeness between pitches corresponds to interpreted proximity between pitches in tonal music. Even though two pitches related by a half step interval are close to each other on a keyboard, they are quite distant in the Spiral Array. And, even though on a keyboard an octave (a perfect eighth) interval spans a greater distance than a major third interval, in the Spiral Array two pitches separated by an octave are closer (in fact, having zero distance) than any other two pitches, such as two separated by a major third interval.

If one imagines that the pitch class spiral is a spring coil, then the distances representing the various interval relationships change depending on how much the spiral is stretched out. The choice of the aspect ratio h/r consequently has implications for all interval representations, not only the perfect fifth and major third. The selection of this parameter will be discussed in greater detail in Appendix A.

In the planar arrangement, intervals could be represented by vectors. In the Spiral Array, intervals could also be represented as vectors, with a little extra work: vectors identical under rotation about the central axis and vertical translation would be deemed to belong to the same equivalence class.

3.3 Representing Chords

This section introduces the modeling of a *center of effect* for higher level tonal constructs in the Spiral Array. Chords and keys are represented by points in space corresponding to centroids of their lower level components. Each centroid is computed as the mathematical sum of its weighted components.

A chord is the composite result, or effect, of its component pitches. A key is the effect of its defining chords. I propose that this effect can be represented spatially. The spatial representation, as well as the effect, cannot be identified only with one of the component elements, but should be a function of their interaction. For example, the C major triad[2] should not be represented by the pitch C alone, but by a combination of its three component pitches, C, E and G.

3.3.1 Description: Center of Triangle

As in the Harmonic Network, triadic chords form triangles. In the Spiral Array geometry, I propose that each chord be represented spatially as a point in the interior of this triangle. Because chords are represented by nonoverlapping triangles in the model, any point inside a triangle is sufficient to uniquely identify the designated chord. This point is not the root, the fifth, or the third, but a convex combination of the three. It represents the center of effect of the chord pitches. See Fig. 3.5 for some examples of chord representations.

---

[2] A *triad* is a chord comprising three pitches: a root, a third, and a fifth. The root and fifth form a perfect fifth. The root and third form a major or minor third for a major or minor triad, respectively. In general, the words chord and triad will be used interchangeably.

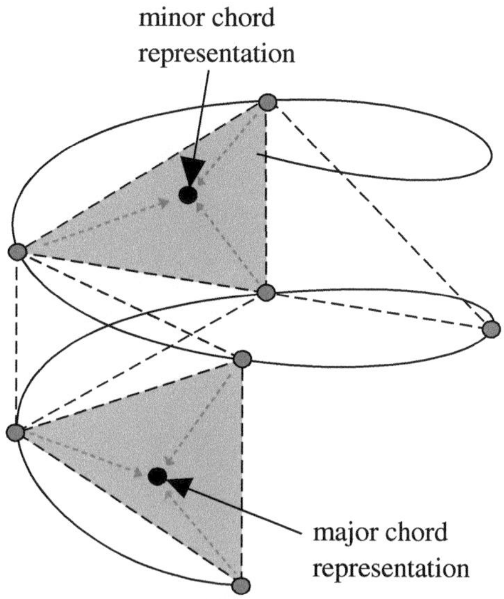

Fig. 3.5 Examples of chord representations: each chord representation is the composite result of its constituent pitches

The three component pitches of the chord form the vertices of the triangle, and the sides of the triangle represent a perfect fifth interval, a major third interval, and a minor third interval. Major and minor triads differ only in their orientation. In the Spiral Array, a major triad can be "flipped" to become a minor triad; in other words, the reflection of a major triad with respect to any one of its sides is a minor triad.

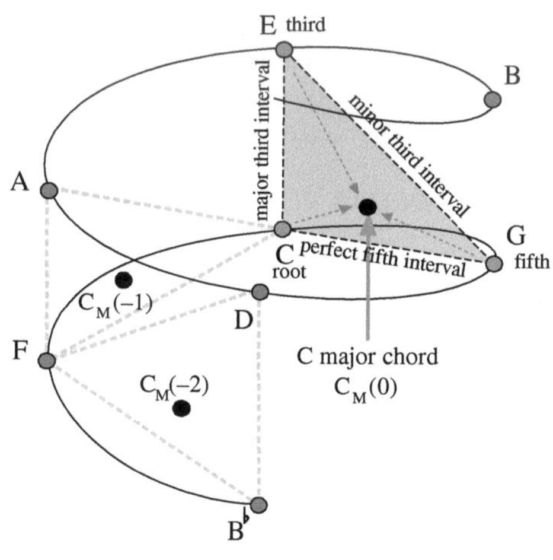

Fig. 3.6 Examples of major chord representations

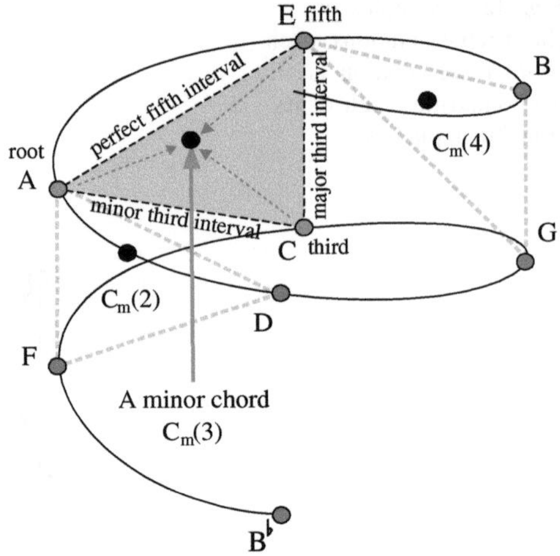

Fig. 3.7 Examples of minor chord representations

3.3.2 Example 1: The C Major Chord Representation

As shown in Fig. 3.6, the C major triad consists of C, E, and G. The root, C, is denoted by $\mathbf{P}(0)$; the fifth, G, is a perfect fifth above C, $\mathbf{P}(1)$; and, the third, E, is a major third, or four perfect fifths, above C, $\mathbf{P}(4)$. The root of the chord, C, is connected to the sides representing a perfect fifth interval and a major third interval. The triad is represented by a point inside the triangle, $\mathbf{C}_{\mathrm{M}}(0)$. The index of the chord corresponds to the index of its root, 0.

Similar representations can be generated for all major triads. These representations also sit on a spiral. This major chord spiral is a little tighter, having a slightly smaller radius, than the outermost pitch spiral, and resides completely inside the outer spiral.

3.3.3 Example 2: The A Minor Chord Representation

As shown in Fig. 3.7, the A minor triad consists of A, C, and E. The root, A, is denoted by $\mathbf{P}(3)$; the fifth, E, is a perfect fifth above A, $\mathbf{P}(4)$; and, the third, C, is a minor third, or three perfect fifths, below A, $\mathbf{P}(3-3) = \mathbf{P}(0)$. The root of the chord, A, is connected to the sides representing a perfect fifth interval and a minor third interval. The triad is represented by a point on the triangle, $\mathbf{C}_{\mathrm{m}}(3)$. The index of the chord corresponds to the index of its root, 3.

Similar chord representations can be generated for all minor triads. Again, these representations sit on a spiral that is a little tighter than the outermost pitch spiral, and resides completely inside that outer spiral.

3.3.4 Mathematical Representation

The chord's representation is generated by a linear combination of its three component pitch positions. The combination is restricted to be convex. Mathematically, this means that the weights on the pitch positions are nonnegative and sum to one. Geometrically, this means that the chord representation resides strictly within the boundaries of the triangle outlined by the triad, as seen in Fig. 3.5. Hence, the chord is represented by a weighted average of the pitch positions: the root, **P**(k), the fifth, **P**(k+1), and the third: **P**(k+ 4), for major triads, or **P**(k−3) for minor triads.

Definition 3 *The representation for a major triad is generated by the convex combination of its root, fifth, and third pitch positions:*

$$\mathbf{C}_{\mathrm{M}}(\mathrm{k}) \stackrel{\mathrm{def}}{=} w_1 \cdot \mathbf{P}(\mathrm{k}) + w_2 \cdot \mathbf{P}(\mathrm{k}+1) + w_3 \cdot \mathbf{P}(\mathrm{k}+4), \tag{3.4}$$

$$\text{where} \qquad w_1 \geq w_2 \geq w_3 > 0 \text{ and } \sum_{i=1}^{3} w_i = 1.$$

The minor triad is generated by a similar combination:

$$\mathbf{C}_{\mathrm{m}}(\mathrm{k}) \stackrel{\mathrm{def}}{=} u_1 \cdot \mathbf{P}(\mathrm{k}) + u_2 \cdot \mathbf{P}(\mathrm{k}+1) + u_3 \cdot \mathbf{P}(\mathrm{k}-3), \tag{3.5}$$

$$\text{where} \qquad u_1 \geq u_2 \geq u_3 > 0 \text{ and } \sum_{i=1}^{3} u_i = 1.$$

The weights, w_i and u_i, on the pitch positions represent the importance of the pitch to the generated chord. For longstanding psychological, physical, and theoretical reasons that are better explained elsewhere, for example in [1, 4], the root is deemed the most important, followed by the fifth, then the third. As such, I limit the weights to be monotonically decreasing from the root, to the fifth, to the third. In order that spatial distance mirrors these relations, there are additional constraints on the aspect ratio, h/r, which will be discussed in Appendix A.2.

3.3.5 Spatial Geometry

As mentioned, the set of points that represents the center of effect for major triads also sits on a spiral, one with a slightly smaller radius than the one for pitches. The chord representations inherit the property of making one full turn in four steps, so that the fourth step lines the current chord vertically above the starting one. Hence, the roots of neighboring major triads are a perfect fifth apart, and the roots of vertical major triad neighbors are a major third apart. For example, for the C major triad, its nearest neighbor a quarter turn up the spiral is also a major triad, the G major triad, and its nearest vertically adjacent major triad neighbor above is the E major triad.

Theorem 1 *The representation for each major chord can be described as a 90° clockwise rotation of the previous major chord followed by a vertical elevation of h units:*

$$\mathbf{C_M}(\mathrm{k}+1) = \mathbf{R} \cdot \mathbf{C_M}(\mathrm{k}) + \mathbf{h}, \text{ where } \mathbf{R} = \begin{bmatrix} 0 & 1 & 0 \\ -1 & 0 & 0 \\ 0 & 0 & 1 \end{bmatrix} \text{ and } \mathbf{h} = \begin{bmatrix} 0 \\ 0 \\ \mathrm{h} \end{bmatrix}. \quad (3.6)$$

The chord representations whose roots form a major third interval are related by a vertical translation of four elevation steps:

$$\mathbf{C_M}(\mathrm{k}+4) = \mathbf{C_M}(\mathrm{k}) + 4 \cdot \mathbf{h}. \quad (3.7)$$

[Proof]: Using definitions for chords as given in Eq. 3.4 and for pitch positions as given in Eq. 3.1, we get

$$\begin{aligned} \mathbf{C_M}(\mathrm{k}+1) &= w_1 \cdot \mathbf{P}(\mathrm{k}+1) + w_2 \cdot \mathbf{P}(\mathrm{k}+2) + w_3 \cdot \mathbf{P}(\mathrm{k}+5) \\ &= w_1 \cdot (\mathbf{RP}(\mathrm{k}) + \mathbf{h}) + w_2 \cdot (\mathbf{RP}(\mathrm{k}+1) + \mathbf{h}) + w_3 \cdot (\mathbf{RP}(\mathrm{k}+4) + \mathbf{h}) \\ &= \mathbf{R} \cdot [w_1 \cdot \mathbf{P}(\mathrm{k}) + w_2 \cdot \mathbf{P}(\mathrm{k}+1) + w_3 \cdot \mathbf{P}(\mathrm{k}+4)] \\ &\quad + (w_1 + w_2 + w_3) \cdot \mathbf{h} \\ &= \mathbf{R} \cdot \mathbf{C_M}(\mathrm{k}) + \mathbf{h}. \end{aligned}$$

The second part of the theorem follows directly from the first. Each step adds a quarter rotation and elevation h, thus, four steps later, the chord is now 4h above where it began. □

Theorem 2 *Similarly, the representation for each minor chord can be described as a 90° clockwise rotation of the previous minor chord followed by a vertical elevation of h units:*

$$\mathbf{C_m}(\mathrm{k}+1) = \mathbf{R} \cdot \mathbf{C_m}(\mathrm{k}) + \mathbf{h}, \text{ where } \mathbf{R} = \begin{bmatrix} 0 & 1 & 0 \\ -1 & 0 & 0 \\ 0 & 0 & 1 \end{bmatrix} \text{ and } \mathbf{h} = \begin{bmatrix} 0 \\ 0 \\ h \end{bmatrix}. \quad (3.8)$$

The representations for chords whose roots form a major third interval are related by a vertical translation of four elevation steps:

$$\mathbf{C_m}(\mathrm{k}+4) = \mathbf{C_m}(\mathrm{k}) + 4 \cdot \mathbf{h}. \quad (3.9)$$

[Proof]: Similar to that of Theorem 1. □

Together, the major and minor chord spirals form a double helix inside the pitch class spiral.

As in the neo-Riemannian transformations described in Sect. 2.2.2, the minor (major) triads neighboring a major (minor) chord each differ from the major (minor) chord by one pitch. Each major triad is flanked by its relative, parallel, and mediant minor triad. Similarly, the major triads neighboring a minor chord are known as the relative, parallel, and submediant major triads.

3.4 Key Representations

An important property of the Spiral Array is that representations of pitches in a given key still occupy a compact neighborhood. In contrast to the Harmonic Network, where periodicities generate an infinite pattern of possible key regions, only one such neighborhood corresponds to the set of (named) pitches in the key. The added dimension in the Spiral Array allows for the spatial representation of a center of effect of the key that is confined within this neighborhood, and that is distinct from any of the member pitches and chords of the key.

3.4.1 The Major Key Representation

Each major chord, together with its right and left neighbor major chords, combine to produce the effect of a major key.[3] See Fig. 3.8 for an example of a major key representation. In music terminology, the chords that define the key are given names, with respect to the key that reflects their function. The center chord is called the

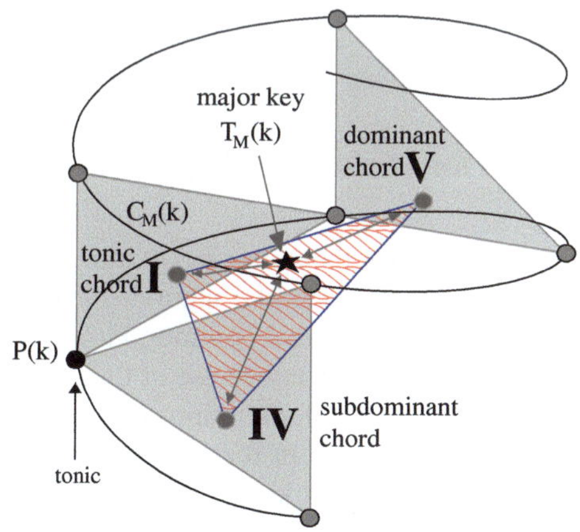

Fig. 3.8 Geometric representation of a major key, the composite effect of its I, V, and IV chords

[3] For a deeper discussion on the connection between chords and keys, see for example [2, 3].

tonic chord (I),[4] the one a quarter turn up the spiral (to the right of the I chord in Fig. 3.8) the dominant (V), and the one a quarter turn down the spiral (to the left of the I chord in Fig. 3.8) the subdominant (IV). I define the representation of the major key as a combination of its I, V, and IV chord representations. For example, the representation of the C major key is generated by the C major, G major, and F major chord representations.

This definition of the center of effect for a key recognizes the bottom-up and top-down nature of the relationship between chords and keys. Chords, depending on their relationships, can generate a key. At the same time, a key gives rise to certain chord relationships with specific functions within the key.

3.4.2 Mathematical Representation

Mathematically, the representation for a major key, $\mathbf{T_M}(k)$, is the weighted average of its tonic triad, $\mathbf{C_M}(k)$, dominant triad, $\mathbf{C_M}(k+1)$, and subdominant triad, $\mathbf{C_M}(k-1)$, representations. As before, the design objective is to have the weights correspond to each chord's significance in the key. Hence, the I chord is given the largest weight, followed by that of the V chord, then the IV chord.

Definition 4 *A major key is represented by a convex combination of its tonic, dominant, and subdominant chords. The weights are restricted to be monotonic and nonincreasing:*

$$\mathbf{T_M}(k) \stackrel{\text{def}}{=} \omega_1 \cdot \mathbf{C_M}(k) + \omega_2 \cdot \mathbf{C_M}(k+1) + \omega_3 \cdot \mathbf{C}_M(k-1), \tag{3.10}$$

$$\text{where} \quad \omega_1 \geq \omega_2 \geq \omega_3 > 0 \text{ and } \sum_{i=1}^{3} \omega_i = 1.$$

3.4.3 The Minor Key Representation

The definition for the minor key is more complicated. This is because there are three types of minor modes: the natural minor, harmonic minor, and melodic minor. In each case, the tonic triad is minor. However, the dominant triad is minor in the case of the natural, major in the case of the harmonic, and could be either major or minor in the case of the melodic minor mode. The subdominant triad is minor in the case of the natural, minor in the case of the harmonic, and major or minor in the case of the melodic minor key. Hence I propose that the center of effect for the minor

[4] I shall use roman numerals to denote chord function within a key. The number indicates the scale degree of the chord's root. For example, I represents the tonic chord. I adopt the convention of denoting major chords by upper case roman numerals, and minor chords by lower case ones. For example, a major chord with the tonic as root is I but a minor chord with the same root is i.

Fig. 3.9 Geometric representation of a minor key, a composite of its tonic (i), dominant (V/v), and subdominant (iv/IV) chords

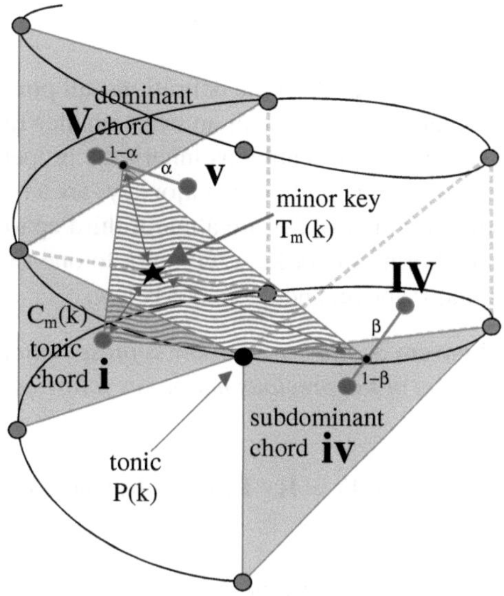

key, $\mathbf{T}_m(k)$, be modeled as a combination of the tonic, $\mathbf{C}_m(k)$, the possible dominant triads, $\mathbf{C}_M(k+1)$ and $\mathbf{C}_m(k+1)$, and the possible subdominant triads, $\mathbf{C}_m(k-1)$ and $\mathbf{C}_M(k-1)$. See Fig. 3.9 for the spatial representation of a minor key.

Definition 5 *The minor key representation is generated by a convex combination of its tonic, dominant (major and minor), and subdominant (minor and major) chords as follows:*

$$\begin{aligned} \mathbf{T}_m(k) &\overset{\text{def}}{=} \upsilon_1 \cdot \mathbf{C}_m(k) \\ &+ \upsilon_2 \cdot [\alpha \cdot \mathbf{C}_M(k+1) + (1-\alpha)\cdot \mathbf{C}_m(k+1)] \\ &+ \upsilon_3 \cdot [\beta \cdot \mathbf{C}_m(k-1) + (1-\beta)\cdot \mathbf{C}_M(k-1)], \qquad (3.11) \\ \text{where} \quad & \upsilon_1 \geq \upsilon_2 \geq \upsilon_3 > 0 \text{ and } \upsilon_1 + \upsilon_2 + \upsilon_3 = 1, \\ \text{and} \quad & 0 \leq \alpha \leq 1, 0 \leq \beta \leq 1. \end{aligned}$$

When the minor key representation is modeled using the tonic minor, the dominant major (and not minor) triad, and the subdominant minor (and not major) triad, $\alpha = 1$ and $\beta = 1$, and the combination of chords generates the effect of a harmonic minor. When one models the minor key representation with only minor triads, $\alpha = 0$ and $\beta = 1$, and the i, v, and iv chords generate the effect of a natural minor. The melodic minor has the composite effect of some combination of all chords represented in the definition.

3.4.4 Spatial Geometry

Again, as a result of the definitions, all points representing the major keys sit on yet another spiral, one with an even smaller radius than the one for major triads. The major key representations inherit the property of a full rotation in four steps so that the tonics of neighboring major keys are a perfect fifth apart, and the tonics of vertical neighbor major keys are a major third apart. For example, the C major key has, for neighboring major keys, G major to one side along the spiral, F major to the other, E major above, and A♭ major below.

Theorem 3 *Each major key representation can be described as a 90° clockwise rotation of the previous major key followed by a vertical elevation of h units:*

$$\mathbf{T}_{\mathrm{M}}(\mathrm{k}+1) = \mathbf{R} \cdot \mathbf{T}_{\mathrm{M}}(\mathrm{k}) + \mathbf{h}, \text{ where } \mathbf{R} = \begin{bmatrix} 0 & 1 & 0 \\ -1 & 0 & 0 \\ 0 & 0 & 1 \end{bmatrix} \text{ and } \mathbf{h} = \begin{bmatrix} 0 \\ 0 \\ \mathrm{h} \end{bmatrix}. \tag{3.12}$$

The representations of keys whose tonics form a major third interval are related by a vertical translation of four elevation steps:

$$\mathbf{T}_{\mathrm{M}}(\mathrm{k}+4) = \mathbf{T}_{\mathrm{M}}(\mathrm{k}) + 4 \cdot \mathbf{h}. \tag{3.13}$$

[Proof]: Using the definition for major keys as given in Eq. 3.10, and invoking Theorem 1,

$$\begin{aligned}
\mathbf{T}_{\mathrm{M}}(\mathrm{k}+1) &= \omega_1 \cdot \mathbf{C}_{\mathrm{M}}(\mathrm{k}+1) + \omega_2 \cdot \mathbf{C}_{\mathrm{M}}(\mathrm{k}+2) + \omega_3 \cdot \mathbf{C}_{\mathrm{M}}(\mathrm{k}) \\
&= \omega_1 \cdot (\mathbf{R} \cdot \mathbf{C}_{\mathrm{M}}(\mathrm{k}) + \mathbf{h}) + \omega_2 \cdot (\mathbf{R} \cdot \mathbf{C}_{\mathrm{M}}(\mathrm{k}+1) + \mathbf{h}) \\
&\quad + \omega_3 \cdot (\mathbf{R} \cdot \mathbf{C}_{\mathrm{M}}(\mathrm{k}-1) + \mathbf{h}) \\
&= \mathbf{R} \cdot [\omega_1 \cdot \mathbf{C}_{\mathrm{M}}(\mathrm{k}) + \omega_2 \cdot \mathbf{C}_{\mathrm{M}}(\mathrm{k}+1) + \omega_3 \cdot \mathbf{C}_{\mathrm{M}}(\mathrm{k}-1)] \\
&\quad + (\omega_1 + \omega_2 + \omega_3) \cdot \mathbf{h} \\
&= \mathbf{R}\mathbf{T}_{\mathrm{M}}(\mathrm{k}) + \mathbf{h}.
\end{aligned}$$

The second part of the theorem follows directly from the first. □

Theorem 4 *Similarly, each minor key can be described as a 90° clockwise rotation of the previous minor key followed by a vertical elevation of h units:*

$$\mathbf{T}_{\mathrm{m}}(\mathrm{k}+1) = \mathbf{R} \cdot \mathbf{T}_{\mathrm{m}}(\mathrm{k}) + \mathbf{h}, \text{ where } \mathbf{R} = \begin{bmatrix} 0 & 1 & 0 \\ -1 & 0 & 0 \\ 0 & 0 & 1 \end{bmatrix} \text{ and } \mathbf{h} = \begin{bmatrix} 0 \\ 0 \\ \mathrm{h} \end{bmatrix}. \tag{3.14}$$

The representations for chords whose roots form a major third interval are related by a vertical translation of four elevation steps:

$$\mathbf{T}_{\mathrm{m}}(\mathrm{k}+4) = \mathbf{T}_{\mathrm{m}}(\mathrm{k}) + 4 \cdot \mathbf{h}. \tag{3.15}$$

[Proof]: Using the definition for minor keys as given in Eq. 3.12, and invoking Theorems 1 and 2,

$$\begin{aligned}
\mathbf{T}_{\mathrm{m}}(\mathrm{k}+1) &= \upsilon_1 \cdot \mathbf{C}_{\mathrm{m}}(\mathrm{k}+1) \\
&\quad + \upsilon_2 \cdot [\alpha \cdot \mathbf{C}_{\mathrm{M}}(\mathrm{k}+2) + (1-\alpha) \cdot \mathbf{C}_{\mathrm{m}}(\mathrm{k}+2)] \\
&\quad + \upsilon_3 \cdot [\beta \cdot \mathbf{C}_{\mathrm{m}}(\mathrm{k}) + (1-\beta) \cdot \mathbf{C}_{\mathrm{M}}(\mathrm{k})] \\
&= \upsilon_1 \cdot (\mathbf{R} \cdot \mathbf{C}_{\mathrm{m}}(\mathrm{k}) + \mathbf{h}) \\
&\quad + \upsilon_2 \cdot [\alpha \cdot (\mathbf{R} \cdot \mathbf{C}_{\mathrm{M}}(\mathrm{k}+1) + \mathbf{h}) + (1-\alpha) \cdot (\mathbf{R} \cdot \mathbf{C}_{\mathrm{m}}(\mathrm{k}+1) + \mathbf{h})] \\
&\quad + \upsilon_3 \cdot [\beta \cdot (\mathbf{R} \cdot \mathbf{C}_{\mathrm{m}}(\mathrm{k}-1) + \mathbf{h}) + (1-\beta) \cdot (\mathbf{R} \cdot \mathbf{C}_{\mathrm{M}}(\mathrm{k}-1) + \mathbf{h})] \\
&= \mathbf{R} \cdot \{\upsilon_1 \cdot \mathbf{C}_{\mathrm{M}}(\mathrm{k}) \\
&\quad + \upsilon_2 \cdot [\alpha \cdot \mathbf{C}_{\mathrm{M}}(\mathrm{k}+1) + (1-\alpha) \cdot \mathbf{C}_{\mathrm{m}}(\mathrm{k}+1)] \\
&\quad + \upsilon_3 \cdot [\beta \cdot \mathbf{C}_{\mathrm{m}}(\mathrm{k}-1) + (1-\beta) \cdot \mathbf{C}_{\mathrm{M}}(\mathrm{k}-1)]\} + (\upsilon_1 + \upsilon_2 + \upsilon_3) \cdot \mathbf{h} \\
&= \mathbf{R} \cdot \mathbf{T}_{\mathrm{m}}(\mathrm{k}) + \mathbf{h}.
\end{aligned}$$

The second part of the theorem follows directly from the first. □

Together, the major key spiral and the minor key spiral form a double helix inside the major/minor chord double helix, which is, in turn, inside the pitch class spiral.

As with the chords, the minor keys neighboring a major key are its relative, parallel, and mediant minor keys; and the major keys neighboring a minor key are its relative, parallel, and submediant major keys.

3.5 An Array of Spirals: A Discussion

The Spiral Array presents a new way to represent and relate musical entities from different hierarchical levels. Its design proposes a new way to generate these entities in the same space. Pitch positions form the basis for both chord and, by inference, key representations. Tonal entities from different hierarchical levels are depicted by the same kind of elements—points in space. This egalitarian representation scheme offers opportunities for comparison and relation among the different elements. The user is no longer limited to comparing like with like—for example, pitches with pitches and keys with keys—and can concurrently consider multiple levels of representations to make cross-level comparisons—such as incorporating into the model distances between pitches and chords, pitches and keys, and chords and keys.

In the tabular arrangement of the Harmonic Network, intervals could be represented by vectors. New rotation-invariant vector representations and similarity metrics would need to be employed in the three-dimensional Spiral Array, as would be the case for a toroid configuration.

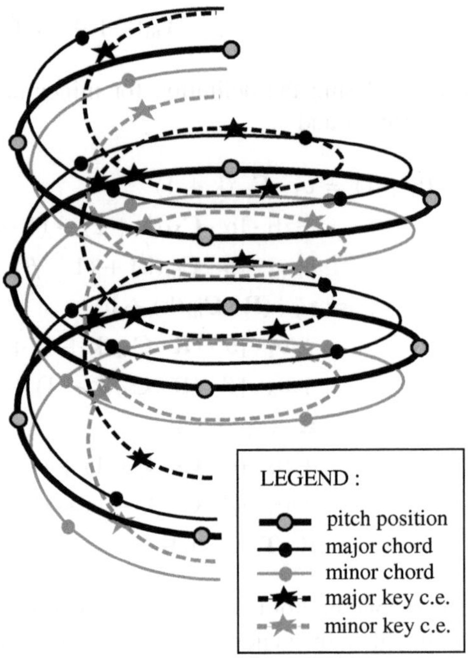

Fig. 3.10 Juxtaposing in the same space the pitch class, major and minor chord, and major and minor key spirals

A strength of the Spiral Array lies in its symmetry, regularity, and transformational invariance across all hierarchical levels. Figure 3.10 juxtaposes pitch, chord, and key representations in the same space. Pitches are represented by positions at every quarter turn of the pitch class spiral. As shown by Theorems 1 and 2, major and minor chords are each represented by positions marking each quarter turn of their respective spirals. As shown by Theorems 3 and 4, the same is true of major and minor key representations. All of these spirals ascend vertically by h units at every quarter turn.

The major/minor chord spirals form a double helix inside the pitch class spiral, and the major/minor key spirals form another double helix inside the major/minor chord double helix.

The name of the model acknowledges the fact that the Spiral Array stems from a spiral configuration of the Harmonic Network which comprises a lattice, an array, of pitch representations. Furthermore, the resulting set of nested helices are themselves an array of spirals. It has been pointed out to me that the coils that make up the Spiral Array model are in effect helices, and not spirals, as the mathematical definition of a spiral refers to a curve emanating from a central point. I use the term spiral the way it is used when referring to the curve of spiral staircases, where the radius stays constant.

3.6 Summary of Definitions

For ease of reference when perusing later chapters, this section provides a compilation of the definitions given in this chapter:

A **pitch** representation:

$$\mathbf{P}(\mathrm{k}) = \begin{bmatrix} x_{\mathrm{k}} \\ y_{\mathrm{k}} \\ z_{\mathrm{k}} \end{bmatrix} = \begin{bmatrix} \mathrm{r}\sin\frac{\mathrm{k}\pi}{2} \\ \mathrm{r}\cos\frac{\mathrm{k}\pi}{2} \\ \mathrm{kh} \end{bmatrix},$$

where k is the index of the pitch along a line of fifths. **P**(0) can be arbitrarily chosen; common choices are C or D.

A **major chord** representation:

$$\mathbf{C}_{\mathrm{M}}(\mathrm{k}) = w_1 \cdot \mathbf{P}(\mathrm{k}) + w_2 \cdot \mathbf{P}(\mathrm{k}+1) + w_3 \cdot \mathbf{P}(\mathrm{k}+4),$$
$$\text{where} \quad w_1 \geq w_2 \geq w_3 > 0 \text{ and } \sum_{i=1}^{3} w_i = 1.$$

k is the index of the root of the major chord, and w_1, w_2, w_3 correspond to the relative significance of the root, fifth, and third of the major triad.

A **minor chord** representation:

$$\mathbf{C}_{\mathrm{m}}(\mathrm{k}) = u_1 \cdot \mathbf{P}(\mathrm{k}) + u_2 \cdot \mathbf{P}(\mathrm{k}+1) + u_3 \cdot \mathbf{P}(\mathrm{k}-3),$$
$$\text{where} \quad u_1 \geq u_2 \geq u_3 > 0 \text{ and } \sum_{i=1}^{3} u_i = 1.$$

k is the index of the root of the minor chord, and u_1, u_2, u_3 correspond to the relative significance of the root, fifth, and third of the minor triad.

A **major key** representation:

$$\mathbf{T}_{\mathrm{M}}(\mathrm{k}) = \omega_1 \cdot \mathbf{C}_{\mathrm{M}}(\mathrm{k}) + \omega_2 \cdot \mathbf{C}_{\mathrm{M}}(\mathrm{k}+1) + \omega_3 \cdot \mathbf{C}_{M}(\mathrm{k}-1),$$
$$\text{where} \quad \omega_1 \geq \omega_2 \geq \omega_3 > 0 \text{ and } \sum_{i=1}^{3} \omega_i = 1.$$

k is the index of the tonic of the major key, and $\omega_1, \omega_2, \omega_3$ correspond to the relative significance of the tonic, dominant, and subdominant chords.

A **minor key** representation:

$$
\begin{aligned}
\mathbf{T}_{\mathrm{m}}(\mathrm{k}) = {} & \upsilon_1 \cdot \mathbf{C}_{\mathrm{m}}(\mathrm{k}) \\
& + \ \upsilon_2 \cdot [\alpha \cdot \mathbf{C}_{\mathrm{M}}(\mathrm{k}+1) + (1-\alpha) \cdot \mathbf{C}_{\mathrm{m}}(\mathrm{k}+1)] \\
& + \ \upsilon_3 \cdot [\beta \cdot \mathbf{C}_{\mathrm{m}}(\mathrm{k}-1) + (1-\beta) \cdot \mathbf{C}_{\mathrm{M}}(\mathrm{k}-1)], \\
\text{where} \quad & \upsilon_1 \geq \upsilon_2 \geq \upsilon_3 > 0 \text{ and } \quad \upsilon_1 + \upsilon_2 + \upsilon_3 = 1, \\
\text{and} \quad & 0 \leq \alpha \leq 1, 0 \leq \beta \leq 1.
\end{aligned}
$$

k is the index of the tonic of the major key, and $\upsilon_1, \upsilon_2, \upsilon_3$ correspond to the relative significance of the tonic, dominant, and subdominant chords, α is the relative weight accorded the major versus the minor dominant chord, and β the relative weight assigned the minor versus the major subdominant chord.

References

1. Bamberger, J.S.: Developing Musical Intuition. Oxford University Press, New York (2000)
2. Schoenberg, A.: Style and Idea. Philosophical Library, New York (1950)
3. Schoenberg, A.: Structural Functions of Harmony. Norton & Co., Inc., New York (1954)
4. Sessions, R.: Harmonic Practice. Harcourt, Brace and Company, New York (1951)

Part III
Key Finding

Chapter 4
The CEG Algorithm (Part I)

Abstract This chapter presents the Center of Effect Generator (CEG) key-finding algorithm based on the Spiral Array model, starting from the idea of the *center of effect* (CE), whereby any sequence of notes maps to a point in the interior of the model. The algorithm is illustrated through a simple melodic example, "Simple Gifts" from Copland's *Appalachian Spring*. Included are discussions on why the model works, and the principles behind the particular set of model parameters used in the test run. At each step, the algorithm reports the three closest keys, ranked by distance; a step-by-step geometric interpretation of the algorithm and a graph showing the evolution of the closest keys accompany the tabular results. A MATLAB version of the CEG code appears in Appendix B. An evaluation of the CEG algorithm follows in Chap. 5; the CEG method is applied to polyphonic music in MIDI (Musical Instrument Digital Interface) format in Chap. 9, and adapted to music audio in Chap. 11.

Since this tonality modeling enterprise was inspired by a question about key, as mentioned in Sect. 1.3, it is only fitting that the first application of the model describes an algorithm for key-finding. The derivation of key from pitch (and rhythm) information has been widely researched in the fields of music perception, artificial intelligence, and cognitive science. The background to some selected previous research in this area is covered in Sect. 4.1.

In this chapter I describe my proposed key-finding algorithm which I call the Center of Effect Generator method, and compare its performance to two other algorithms, namely, Longuet-Higgins and Steedman's Shape Matching Algorithm, and Krumhansl and Schmuckler's Probe Tone Profile Method. The CEG method's average performance surpasses that of the two previous models on the Bach *Well-Tempered Clavier Book I* fugue subjects test set, and its performance is close to optimal. I also briefly discuss the issue of model validation. The application of the

---

This chapter is a minor revision of the first part of Finding Keys (Chapter 5) of "Towards a Mathematical Modeling of Tonality" by Elaine Chew, an MIT PhD dissertation, Cambridge, Massachusetts (2000) https://dspace.mit.edu/handle/1721.1/9139

E. Chew, *Mathematical and Computational Modeling of Tonality*, International Series in Operations Research & Management Science 204,
DOI: 10.1007/978-1-4614-9475-1_4,

CEG method to denser musical textures—for example, musical passages with more than one melodic line—will be discussed in later chapters.

4.1 Introduction to the CEG Key-Finding Method

This section describes a key-finding algorithm called the Center of Effect Generator (CEG) based on the Spiral Array.

Analyzing the key of a melody poses many challenges. In many respects, it can be a far more difficult problem than attempting to discern the key of a large-scale orchestral piece, or even a piece of moderately dense texture, such as a two-hand piano piece. In the case of a single melodic line, one must make informed decisions about its key based on sparse information. Furthermore, there could be more than one equally valid answer, in which case a list of the most likely key candidates may be a more appropriate answer than one definite key. The CEG algorithm produces a ranked list of possible keys, and a distance (indicator of fit) for each of the keys.

CEG is not only an acronym for the Center of Effect Generator, the letters also correspond to the names of the pitches in the C major triad. CG forms a perfect fifth interval and CE forms a major third interval, important relationships in the Spiral Array model. Because the Spiral Array model uses distance as a measure of perceived closeness, the CEG algorithm uses the Spiral Array to reframe the problem of key recognition as a computationally simple one of finding a distance-minimizing key.

4.1.1 The Center of Effect

The algorithm uses pitch and duration information to generate a *center of effect*:(CE) This CE represents, spatially in the model, the tonal space generated by the pitches sounded. In the model, the CE is the mathematical sum of the pitches, a composite of their individual positions. In fact, the key representations in the Spiral Array are also CEs, the only difference being that they are generated from a specific combination of pitches.

The CE generated by the melody's pitches can then be compared to the framework of key representations in the Spiral Array. The quality of the melody's fit to a key is measured by proximity to the candidate key. All the candidate keys can be ranked by their distance to the CE, when accounting for tonal ambiguities.

4.1.2 Why the Algorithm Works

The collection of pitches in a given key defines a compact space in the Spiral Array. A musical passage in that key employs pitches from this collection. Thus, the pitches from the passage also occupy the same compact space in the Spiral Array. For a melody, as pitches are sounded, one gets a better idea of the pitch collection used by the melody, and by inference, its key. Analogously, in the model, as the music progresses, one gets a better idea of the tonal space occupied by the pitches.

As the number of pitches increases, the geometrical shape defined by the pitch positions, having them as its vertices, becomes more complex. Instead of using this complex shape to identify the key, the algorithm reduces the pitch information down to a single point, the center of effect. In this manner, the pitches combine to create an object in space—a point which is the composite sum of the pitch positions.

Since keys are also defined as points in space, it is then simple to compute the distance between the CE and the key, and nearby keys, to determine which key is closest to the CE. Thus the mathematical sum of pitches affords parsimonious descriptions of, and comparisons between, different pitch collections.

However, the CEG algorithm does more than simply compare pitch collections. By definition, the key representations favor triadic pitch configurations, and also tonic-dominant and tonic-subdominant relationships. Thus, not all pitches are weighted equally; and the key representation is a nonlinear combination of its pitch collection. This is a specific definition of key representations, one that incorporates different levels of hierarchical structure. The nonlinear combination of the pitch collection incorporates particular pitch-to-key and interval-to-key preferences described in Appendix A, which is devoted to discussions on the calibration of the model. Thus, by comparing the CEs to these key representations, by design, certain pitch relations will prevail.

The algorithm is best explained by an example. The following section applies the CEG algorithm to the melody "Simple Gifts."

4.2 Key-Finding Example Using "Simple Gifts"

An unfolding melody generates a key by the consecutive sounding of pitches in the key, producing interval relations that serve as cues to the tonal space of the melody. As more pitches are heard, the listener mentally (maybe unconsciously) forms an idea of the extent of this tonal space and creates a hierarchy based on pitch stability.

Correspondingly, in the Spiral Array, as more pitches are sounded and mapped onto their respective spatial representations, one gets a better definition of the analogous space occupied by the key in the model. This information can be summarized by a sum of the pitch positions. I have chosen to weight each pitch position by its duration since pitch duration is one of the factors that contribute to the cognition of pitch importance. Note that other plausible weighting schemes may incorporate metrical information (strength of beat), and in the case of polyphonic music, harmonically salient bass notes.

Fig. 4.1 "Simple Gifts"

4.2.1 The Melody

An example will illustrate how the CEG method spatially models the generating of a key by a melody. The melody used is the Shaker tune, as it appears in Copland's symphonic suite *Appalachian Spring* (1945) [2]. The original melody appears in several variations in the piece. It is notated in Fig. 4.1, for convenience, in the key of F major.

4.2.2 Model Parameters

Appendix A provides a progressive description of the model constraints and the parameter solutions satisfying a set of principles of tonal cognition. The model is run with the final set of parameters displayed in Table A.9 of the appendix: the aspect ratio is set to $h = \sqrt{2/15}$ ($r = 1$); pitch weights for major chords and the chord weights for the major and minor keys are [0.6025, 0.2930, 0.1145]; and the pitch weights for minor chords are [0.6011, 0.2121, 0.1868]. This particular assignment ensures that conditions B and C are met, and A is close to satisfied:

Condition A: The key nearest an isolated pitch, $\mathbf{P}(\mathrm{k})$, should be a major key of the same name, $\mathbf{T}_{\mathrm{M}}(\mathrm{k})$.

Condition B: A half step interval should be interpreted as the leading tone and tonic, i.e., scale degrees $\hat{7}$ and $\hat{1}$, of the major key.

Condition C: The lower and upper pitches of a perfect fourth interval should be interpreted as the dominant, $\hat{5}$, and tonic, $\hat{1}$, of the major key.

For the chosen weights, the key closest to an isolated pitch, $\mathbf{P}(\mathrm{k})$, is the minor key, $\mathbf{T}_{\mathrm{m}}(\mathrm{k})$, instead of the major key of the same name, which is also plausible. Additionally, the choice of weights result in a major chord being closest to its root, followed by its fifth, then its third (Theorem 7 of Appendix A); and, a minor chord being closest to its root, followed by its fifth, then its third (Theorem 9 of Appendix A).

4.2.3 The Center of Effect Generator

At any point in time, the CEG method can generate a CE that summarizes the tonal space generated by the pitches sounded thus far. Define a step to be a pitch event. At each step, each of the pitches from the beginning to the present is weighted (multiplied) by its proportional duration, and the center of effect is generated by the sum of these weighted pitch positions. Note that this definition is a general one. If desired, the "beat-in-bar" can also be incorporated into this summation by giving important beats greater weight.

However, in this instance of the algorithm, I use pitch and duration information to generate this center. And, a CE is generated at each pitch event. If the ith note is represented in the Spiral Array by pitch position p_i and has duration d_i, then the aggregate center at the ith pitch event pitch is defined as

$$c_i \stackrel{\text{def}}{=} \frac{1}{D_i}\sum_{j=1}^{i} d_j \cdot p_j, \qquad \text{where } D_i = \sum_{j=1}^{i} d_j.$$

The CEG method updates its decision with each note or pitch event. The distance from the key representations to c_i is calculated and ranked as

$$\{t^i_{(1)}, t^i_{(2)}, t^i_{(3)}, \ldots\},$$

where $t^i_{(1)}$ is the key closest to c_i, $t^i_{(2)}$ the key second closest to c_i, and so on.

4.2.4 Output Keys and Distances

The CEG algorithm is applied to "Simple Gifts". Table 4.2 documents the likely keys ranked by distance. Each row documents the results up to and including the pitch event at that time. In the "Durations" column, the letters represent note values as given in Table 4.1.

4.2.5 Step-by-Step Geometry

The First Few Steps

At the sounding of the first pitch, C, the first CE is generated: $c_1 = \mathbf{P}(0)$. As shown in Fig. 4.2, c_1 is close to F minor, C major, and F major, and almost equidistant from all three. More precisely, $t^1_{(1)}$ = F minor (0.5941), $t^1_{(2)}$ = C major (0.5998), and $t^1_{(3)}$ =

Table 4.1 Translation of symbols representing note values

Symbol	Derivation	Note value	Assigned weight
b	semi-breve	whole note	1.0
m	minim	half note	0.5
dc	dotted crotchet	dotted quarter note	0.375
c+s	crotchet+semi-quaver	quarter note + sixteenth	0.3125
c	crotchet	quarter note	0.25
dq	dotted-quaver	dotted eighth-note	0.1875
q	quaver	eighth-note	0.125
s	semi-quaver	sixteenth-note	0.0625

F major (0.6152), where the numbers in the brackets denote the distance of the first CE, c_1, from the respective key representations.

At the onset of the second pitch, F, which happens to be of duration equal to the first, the center now shifts to the midpoint of C and F, c_2. Observe in Fig. 4.2, that this new center is now closest to F major and F minor, almost equidistant from either one. This transition from pitch C to pitch F constitutes a rising perfect fourth interval. Section A.8 in Appendix A explains the implication of such a transition and gives two examples in Fig. A.14—one in a minor key, and the other in a major key. Thus the CEG algorithm's choices of F major and F minor agree with the tonal cognition implications. In this case, $t^2_{(1)}$ = F major (0.0742), $t^2_{(2)}$ = F minor (0.0742).

With the next pitch, F, the center inches closer toward the F major key, preferring this over the F minor. Now, $t^3_{(1)}$ = F major (0.0940), and $t^2_{(2)}$ = F minor (0.0982).

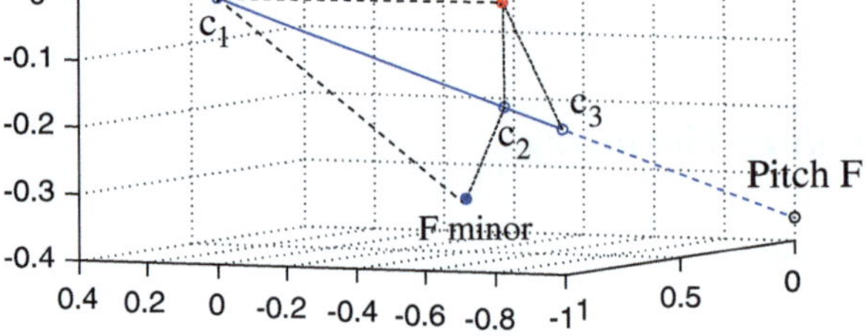

Fig. 4.2 Generating centers as "Simple Gifts" unfolds

Table 4.2 Key selection for "Simple Gifts" at each pitch event

Melody		Key choice		
Pitches	Durations	First	Second	Third
C	c	f (0.5941)[a]	C (0.5998)	F (0.6152)
F	c	F (0.0742)	f (0.0742)	B♭ (0.8424)
F	q	F (0.0940)	f (0.0982)	B♭ (0.7117)
G	q	F (0.0092)	f (0.0451)	c (0.5263)
A	q	F (0.0118)	f (0.1673)	C (0.6288)
F	q	F (0.0216)	f (0.1603)	B♭ (0.5830)
A	q	F (0.0669)	f (0.2872)	B♭ (0.6861)
B♭	q	F (0.0661)	f (0.2454)	B♭ (0.4499)
C	c	F (0.0102)	f (0.1561)	B♭ (0.5836)
C	q	F (0.0044)	f (0.1374)	c (0.5758)
B♭	q	F (0.0094)	f (0.1194)	B♭ (0.4857)
A	c	F (0.0422)	f (0.2476)	B♭ (0.5941)
G	q	F (0.0455)	f (0.2502)	C (0.5168)
F	q	F (0.0399)	f (0.2344)	B♭ (0.5483)
G	c	F (0.0672)	f (0.2617)	C (0.4024)
C	c	F (0.0508)	f (0.2257)	C (0.3541)
G	c	F (0.1008)	C (0.2521)	c (0.2690)
C	c	F (0.0935)	C (0.2319)	c (0.2491)
G	q	F (0.1201)	C (0.1971)	c (0.2129)
A	q	F (0.1156)	C (0.2137)	c (0.2526)
G	q	F (0.1411)	C (0.1834)	c (0.2202)
E	q	F (0.1597)	C (0.1645)	c (0.2292)
C	c	F (0.1536)	C (0.1571)	c (0.2191)
C	c	F (0.1521)	C (0.1545)	c (0.2140)
F	c	F (0.1076)	C (0.2062)	c (0.2519)
F	q	F (0.0898)	C (0.2327)	f (0.2592)
G	q	F (0.1068)	C (0.2058)	c (0.2436)
A	q	F (0.1049)	C (0.2185)	c (0.2723)
F	q	F (0.0892)	C (0.2434)	f (0.2731)
A	q	F (0.0903)	C (0.2570)	f (0.2910)
B♭	q	F (0.0847)	C (0.2655)	f (0.2760)
C	c	F (0.0828)	C (0.2547)	f (0.2645)
C	q	F (0.0827)	C (0.2505)	f (0.2599)
B♭	q	F (0.0780)	f (0.2472)	C (0.2586)
A	c	F (0.0799)	f (0.2784)	C (0.2808)
G	q	F (0.0915)	C (0.2573)	f (0.2899)
F	q	F (0.0807)	f (0.2757)	C (0.2779)

(continued)

Table 4.2 (Continued)

Melody		Key choice		
Pitches	Durations	First	Second	Third
G	c	F (0.1044)	C (0.2355)	c (0.2814)
C	c	F (0.1015)	C (0.2272)	c (0.2722)
A	c	F (0.1029)	C (0.2479)	f (0.3143)
G	c	F (0.1249)	C (0.2125)	c (0.2765)
F	c	F (0.1031)	C (0.2455)	c (0.3011)
F	c	F (0.0850)	C (0.2788)	f (0.2836)
F	m	F (0.0580)	f (0.2458)	C (0.3452)

[a]Numbers in brackets indicate the distance from the key to the generated center

The "Random Walk" Around *F* Major

Examining the contents of Table 4.2, one can see that the rising fourth between the first and second pitch events establishes the melody as being in F major. The generated centers continue to hover around F major, moving farthest away from it when the melody dwells in the dominant area. This happens around the GCGCGAGEC sequence in the third and fourth bars shown in Fig. 4.1.

Figure 4.3 shows a bird's-eye view of how the centers generated by the melody's notes, $\{c_i\}$, approach F major, then perform a "random walk" around it. The minimum

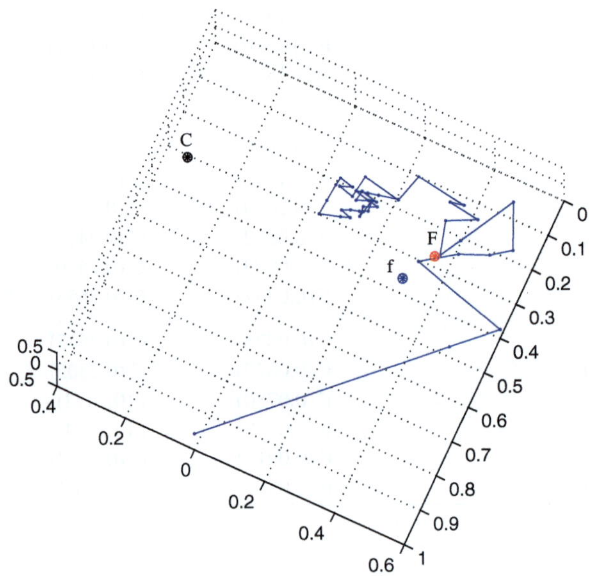

Fig. 4.3 An application of the CEG algorithm: a bird's eye view of the path traced by the CEs, $\{c_i\}$, as "Simple Gifts" unfolds, establishing its affiliation to F major

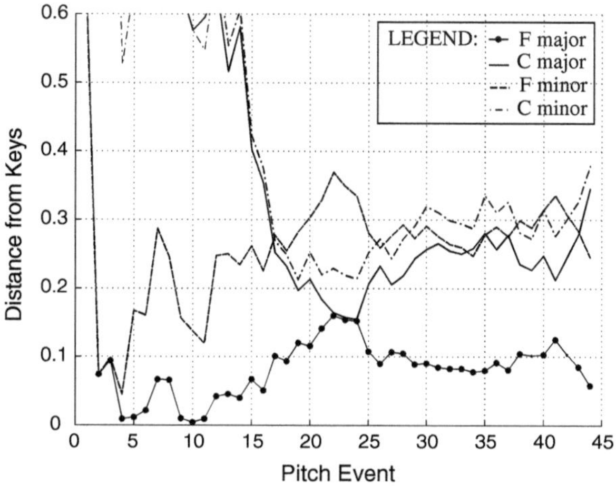

Fig. 4.4 Distance to various keys as "Simple Gifts" unfolds

distance is achieved at pitch event $i = 10$, although it comes extremely close at pitch event $i = 4$. Relating this to the numbers in Table 4.2, the distance from c_i to F major is 0.0092 when $i = 4$, and 0.0044 when $i = 10$. Compare this to the initial distances of approximately 0.6 each from c_i to F minor, C major and F major.

Figure 4.4 shows the exact distances from the four closest keys (F major, C major, F minor, and C minor), at each successive pitch event. Observe in the graph that F major quickly establishes itself as the closest key. However, between pitch events $i = 22$ and 24, C major (the dominant of F) vies with F major for preeminence. The melody dwells on the dominant key area from $i = 19$ to 24, outlining the C major triad from $i = 21$ to 24. Thus, this behavior in the model is in accordance with listener cognition, and is a desired trait.

In the following chapter, the CEG algorithm will be evaluated against two fundamental methods for key finding, namely Longuet-Higgins and Steedman's Shape Matching Algorithm (SMA) [4], and Krumhansl and Schmuckler's Probe Tone Profile Method (PTPM) [3]. The evaluation will be conducted using the classic test set of the fugue subjects from Bach's *Well-Tempered Clavier* Book I [1].

References

1. Bach, J.S.: The Well-Tempered Clavier-Part I, BWV 846–869. Henle Urtext Edition, Munich (1997)
2. Copland, A.: Appalachian Spring. Boosey & Hawkes, New York (1945)
3. Krumhansl, C.L.: Cognitive Foundations of Musical Pitch. Oxford University Press, New York (1990)
4. Longuet-Higgins, H.C., Steedman, M.J.: On interpreting Bach. In: Meltzer, B., Michie, D. (eds.) Machine Intelligence 6, pp. 221–241. Edinburgh University Press, Edinburgh (1971)

Chapter 5
The CEG Algorithm (Part II): Validation

Abstract In this chapter, I compare the performance of the Center of Effect Generator (CEG) key-finding method, based on the Spiral Array model, to that of two classic key-finding algorithms, namely, Krumhansl and Schmuckler's Probe Tone Profile Method (PTPM), and Longuet-Higgins and Steedman's Shape Matching Algorithm (SMA). The three algorithms are tested using the fugue subjects of Bach's *Well-Tempered Clavier*; the evaluation criterion being the number of pitch events needed to correctly determine the key of the fugue. The CEG required on average 3.75 pitches, the PTPM 5.25, and the SMA 8.71. If one considered only the fugue subjects in which the tonic-dominant rule was not applied by any of the three algorithms, the CEG required 3.57 pitches, the PTPM 4.79, and the SMA 8.21. The MATLAB code and details of each test run are given in Appendix B. Here, I present an analysis of the results of the comparison, and discuss the issue of model validation. A MATLAB version of the CEG code appears in Appendix B. The CEG method is applied to polyphonic music in MIDI (Musical Instrument Digital Interface) format in Chap. 9, and adapted to music audio in Chap. 11.

5.1 Model Validation

How does one know when a key-finding algorithm generates the correct answer, or agrees with listener cognition? For a given piece of music, how does a listener decide what is the correct key or when the musical passage deviates from this established key?

In music theoretic work, researchers often refer to their own musical experiences and insights for benchmark musical analyses. Individual analysis was employed

This chapter is a revision of parts of Finding Keys (Chap. 5) in "Towards a Mathematical Modeling of Tonality" by Elaine Chew, an MIT PhD dissertation, Cambridge, Massachusetts (2000) https://dspace.mit.edu/handle/1721.1/9139.

E. Chew, *Mathematical and Computational Modeling of Tonality*,
International Series in Operations Research & Management Science 204,
DOI: 10.1007/978-1-4614-9475-1_5, © Springer Science+Business Media New York 2014

by Lewin to devise his "Generalized Musical Intervals and Transformations" [9], by Lerdahl and Jackendoff in their "Generative Theory of Tonal Music" [8], and by Schenker in his theory of a fundamental structure underlying musical pieces (see [5]). Outside of music theory, in the field of theoretical linguistics, the researcher refers to his/her own experience to determine if a sentence is syntactically correct.

With the influence of the fields of artificial intelligence and psychology music, there has been a movement toward empirical approaches to music research, such as Temperley's corpus-based studies in his book, "The Cognition of Basic Musical Structures" [11]. One systematic way to identify and verify the answer may be through listener annotation; another is through the composer's own documentation of the intended key, for example, as listed in the title. In general, there is agreement amongst informed listeners concerning the keys generated, and the chords employed, in a piece of music. The precise point of modulation, when a new key is established and the previous one finished, is subject to more debate, as will be discussed in Chap. 6.

5.1.1 Test Data

A classic test set that has been invoked in the testing of key-finding algorithms is the fugue subjects of Bach's *Well-Tempered Clavier* (WTC) [2, 3]. Bach wrote a prelude and fugue pair for every key, twelve major and twelve minor, totaling twenty-four, in each Book, I and II, of the WTC, resulting in a total of forty-eight pairs of preludes and fugues.

I use the CEG algorithm to analyze the keys of the twenty-four fugue subjects from Book I of the WTC, comparing these results with other algorithms' findings. Later in this book, larger polyphonic test sets–score-based, MIDI (Musical Instrument Digital Interface), and audio–will be employed in algorithmic testing. For clarity and historical reasons, the tests here begin with the WTC score-based melodic data.

Analyzing a single melodic line is itself a challenging task because of the limited information embodied in a sequence of individual note events. Furthermore, some melodies can be harmonized in several different ways as evidenced by Bach himself in his chorales [1]. The subjects of fugal compositions are chosen for their contrapuntal potential. They are melodies that can be transformed in many different ways—for example, inverted, augmented (stretched in time), and presented in diminution (compressed time-wise)—and yet harmonize when one transformation is overlapped with another. Often, for this reason, fugue subjects are melodies that can be harmonized in many different ways, and hence may frequently be tonally ambiguous.

One advantage of employing fugue subjects as test data for key-finding is that their succinctness limits the degree of tonal meandering, making it more likely that the excerpt sticks to the goal of establishing the intended key. Nevertheless, some fugue subjects may establish a key, then move to another key area, say the dominant. In response to key ambiguities, algorithms can list the most likely key candidates, recognizing that the list may evolve over time.

The numerical details of the CEG's output are presented in Appendix B, comparison results follow in Sect. 5.2, and detailed analyses and discussion of the results are presented in Sect. 5.3.

5.2 Comparing Key-Finding Algorithms

In this section, I compare the CEG algorithm with two classic key-finding methods in the music cognition literature. These are Longuet-Higgins and Steedman's Shape Matching Algorithm [10], and Krumhansl and Schmuckler's Probe Tone Profile Method, see [7], for key-finding. The benchmark examples common to the two are the fugue subjects from Bach's forty-eight preludes and fugues of his *Well-Tempered Clavier*, Book I [2] and Book II [3].

5.2.1 Method 3: The Shape Matching Algorithm

Longuet-Higgins and Steedman's Shape Matching Algorithm (SMA) [10] makes use of the fact that pitches belonging to a major key are located on the Harmonic Network in such a way as to form a distinct and compact shape. The shape remains the same for all major keys, the only difference being the pitch collection inside the shape. Pitches belonging to a minor key formed a different unique and compact shape. Hence each key is represented by one of two shapes, and the key identity is given by the location of the shape. Figure 5.1 shows one example each of major and minor key shapes.

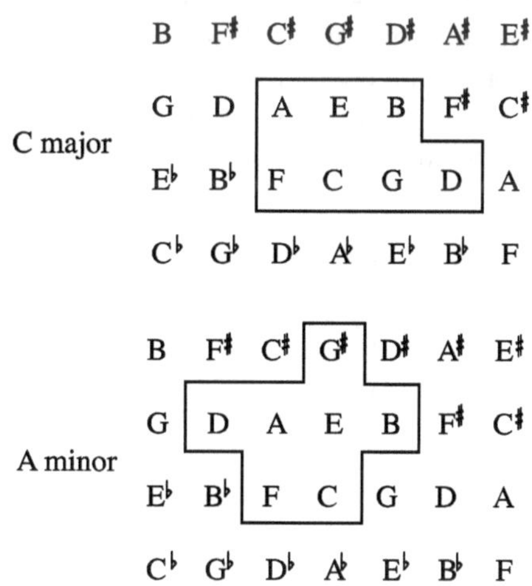

Fig. 5.1 Examples of shapes outlining a major key, C major, and a minor key, A minor, in the Harmonic Network

The SMA works by a process of elimination. With each pitch event, shapes not covering the pitches sounded are eliminated, together with the keys they represent. This process is repeated until only one key remains.

If the music sample to be analyzed has ended and multiple solutions remain, the tonic-dominant rule is invoked. If all keys have been eliminated, the algorithm takes one step back, then invokes the tonic-dominant rule. The tonic-dominant rule simply assumes that the first pitch in the melody is most likely to be the tonic, and next most likely to be the dominant, of the actual key. Hence, if either of these are candidates for selection, they are to be chosen in that order.

The algorithm was tested on all forty-eight fugue subjects in Bach's *Well-Tempered Clavier* (WTC) [2, 3]. The intended key was found in 26 of the 48 fugue subjects. Of the remaining 22, the keys in 17 of the fugues were found by invoking the tonic-dominant rule.

5.2.2 Method 2: The Probe Tone Profile Method

In [6], Krumhansl and Kessler created probe tone profiles based on listener judgements of how well each of the twelve pitch classes fit into a given key context. A tonal context is established, for example, by sounding the I-IV-V-I chord progression in the key of C major; an individual tone, say E♭, is then sounded, and the listener asked to rate, on a scale of one to seven, how well the tone fit in that context, where seven is an excellent fit, and one a poor fit. Figure 5.2 shows the average listener ratings of how well each of the twelve tones in a chromatic scale fit in the C major and C minor contexts, respectively.

Krumhansl and Schmuckler, see [7], later used these profiles in their Probe Tone Profile (PTPM) key-finding method. In the PTPM, the cumulative durations of each of the twelve pitch classes of the test data are recorded in an input vector, and the correlation coefficient is calculated between the vector and the probe tone profiles of each of the twenty-four major and minor keys. The key profile with the highest correlation reveals the key of the test data. Key candidates are ranked by likelihood;

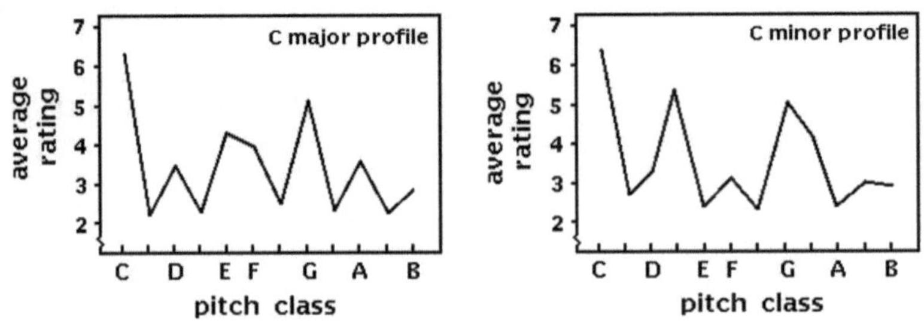

Fig. 5.2 Krumhansl and Kessler's probe tone profiles for C major and C minor (from [6])

and the output of the PTPM is a vector containing numerical values rating the strength of each key. Again, the tonic-dominant rule is invoked when there is a tie.

For evaluation, the PTPM analyzes the "melody-to-date" at each pitch event. With prior knowledge of the intended key, the first instance at which the intended key ranks first in strength is selected to be the point at which the key is determined. This same stopping criterion was invoked in a comparison between the PTPM and SMA. Considering only the fugue subjects (from Books I and II) in which the tonic-dominant rule was not applied, the PTPM algorithm required on average 5.11 pitches, and the SMA 9.42 pitches.

5.2.3 Method 1: The CEG Algorithm

The Spiral Array comprises an array of helices representing pitches, chords, and keys in the same three-dimensional space; each higher level entity is generated from a convex combination of its lower level components, chords from their component pitches, and keys from their defining chords (see description in Chap. 3). The model can be calibrated to concur with principles of tonal cognition, as shown in Appendix A. Figure 5.3 shows the major and minor key helices generated from their component chords; for clarity, the helices are shown on two separate diagrams, in reality they exist in the same space.

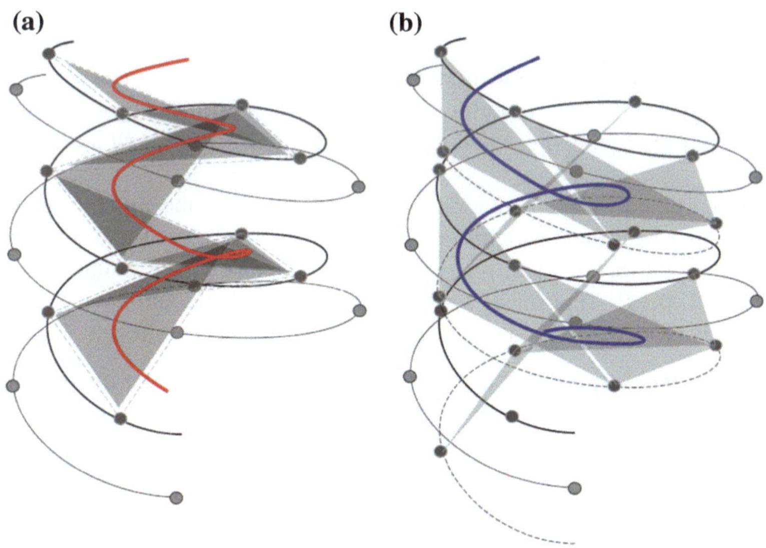

Fig. 5.3 Major key and minor key helices generated by their defining chords, which are in turn generated by their component pitches, in the Spiral Array. **a** Innermost helix represents major keys **b** innermost helix represents minor keys

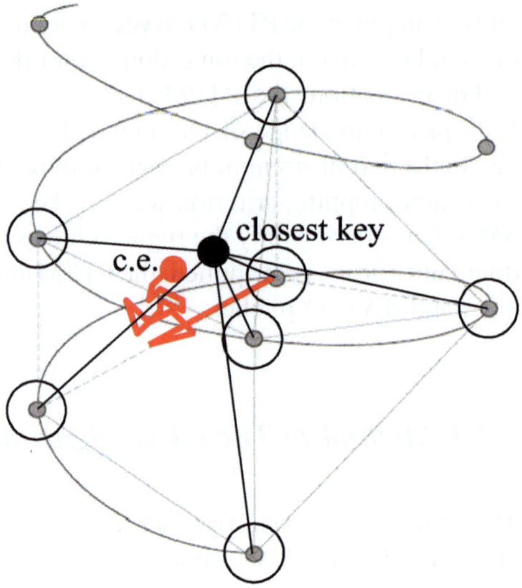

Fig. 5.4 Illustration of the CEG key-finding algorithm: shown in the diagram are the evolving CE, the closest key representation, and the convex hull of the pitches of that key

Not only can higher level components be generated as the weighted center of their member pitches, any collection of pitches can be mapped to a point in space that is the *center of effect* (CE) of the pitch representations. This is the central idea of the CEG algorithm. The twelve pitch classes of the input stream are mapped to their corresponding pitch positions, and weighted by their proportional cumulative durations to get the CE of the input. A nearest-neighbor search for the closest key representation in space reveals the key of the excerpt. Figure 5.4 shows the process: the CE evolves with each incremental step, and the closest key is given as the answer. The algorithm can also give as output a list of likely keys, ranked by closeness to the CE. The CEG method is presented and described in detail through a case study in Chap. 4.

In this particular implementation of the Spiral Array, the aspect ratio $h/r = \sqrt{2/15}$, where $r = 1$. Pitch weights for major/minor chords (w, u), and chord weights for major/minor keys (ω, υ) are restricted to be the same, and are assigned the values [0.536, 0.274, 0.19]. This set of assignments ensures that conditions A and B are met, and C is close to satisfied, where their definitions are as follows:

Condition A: The key nearest an isolated pitch, $\mathbf{P}(k)$, should be a major key of the same name, $\mathbf{T}_M(k)$.

Condition B: A half-step interval should be interpreted as the leading tone and tonic, i.e., scale degrees $\hat{7}$ and $\hat{1}$, of the major key.

Condition C: The lower and upper pitches of a perfect fourth interval should be interpreted as the dominant, $\hat{5}$, and tonic, $\hat{1}$, of the major key.

For these weights, the lower and upper pitches of a perfect fourth interval is interpreted as the dominant, $\hat{5}$, and tonic, $\hat{1}$, of the minor key, instead of the major key, which is also plausible. Additionally, the choice of weights results in a major chord being closest to its root, followed by its fifth, then its third (Theorem 7); and, a minor chord being closest to its root, followed by its fifth, then its third (Theorem 9). Appendix A progressively lays out the calibration criteria for the model.

Similar to the PTPM, the CEG considers pitch and duration information, mapping pitches to their corresponding pitches in the pitch class spiral, then weighting each point with the proportional duration to get the center of effect (CE) of the test data. In contrast to the PTPM, which takes as input a vector of pitch names and corresponding durations, the CEG's input is a point in space, the CE summarizing the pitch and duration information.

To align the CEG against the SMA and PTPM in the ensuing tests, for each incremental pitch event, the model gives as output a vector of possible keys ranked by likelihood, and a vector of distances between each key representation and the aggregate position. Note that because the model has been designed so that close relationships are indicated by spatial proximity, a low distance value represents a close match. For more meaningful comparisons, I will determine the number of pitches to key discernment in the same way as Krumhansl and Schmuckler. The first occurrence of the intended key is recorded as the pitch event at which the key is correctly determined. The key selected at the first note, even if it is the correct key, is discarded as it is not a meaningful correct choice. The results of the comparison are reported in Table 5.1—the PTPM and SMA results are as reported in [7].

5.3 Detailed Analysis of Comparison Results

The following conclusions are gathered from observations of model performance on the fugue subjects of the WTC Book I as given by Table 5.1, and documented in Appendix B. In this section, I conduct a case-by-case examination of each algorithm's performance, and summarize the strengths and weaknesses of each algorithm.

5.3.1 Results Summary

For the fugue subjects in Book I of the WTC, the CEG required on average 3.75 pitches, the PTPM 5.25, and the SMA 8.71. If one considered only the fugue subjects in which the tonic-dominant rule was not applied by any of the three algorithms, the CEG required 3.57 pitches, the PTPM 4.79, and the SMA 8.21. Details of each test run are given in Appendix B.

Given a melody, a hypothesis of its key based on its first pitch is not a meaningful one. The reliability of a hypothesis based on two pitch events remains questionable. As a general rule of thumb, on average, the minimum number of pitches required to form an opinion of the key is three. The CEG algorithm required, on average, 3.75 steps to determine the key of the 24 fugue subjects. This is an average performance that is close to optimal.

Table 5.1 Comparison of key-finding algorithms on Bach's WTC I fugue subjects

	WTC Book 1	Steps to key		
	Fugue subjects	CEG[a]	PTPM	SMA
1	C major	2	2	16[†]
2	C minor	5	5	5
3	C♯ major	6	7	16
4	C♯ minor	3	3	4
5	D major	2	2	15[†]
6	D minor	3	3	8
7	E♭ major	2	6	11[†]
8	D♯ minor	2	6	12[†]
9	E major	14	12[†]	11
10	E minor	3	2	7[†]
11	F major	4	10	6
12	F minor	3	15	4[†]
13	F♯ major	3	2	8
14	F♯ minor	7	18	5[†]
15	G major	2	2	15
16	G minor	3	3	4
17	A♭ major	3	2	7[†]
18	G♯ minor	5	5	5
19	A major	2	4	7
20	A minor	5	5	5
21	B♭ major	4	4	14
22	B♭ minor	2	3	6[†]
23	B major	2	11	11
24	B minor	3	3	7
	Average	3.75	5.25	8.71
		(3.57)[b]	(4.79)	(8.21)

[a] Numbers generated using $h = \sqrt{2/15}$ ($r = 1$), and all weights set to [0.536, 0.274, 0.19]
[†] Tonic-dominant rule invoked
[b] Numbers in brackets denote average for fugue subjects in which the tonic-dominant rule was not applied by any method

5.3.2 Case 1: Same Performance by All Three Algorithms

The fugue subjects shown in Fig. 5.5, Nos. 2, 18, and 20, required the same number of steps in all three algorithms to determine their keys. Each algorithm took five steps for each fugue subject. It is interesting to note that all three examples are in the minor mode, and Fugue Nos. 18 and 20 have essentially the same pitch relations in the first five notes—degrees $\hat{1}, \hat{7}, \hat{1}, \hat{2}, \hat{3}$ of their respective minor scales.

In each example, the fifth pitch is the one that informs the algorithms that the melody is unequivocally in the minor mode. Even a human listener would not be able to determine if the melody was in a major or minor key given only the first four pitches—scale degrees $\hat{1}, \hat{7}, \hat{1}, \hat{5}$ in the case of Fugue No. 2; and, scale

Fig. 5.5 Fugue subjects for which all three algorithms performed equally well, requiring the same number of steps to determine the key

degrees $\hat{1}$, $\hat{7}$, $\hat{1}$, $\hat{2}$ in the case of Fugue Nos. 18 and 20. It is the fifth pitch event—$\hat{6}$ in Fugue No. 2; and, $\hat{3}$ in Fugues 18 and 20—each of which would have been a half step higher in the corresponding major scale that tips the balance in favor of the minor mode. Thus, all three algorithms agree with the listener's perception.

5.3.3 Case 1a: Almost Same Performance by All Three Algorithms

For the subjects of Fugue Nos. 4 and 16, shown in Fig. 5.6, the CEG and PTPM took three steps, and the SMA took four steps, to determine the intended keys. In these examples, again both in the minor mode, the scale degrees of the first three notes are: $\hat{1}$, $\hat{7}$, $\hat{3}$ in Fugue No. 4; and, $\hat{5}$, $\hat{6}$, $\hat{1}$ in Fugue No. 16.

In this case, both the CEG (by design) and PTPM (by inference from listener decisions) took into account interval relations important to a triad—perfect fifths and minor thirds. These intervals are formed by $\hat{1} - \hat{3}$ in Fugue No. 4, and $\hat{1} - \hat{5}$ in Fugue No. 16.

Fig. 5.6 Fugue subjects for which all three algorithms performed nearly equally well, requiring almost the same number of steps to determine the key

If one considered only the pitch collection up to step three, Fugue No. 16 could be in C minor (in which case the pitches would be scale degrees $\hat{2}$, $\hat{3}$, $\hat{5}$) or G minor (where the same pitches would be scale degrees $\hat{5}$, $\hat{6}$, $\hat{1}$). Similarly, the first three pitch events of Fugue No. 4 could be in E♯ minor if enharmonic equivalents are allowed (in which case the pitches would be scale degrees $\hat{6}$, $\hat{5}$, $\hat{7}$) or C♯ minor (where the same pitches would be scale degrees $\hat{1}$, $\hat{7}$, $\hat{3}$). Thus, the SMA is unable to make a decision between the competing minor keys until the appearance of the second half-step, scale degrees $\hat{3}$–$\hat{2}$ of C♯ minor in Fugue No. 4, and scale degrees $\hat{1}$–$\hat{7}$ of G minor in Fugue No. 16.

5.3.4 Case 2: SMA Worse than CEG and PTPM

I divide the fugue subjects in which the SMA performed markedly worse than the CEG and PTPM into two sets. In the first set, the SMA invoked the tonic-dominant rule, and in the second set, the SMA eventually chose the intended key.

In the first set, where the SMA invoked the tonic-dominant rule, the two fugue subjects, shown in Fig. 5.7, begin with a lower tetrachord (the first four scale degrees) of the intended major scale. The CEG and PTPM immediately chose the appropriate key based on the first two pitches, $\hat{1} - \hat{2}$. This implies that, given two pitches related by a whole-step interval in isolation, both algorithms assign the $\hat{1}, \hat{2}$ scale degree interpretation to the two pitches, which is intuitively plausible and happens to be correct.

Using the SMA, the pitches in fugue subject No. 1 could be in C major, which corresponds to a scale degree $\hat{1}$, $\hat{2}$, $\hat{3}$, $\hat{4}$ interpretation, or F major, which corresponds to a scale degree $\hat{5}$, $\hat{6}$, $\hat{7}$, $\hat{1}$ interpretation; and the pitches in fugue subject No. 5, could be in D major, which implies a scale degree $\hat{1}$, $\hat{2}$, $\hat{3}$, $\hat{4}$ interpretation, or G major, which is a scale degree $\hat{5}$, $\hat{6}$, $\hat{7}$, $\hat{1}$ interpretation. In either case, the SMA is unable to choose between the two candidate keys for the entire length of the fugue subjects because the leading-tone ($\hat{7}$), which would break the tie, is missing.

We now move on to the second set, examples in which the SMA eventually chose the intended key. In the first part of this second set, the SMA required seven and

Fig. 5.7 Fugue subjects in which the SMA required the tonic-dominant rule to break the tie, but the CEG and PTPM performed well

eight steps, respectively, for the two minor key examples, subjects of Fugue Nos. 24 and 6 shown in Fig. 5.8, compared to the CEG's and PTPM's three. Again, the SMA suffered from an inability to disambiguate between keys using the same pitch subsets. Fugue subject No. 24 could be in D major or B minor until the A♯ pitch event appears at step seven, unequivocally demonstrating that the melody is in the key of B minor; and, fugue subject No. 6 could be in C major, F major, D minor, or A minor until the C♯ appears at step eight, determining without a doubt that the key is D minor.

In contrast, the CEG algorithm based on the Spiral Array model recognizes and prioritizes the B minor triad in the opening of Fugue No. 24, and the ascending scale figure at the start of Fugue No. 6, as significant clues to the key. The PTPM's comparisons to listener judgements comes to the same conclusion.

In the second part of the second set, the SMA required fourteen steps versus the CEG's and PTPM's four to determine the key of Fugue No. 21; and, fifteen steps versus two in Fugue No. 15; both fugue subjects are shown in Fig. 5.9. The SMA is unable to resolve the ambiguities until step fourteen, when it chose B♭ major in favor of its competitor, F major, in Fugue No. 21; and, at step fifteen, when it chose G major in favor of the alternative, D major, in Fugue No. 15. Both the CEG and PTPM decided at step two, in Fugue No. 15, that the whole step interval between the first two pitches implied G major. The two algorithms again agreed at step four, in Fugue No. 21, that the key was B♭ major and not F major.

Fig. 5.8 Fugue subjects in which the SMA eventually found the key, but the CEG and PTPM performed well (Part 1)

Fig. 5.9 Fugue subjects in which the SMA eventually found the key, but the CEG and PTPM performed well (Part 2)

Fugue No. 9 in E major

Fig. 5.10 Fugue subject in which the CEG performed worst

5.3.5 Case 3a: Worst Case for the CEG Algorithm

In one instance, the CEG algorithm performed worse than the PTPM and SMA. This was for the subject of Fugue No. 9, shown in Fig. 5.10. The CEG algorithm took fourteen steps, the SMA only eleven, and the PTPM invoked the tonic-dominant rule at step twelve, to determine the intended key.

The first pitch in the sequence is E, and thus closest to the key of E major; as a solution at step one is ignored, the CEG algorithm continues to the next note, *F*♯. Because each pitch is weighted by its duration, the long duration (longest in the sequence) of the pitch F♯ tips the balance towards a B major/minor tonality. Ignoring the first note, the next four could be a variation on the opening to Bach's Minuet in G, transposed to B major. Thus, B major is not an unreasonable choice. The influence of the B major tonality is so strong, and the ensuing notes of such short duration (thus having lesser influence) that it took fourteen steps to cancel out the effect of the extended F♯ pitch.

The PTPM was probably also affected by taking into consideration pitch duration, which happens to be a hindrance rather than help in this case. The SMA took almost as long to find the key for different reasons. Step eleven is pitch A, which finally allows the SMA to choose E major over B major.

5.3.6 Case 3b: Best Cases for the CEG Algorithm

The CEG algorithm outperformed the SMA and PTPM in several occasions. These examples have been divided into two sets. The first set gives the fugue subject in which the CEG algorithm chose the intended key in two steps, and both the SMA and PTPM found the key in eleven. The second set is a collection of fugue subjects in which the CEG outperformed the PTPM, which in turn performed better than the SMA. Some other examples in which the CEG performed best have been categorized under other scenarios where the SMA or PTPM performed worst.

The first, shown in Fig. 5.11, is the subject of Fugue No. 23 in B major, and it begins with the scale degrees $\hat{1}$, $\hat{7}$. By design, the CEG algorithm assigns these two pitches the correct scale degrees. Both the SMA and PTPM wait until step eleven, when all seven pitches in the key have been sounded, to decide on B major in favor of F♯ major.

Fig. 5.11 Fugue subjects in which the CEG performed best (Part 1)

Fig. 5.12 Fugue subjects in which the CEG performed best (Part 2)

In the first two examples in the second set, Fugue Nos. 19 and 22 shown in Fig. 5.12, the CEG algorithm chose the correct key after two steps. As with Fugue No. 23 in the first set, the first two pitches in fugue subject No. 19 are correctly interpreted to be $\hat{1}$, $\hat{7}$. The PTPM required step four, with the added $\hat{3}$, $\hat{1}$ pitches to assess the key, and the SMA needed seven steps and the appearance of six out of the seven pitches of the key, especially pitch E, to choose A major in favor of F♯ minor.

In Fugue No. 22, the CEG algorithm arrived at the correct decision in two steps rather fortuitously. One of the desired conditions when choosing weights for the model was that a perfect fourth interval in such a situation would be interpreted as $\hat{1}$, $\hat{5}$ of B♭ major. However, the particular set of weights used in this implementation of the Spiral Array did not satisfy this condition, preferring instead the minor mode. The PTPM made the correct choice at step three, and the SMA required the tonic-dominant rule to break the tie between D♭ major and B♭ minor.

In Fugue No. 3, the CEG required only six steps to ascertain the key. To a listener, the first five pitches could as likely be in F♯ major as C♯ major, but the sixth pitch, E♯, strongly suggests the C♯ major triad, making C♯ major the closest key, which is a correct answer. The PTPM required one additional step to come to the same conclusion. The SMA, however, needed the B♯ at step sixteen to choose C♯ major in favor of F♯ major.

In Fugue Nos. 7 and 8, no such defining moment occurred for the SMA, and the tonic-dominant rule had to be invoked. Because the CEG algorithm prioritizes triadic relations, the first two pitches in fugue subjects No. 7 and No. 8 were assigned, correctly, scale degrees $\hat{5}, \hat{3}$ and $\hat{1}, \hat{5}$, respectively. The PTPM, in both cases, took until step six to choose the appropriate key, waiting for the fourth and third scale degrees respectively.

5.3.7 Case 4a: Worst Cases for the PTPM Algorithm

The PTPM performed worse than the SMA and CEG on several occasions. I analyze the results in two parts. The first part addresses an example in which both the CEG and SMA arrived at the correct key solution before the PTPM. The second part consists of examples in which the SMA invoked the tonic-dominant rule.

In the first example, Fugue subject No. 11, shown in Fig. 5.13, the CEG algorithm chooses F major at step four, a reasonable point at which to make this decision since the previous three pitches could just as likely be in C major, amongst other alternatives. The SMA waits for the leading tone, pitch E (without the ♭), to choose F major over B♭ major. The PTPM required ten steps to select F major as the key, probably because of the brevity of the tonic, pitch F, lowered the correlation coefficient with the F major probe tone profile.

Both examples in part two, Fugue Nos. 12 and 14, shown in Fig. 5.14, employ chromatic pitches not in the scale. As a result, the PTPM required many pitch events, fifteen and eighteen for fugue subjects Nos. 12 and 14, respectively, to occur before choosing the correct key. The SMA is stumped early in the melody, after only four and five pitch events for the pair of fugue subjects, having eliminated all key options, and invokes the tonic-dominant rule. Because the CEG algorithm collapses pitch information into a single point, it is a relatively more robust algorithm when applied to melodies with chromatic pitches, especially in the case of Fugue No. 14, where it

Fig. 5.13 Fugue subjects in which the PTPM performed worst (Part 1)

Fig. 5.14 Fugue subjects in which the PTPM performed worst (Part 2)

Fig. 5.15 Fugue subjects in which the PTPM performed best

finds the key in seven steps despite the non-key pitch A♯ at steps five and seven. It finds the key in three steps for Fugue No. 12, before the non-key pitches are found.

5.3.8 Case 4b: Best Cases for the PTPM Algorithm

In the next three fugue subjects, shown in Fig. 5.15, where the PTPM algorithm performed best of the three, the PTPM took two steps to ascertain the key, while the CEG algorithm took three steps. In Fugue No. 13, the SMA took eight steps; and, in Fugue Nos. 10 and 17, it invoked the tonic-dominant rule.

In all three cases, the PTPM chose the correct key in two steps because it selected the minor mode when given two pitches a minor third apart, i.e., imposing a $\hat{1} - \hat{3}$ interpretation, and the major mode in the $\hat{1} - \hat{5}$ or $\hat{5} - \hat{1}$ interpretation when given two pitches a perfect fifth or perfect fourth apart, respectively. The CEG algorithm did the opposite—choosing the major mode when given two pitches a minor third apart, i.e. implying a $\hat{3} - \hat{5}$ interpretation, and the minor mode in the $\hat{1} - \hat{5}$ or $\hat{5} - \hat{1}$ interpretation when given two pitches a perfect fifth or perfect fourth apart, respectively—this error is quickly corrected in the following step. The SMA required step eight, pitch B, in Fugue No. 13 to select F♯ major over C♯ major; and, required the tonic-dominant rule in the other two cases.

5.4 Discussion: Strengths and Weaknesses

The Shape Matching Algorithm

The SMA associates each key with only its pitch collection, without prioritizing triadic relations. It is highly dependent on pitches that eliminate competing key candidates. For example, this critical pitch is often the leading-note, scale degree $\hat{7}$, of the key with more sharps (or fewer flats) or the subdominant, scale degree $\hat{4}$, of the key with fewer sharps¡ (or more flats) when disambiguating between major keys a perfect fifth apart, i.e., keys differing by only one sharp or flat.

The SMA is unable to parse fugue subjects with chromaticism, non-key pitches that may act as passing tones between diatonic key pitches. In these cases, one or more non-scale pitches eliminates all key candidates, and the tonic-dominant rule has to be invoked.

Only in one case does the SMA perform better than the PTPM and CEG without invoking the tonic-dominant rule. This happened in the scalic subject of Fugue No. 9, where the PTPM and CEG were stumped by the long duration of the second note on the supertonic, second degree of the scale, but the SMA benefitted from the semi-quaver run that ensued that outlined the pitches of the scale. In another instance, Fugue No. 11, the SMA's performance, selecting the key in six steps, was between that of the CEG (four steps) and the PTPM (ten steps).

In all other cases, either the tonic-dominant rule was invoked, or the SMA took more steps than the other two to choose the intended key. Thus, in general, the SMA serves as a lower bound for the three key-finding algorithms.

The Probe Tone Profile Method

The PTPM captures listeners' judgements about pitch-key relationships. As such, it models the intuitions of a sample of listeners.

The probe tone profiles of the PTPM method incorporate some of the design principles of the Spiral Array. The prominence of the tonic, followed by the fifth, then the third in the contour of the pitch profile serves a similar function to the Spiral Array's more significant weights on the tonic, followed by the fifth, then the third of each triad, combined with the closeness of the tonic pitch to the corresponding key.

Some secondary interval-key preferences can be inferred from the PTPM's performance. When given two pitches a perfect fifth or a perfect fourth apart, the PTPM selects the major key of the lower or upper pitch, respectively; when given two pitches a minor third apart, the PTPM treats them as the lower half of a minor triad;

and, when given two pitches a major second apart, the PTPM selects the major key of the lower pitch. These preferences are similar, but not identical, to the kinds of preferences built into the Spiral Array model.

The PTPM suffers from not using the relatively rarer half-step intervals as cues to the key. It also does not respond as well to non-scale chromatic pitches as shown in its worst-case scenario discussed in the previous section.

The Center of Effect Generator

By definition, the Spiral Array accounts for triadic relations (in particular, the perfect fifth and major/minor third interval relations) between pitches, and fifth relations among chords. As a result, the CEG algorithm prioritizes these relations when assessing pitch information.

Because of the way pitch and duration information is reduced to a point in space, the CEG is more robust to "noise" in the form of non-scale tones, and thus performs better than the PTPM and the SMA in the presence of chromatic pitches. For both the CEG and PTPM, the use of the duration information can sometimes be a hinderance, as was the case in Fugue No. 9.

By design, this particular implementation of the CEG method places greatest emphasis on the half-step, which is interpreted to be a $\hat{7} - \hat{1}$ transition, and the tonic of the key—conditions that served the algorithm well in the key-finding experiment. The algorithm also, to some extent, accounts for the fact that a perfect fourth or perfect fifth interval results in initially selecting the minor key of the upper or lower pitch, respectively, whereas the PTPM prefers the major key. Also distinct from the PTPM, the CEG interprets two pitches a minor third apart as being the upper half of a major triad. Most of the time this works well, except when the minor third is actually the bottom of a minor triad as in the E minor Fugue (No. 10).

In general, the CEG performed better than the PTPM and SMA, taking an average of 3.75 notes to determine the key, compared to 5.25 (PTPM) and 8.71 (SMA). Intuitively, one would expect that the absolute minimum number of pitches required to form an opinion of the key in the Spiral Array is three. Geometrically speaking, in the Spiral Array space, to determine the key space of a sample of music, one pitch gives only a point, two form a line, with three we get a 2-simplex (triangle), and four we get a 3-simplex (a tetrahedron). The 3-simplex would be the first to get closer to a volume in space, while the right 2-simplex may give important cues as to the key space. Thus, the CEG algorithm's 3.75 average—3.57 when considering the subset where no tonic-dominant rule was invoked by the other two algorithms—is on the expected threshold at which sufficient information has transpired to determine key.

5.5 Evaluation Method Discussion

Finding a way to measure key-finding performance is a challenging problem, and the use of the number of pitches required to achieve a "right" key answer as a yardstick is not without its flaws. There are two basic concerns regarding the use of this statistic to measure algorithm performance.

The first instance in which the intended key becomes the top ranked one is not necessarily the point at which a key has been strongly established; there is also no guarantee that the algorithm will stick to the key thereafter, and perhaps rightly so, as will be discussed in the following paragraphs. Consider the cases, of which there are not a few, when the number of steps to key is two. Certainly, one may form an opinion, even a strong opinion, about the key given two pitch events. But, it is difficult to justify a key decision after only two pitches because there are so many ways in which the key can deviate from this expectation even at the very next note. At step two, a hypothesis of the key is simply that, a hypothesis to be verified or disqualified as more information accrues.

A large number of required pitch events does not necessarily portend poor model performance. Often, a melodic line can remain ambiguous, deliberately non-committal, for a while, until a strategically placed note confirms or denies a key. This is particularly true for the minor key Fugues: No. 2 in C minor, No. 18 in G♯ minor, and No. 20 in A minor. In all three cases, each algorithm determined the intended key after five notes, so the variance is zero. If an example of this type occurred that required twenty-eight notes, the result could skew the statistics for each algorithm.

Once a key is established, the music may modulate to another key. In fact, the art of making such transitions forms an important hallmark of a good composer. Thus, the key stated in a title may not be the one the ear interprets throughout a piece, nor might it be the most prevalent one. By choosing to use fugue subjects, which are relatively brief and intended to establish the key at the outset of the piece, many of the problems of using the title key as the ground truth might be averted. Despite the limited time in which to establish and vary a key in the statement of a fugue subject, an occasional one may still do so, for example fugue subject No. 7 starts by quickly establishing E♭ major as the key, then rapidly moving on to B♭ major. See [4] for further discussion of when the title key may not be the key of an excerpt from the piece.

The tonic-dominant rule, used by both the SMA and PTPM, is based on the observation, external to the algorithms, that many melodies begin on the tonic or the dominant of their intended key. Hence the first pitch that is sounded is highly likely to be the tonic or dominant of the key. In the CEG method, I have set up the model to conform to a small number of basic conditions governing relationships between tonal elements. One of the conditions, the proximity of the key to its tonic, effectively enacts one part of the tonic-dominant rule by assuming that the first pitch event is the tonic. Another condition, the interpretation of perfect fifth or perfect fourth intervals as $\hat{1} - \hat{5}$ or $\hat{5} - \hat{1}$ transitions or pairings, while distinct from the tonic-dominant rule in that it is designed to capture interval-key relations rather than single pitch-key relations, is similar in spirit to the tonic-dominant rule in that it acknowledges the importance of the tonic and dominant. Note that the Spiral Array could also be readily set up with a different set of conditions.

Finally, the goal of the SMA differs from the PTPM and CEG in significant ways. The SMA behaves conservatively, attempting to establish beyond doubt the identity of the unique key generated by the musical segment. In contrast, the PTPM and CEG would venture a guess for the most likely key from the very first pitch. Pitch, duration, and meter work together to establish a tonality. Both the PTPM and CEG use duration information in their calculations, while the SMA does not. Hence, in some respects, the number of steps to achieve the different goals may not necessarily be a fair measure of algorithmic performance.

References

1. Bach, J.S.: 371 Harmonized Chorales and 69 Chorale Melodies with Figured Bass. (Revised, corrected, edited, and annotated by Riemenschneider, A.) G. Schirmer Inc., New York (1941)
2. Bach, J.S.: The Well-Tempered Clavier—Part I, BWV 846–869. Henle Urtext Edition, Munich (1997)
3. Bach, J.S.: The Well-Tempered Clavier—Part II, BWV 870–893. Henle Urtext Edition, Munich (2007)
4. Chuan, C.-H., Chew, E.: Creating ground truth for audio key-finding: when the title key may not be the key. In: Proceedings of the International Conference on Music Information Retrieval (2012)
5. Jonas, O.: Introduction to the Theory of Heinrich Shenker: the Nature of the Musical Work of Art. Longman, New York (1982)
6. Krumhansl, C.L., Kessler, E.J.: Tracing the dynamic changes in perceived tonal organisation in a spatial representation of musical keys. Psychol. Rev. **89**(4), 334–368 (1982)
7. Krumhansl, C.L.: Cognitive Foundations of Musical Pitch. Oxford University Press, New York (1990)
8. Lerdahl, F., Jackendoff, R.: A Generative Theory of Tonal Music. MIT Press, Cambridge (1983)
9. Lewin, D.: Generalized Musical Intervals and Transformations. Yale University Press, New York (1987)
10. Longuet-Higgins, H.C., Steedman, M.J.: On interpreting Bach. In: Meltzer, B., Michie, D. (eds.) Machine Intelligence 6, pp. 221–241. Edinburgh University Press, Edinburgh (1971)
11. Temperley, D.: The Cognition of Basic Musical Structures. MIT Press, Cambridge (2001)

Part IV
Segmentation

Chapter 6
Determining Key Boundaries

Abstract Computer models for determining key boundaries are important tools for computer analysis of music, computational modeling of music cognition, content-based categorization and retrieval of music information and automatic generating of expressive performance. This chapter describes a Boundary Search Algorithm (BSA) for determining points of modulation in a piece of music using a geometric model for tonality called the Spiral Array. For a given number of key changes, the computational complexity of the algorithm is polynomial in the number of pitch events. We present and discuss computational results for two selections from J.S. Bach's *A Little Notebook for Anna Magdalena*. Comparisons between the choices of an expert listener and the algorithm indicates that in human cognition, a dynamic interplay exists between memory and present knowledge, thus maximizing the opportunity for the information to coalesce into meaningful patterns.

This chapter proposes a method for segmenting a piece of music into its respective key areas using the Spiral Array model. The *key* of a musical piece imparts information about the principal pitch set of the composition. More importantly, it identifies the most stable pitch, the *tonic*, and its hierarchical relations to all other pitches in that pitch set. The *tonic* is sometimes referred to as the *tonal center*, or simply as "doh." Once the key is established, ensuing notes are heard in reference to the tonic. This reference point allows the listener to navigate through the soundscape generated by the piece. The computational modeling of the human ability to determine the key of a melody is a well-studied problem and numerous methods have been proposed for mimicking this cognitive ability. See, for example, [3, 5, 9, 11, 13, 15–18], and [1, 14] in [8].

Modulation or the change of key is a common compositional device. When the key changes, each pitch assumes a different role in relation to the new tonal center.

This chapter is a minor revision of the work, "The Spiral Array: An Algorithm for Determining Key Boundaries" by E. Chew, published in Anagnostopoulou, C., Ferrand, M., Smaill, A. (eds.): Music and Artificial Intelligence, LNCS/LNAI **2445**, pp. 18–31. Springer-Verlag, Heidelberg (2002).

E. Chew, *Mathematical and Computational Modeling of Tonality*,
International Series in Operations Research & Management Science 204,
DOI: 10.1007/978-1-4614-9475-1_6,

This shift in paradigm transforms the tonal context in which the notes are heard. One way in which the human mind organizes music information into coherent chunks is by way of its key contexts. Conversely, key changes often occur at structural boundaries. The skill with which a composer moves persuasively or abruptly from one context to another constitutes an important feature of her style. A performer may choose to highlight unusual modulations through changes in the dynamics or tempo. These are but a few examples to illustrate the importance of developing robust and computationally viable methods for determining key boundaries. Computer modeling of the cognitive task of determining key boundaries is potentially an exponentially hard problem, and very little work has been done in this area.

By using the Spiral Array and by incorporating music knowledge in modeling the problem, this chapter proposes a computationally efficient, knowledge-based way to determine points of modulation in a piece of music. The Spiral Array is a spatial arrangement of pitch classes, chords, and keys proposed by Chew [4] as a geometric model for *tonality*, the system of musical principles underlying tonal music. According to Bamberger [2], *Tonality and its internal logic frame the coherence among pitch relations in the music with which (we) are most familiar.* The perception of key results from this system of hierarchical relations amongst pitches.

The Spiral Array is derived from the Harmonic Network (described in various contexts in [6, 7, 10–12]), an arrangement of pitch classes on a lattice. Chord pitches and pitches in a key map to compact sets on this grid. The Spiral Array preserves the clustering of tonally important pitch sets. In addition, its 3D configuration allows higher level musical entities to be defined and embedded in the interior of the structure. A generating principle in the Spiral Array is that higher level musical entities can be represented unambiguously as convex combinations of their lower level components. Spatial coordinates representing each key were generated from weighted averages of its I, V, and IV chord representations, which were in turn generated from each chord's root, fifth and third pitch representations.

In the Spiral Array, musical information is condensed and summarized by 3D coordinates that represent the *center of effect* (CE) of the pitches. This idea was extended in the CEG algorithm [5] to generate CEs that mapped the establishing of key over time to spatial trajectories. The distance from the moving CE to the key representations serves as a likelihood indicator that the piece is in each of those keys. The distance-minimizing key is elected as the most likely solution. It was shown that the CEG algorithm offers a competitive and viable means to identify the key of a melody. See Chaps. 4 and 5 for a description and an evaluation of the CEG algorithm.

In this chapter, we show that determining the points of modulation can be recast as a problem of finding the distance-minimizing boundaries for the different key contexts using the Spiral Array framework. For this reason, we call the algorithm the Boundary Search Algorithm (BSA). A straightforward implementation of the algorithm yields an $O(n^m)$ performance time, where n is the number of event onsets and m (≥ 2) is the number of boundaries. With more information, a listener can assess the key identity with more confidence, and correspondingly, the algorithm works best when n is large and m is small.

Given a piece of music, exhaustively enumerating all segmentation possibilities based on key boundaries is a formidable and exponential task. We show that by integrating musical knowledge, one can drastically reduce the search space and propose a tractable polynomial-time solution to the problem of determining key boundaries for the two Bach examples addressed in this chapter. Realistically speaking, the number of key changes, m, is typically a small number in relation to the number of event onsets, n. The two Bach examples each contain one point of departure and one return to a primary key area, that is to say, $m = 2$. The problem is further constrained by the fact that the first and third (last) key areas must be identical.

Apart from computational tractability, the benefits of a knowledge-based approach to determining key boundaries also include more accurate models for computer analysis and tracking of music, a better understanding of the perception of musical structure, more realistic computer-generated expressive performances, and more robust algorithms for content-based music information retrieval.

Comparisons between a human expert's perception of key boundaries and the algorithm's choices indicate that in human processing of music information, a dynamic interplay exists between past (memory) and present knowledge, thus maximizing the opportunity for the information to coalesce into meaningful patterns. Also, the human's perception of key boundaries is influenced by phrase structure.

6.1 A Mathematical Model for Tonality

Any computer model for processing music information must begin with a representation that converts content information into numerical data for analysis. The representation we use is the Spiral Array model first introduced in [4]. The Spiral Array grew out of a 3D configuration of the Harmonic Network to eliminate periodicity and redundancy (see Fig. 6.1). The use of this geometric model to determine the key of a musical passage with a likelihood score based on spatial proximity is outlined in [5] and in Chap. 4. We briefly describe how key representations are generated in this structure.

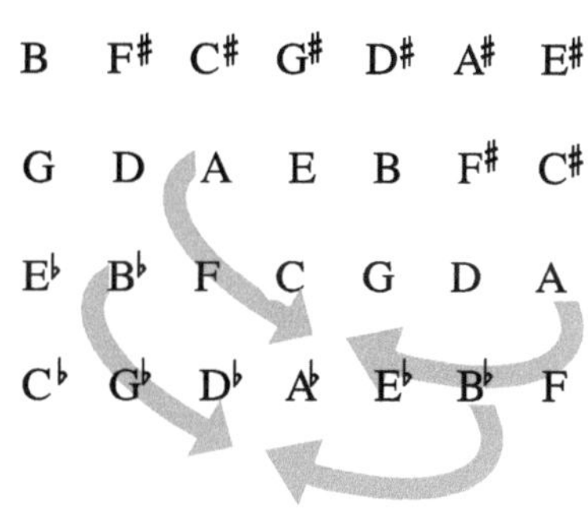

Fig. 6.1 Rolling up the harmonic network to eliminate redundancy

6.1.1 Placement of Pitch Classes, Chords and Keys

In the Spiral Array, pitch classes are represented by spatial coordinates along a spiral. Each pitch class is indexed by its number of perfect fifths from an arbitrarily chosen reference pitch, C (set at position [0, 1, 0]). An increment in the index results in a quarter turn along the spiral. Four quarter turns places pitch classes related by a major third interval in vertical alignment with each other. Strict enharmonic equivalence would require that the spiral be turned into a toroid. The spiral form is assumed so as to preserve the symmetry in the distances among pitch entities.

Definition 6.1 The radius of the cylinder, r, and the height gain per quarter turn, h, uniquely define the position of a pitch representation, which can be described as

$$\mathbf{P}(\mathrm{k}) \stackrel{\text{def}}{=} \begin{bmatrix} x_\mathrm{k} \\ y_\mathrm{k} \\ z_\mathrm{k} \end{bmatrix} = \begin{bmatrix} \mathrm{r}\sin\frac{\mathrm{k}\pi}{2} \\ \mathrm{r}\cos\frac{\mathrm{k}\pi}{2} \\ \mathrm{kh} \end{bmatrix}.$$

Chord representations are defined uniquely as convex combinations of their component pitches. Triads map to nonoverlapping triangles on the Spiral Array. Hence, each triangle can be uniquely represented by the convex combination of its component pitches. The weights are constrained to be monotonically decreasing from the root, to the fifth, to the third to mirror the relative importance of each pitch to the chord.

Definition 6.2 The representation for a major triad is generated by the convex combination of its root, fifth and third pitch positions:

$$\mathbf{C}_\mathrm{M}(\mathrm{k}) \stackrel{\text{def}}{=} w_1 \cdot \mathbf{P}(\mathrm{k}) + w_2 \cdot \mathbf{P}(\mathrm{k}+1) + w_3 \cdot \mathbf{P}(\mathrm{k}+4),$$

$$\text{where} \quad w_1 \geq w_2 \geq w_3 > 0 \text{ and } \sum_{i=1}^{3} w_i = 1.$$

The minor triad is generated by a similar combination,

$$\mathbf{C}_\mathrm{m}(\mathrm{k}) \stackrel{\text{def}}{=} u_1 \cdot \mathbf{P}(\mathrm{k}) + u_2 \cdot \mathbf{P}(\mathrm{k}+1) + u_3 \cdot \mathbf{P}(\mathrm{k}-3),$$

$$\text{where} \quad u_1 \geq u_2 \geq u_3 > 0 \text{ and } \sum_{i=1}^{3} u_i = 1.$$

Major key representations are generated as convex combinations of their I, V, and IV chords. Figure 6.2 shows an example of a major key representation. The weights are constrained to be monotonically nonincreasing from I to V to IV.

Definition 6.3 A major key is represented by a convex combination of its tonic, dominant and subdominant chords. The weights are restricted to be monotonic and

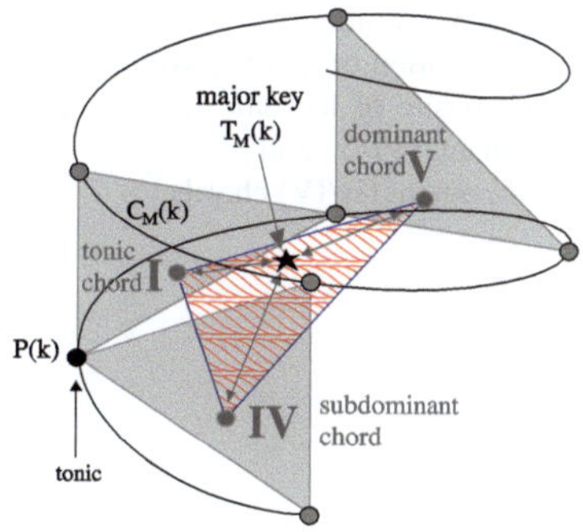

Fig. 6.2 Geometric representation of a major key, a composite of its tonic (I), dominant (V), and subdominant (IV) chords

nonincreasing.

$$\begin{aligned}\mathbf{T}_\mathrm{M}(\mathrm{k}) &\stackrel{\mathrm{def}}{=} \omega_1 \cdot \mathbf{C}_\mathrm{M}(\mathrm{k}) + \omega_2 \cdot \mathbf{C}_\mathrm{M}(\mathrm{k}+1) + \omega_3 \cdot \mathbf{C}_\mathrm{M}(\mathrm{k}-1),\\ \text{where} &\quad \omega_1 \geq \omega_2 \geq \omega_3 > 0 \text{ and } 0 \leq \alpha \leq 1, 0 \leq \beta \leq 1.\end{aligned}$$

The minor key representation is a combination of the i, (V/v) and (iv/IV) chords. The additional parameters α and β model the relative importance and usage of V versus v and iv versus IV chords, respectively, in the minor key.

Definition 6.4 The minor key representation is generated by a convex combination of its tonic, dominant (major or minor), and subdominant (major or minor) chords as follows:

$$\begin{aligned}\mathbf{T}_\mathrm{m}(\mathrm{k}) &\stackrel{\mathrm{def}}{=} \upsilon_1 \cdot \mathbf{C}_\mathrm{m}(\mathrm{k})\\ &\quad + \upsilon_2 \cdot [\alpha \cdot \mathbf{C}_\mathrm{M}(\mathrm{k}+1) + (1-\alpha) \cdot \mathbf{C}_\mathrm{m}(\mathrm{k}+1)]\\ &\quad + \upsilon_3 \cdot [\beta \cdot \mathbf{C}_\mathrm{m}(\mathrm{k}-1) + (1-\beta) \cdot \mathbf{C}_\mathrm{M}(\mathrm{k}-1)],\\ \text{where} &\quad \upsilon_1 \geq \upsilon_2 \geq \upsilon_3 > 0 \text{ and } \upsilon_1 + \upsilon_2 + \upsilon_3 = 1,\\ \text{and} &\quad 0 \leq \alpha \leq 1, 0 \leq \beta \leq 1.\end{aligned}$$

See Fig. 6.3 for the spatial representation of a minor key.

For the two examples in this chapter, the key representations in the Spiral Array were generated by setting the weights (w, u, ω, υ) to be the same and equal to [0.516, 0.315, 0.168], $\mathrm{h} = \sqrt{2/15}$ ($\mathrm{r} = 1$). α is set to 1 and β to 0. These weights model approximately the perceived interval relations: pitches related by a distance of a perfect fifth are no farther than those a major third apart; two pitches an interval of a half-step apart generate a center that is closest to the key of the upper pitch; and, the coordinates of a pitch class are closest to the key representation of the same name.

The Spiral Array turns the original 2D lattice of pitch classes into a 3D configuration that eliminates periodicity. By moving to a higher dimensional space, the Spiral Array allows the definition of higher level musical entities as coordinates in the same space as the pitch classes, blending both discrete and continuous space.

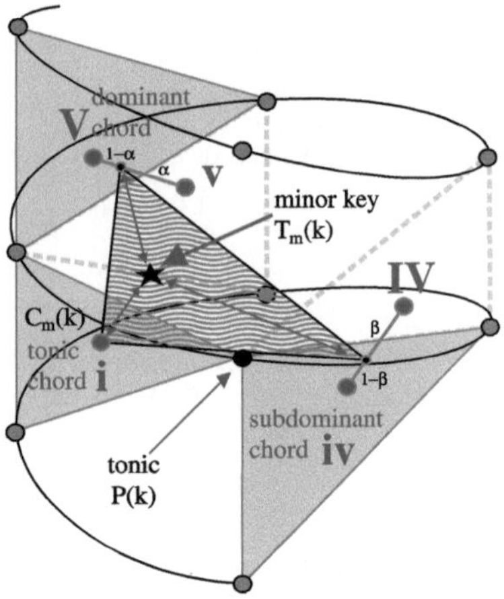

Fig. 6.3 Geometric representation of a minor key, a composite of its tonic i, dominant (V/v), and subdominant (iv/IV) chords

6.2 An Algorithm for Finding Key-Change Boundaries

The idea behind the Boundary Search Algorithm (BSA) is a simple one. The BSA collapses each set of pitch and duration information down to a spatial point, a center of effect (CE). This point is compared to the key representations in the Spiral Array Model. The better the pitch collection adheres to the pitch classes of the key in the proportions determined by their tonal relations, the closer the CE will be to the key representation.

When two or more key areas exist in a piece, the key areas are delineated by modulation boundaries. The problem of finding the point(s) of modulation can then be posed as a problem of determining the placement of boundaries so as to minimize the distances between the respective CEs and their closest keys. This section presents a formal description of the BSA algorithm. Section 6.2.1 describes how music information is condensed to a spatial coordinate for comparison to a key object in the Spiral Array. Section 6.2.2 gives the problem statement the solution to which will determine the optimal choice of modulation boundaries.

6.2.1 The Center of Effect and Distance to Key

A musical passage consists of a collection of pitch events, each comprising pitch and duration information. If $\mathbf{p}_i$ represents the position of the ith note's pitch class in the Spiral Array, and d_i represents its duration, the collection of N notes can be written as $\{\mathbf{p}_i, d_i): i = 1 \ldots N\}$. The center of effect of this pitch collection is defined as

$$\mathbf{c} \stackrel{\text{def}}{=} \sum_{i=1}^{N} \frac{d_i}{D} \cdot \mathbf{p}_i, \qquad \text{where } D = \sum_{i=1}^{N} d_i$$

If $\mathbf{T}$ is a finite collection of keys represented in the Spiral Array, both major and minor, that is to say

$$\mathbf{T} = \{\mathbf{T}_{\mathrm{m}}(\mathrm{k})\,\forall\,\mathrm{k}\} \cup \{\mathbf{T}_{\mathrm{M}}(\mathrm{k})\,\forall\,\mathrm{k}\},$$

then the most likely key of the passage is given by

$$\arg\min_{T \in \mathbf{T}} \|\mathbf{c} - T\|,$$

which can be readily implemented using a nearest-neighbor search. The likelihood that the pitch collection is in any given key is indicated by its proximity to that key. The minimizing distance, also a likelihood indicator, is given as

$$d^{\min} = \min_{T \in \mathbf{T}} \|\mathbf{c} - T\|.$$

6.2.2 Problem Statement and Algorithm

Suppose that m boundaries have been chosen as given by $(\mathrm{B}_1, \ldots, \mathrm{B}_m)$. Let time 0 be B_0, and the end of the passage be $\mathrm{B}_{m+1} = \mathrm{L}$. The note material between any two boundaries $(\mathrm{B}_i, \mathrm{B}_{i+1})$ generates a CE given by $\mathbf{c}_{(\mathrm{B}_i, \mathrm{B}_{i+1})}$. The most likely key has a distance $d^{\min}_{(\mathrm{B}_i, \mathrm{B}_{i+1})}$ from this CE.

Our hypothesis is that the best boundary candidates will segment the piece in such a way as to minimize the distances $d^{\min}_{(\mathrm{B}_i, \mathrm{B}_{i+1})}$, $i = 0, \ldots, m$,. Hence, the goal is to find the boundaries $(\mathrm{B}_1, \ldots, \mathrm{B}_m)$ that minimize the sum of these distances. Let n_i be the number of pitch events from the beginning up to the boundary B_i, with the appropriate boundary value of $n_0 = 0$. The optimal boundaries, $(\mathrm{B}_1^*, \ldots, \mathrm{B}_m^*)$, are then given by the solution to the following system:

$$\begin{aligned}
\min \quad & \sum_{i=0}^{m} d^{\min}_{(\mathrm{B}_i, \mathrm{B}_{i+1})} \\
\text{s.t.} \quad & d^{\min}_{(\mathrm{B}_i, \mathrm{B}_{i+1})} = \min_{T \in \mathbf{T}} \|\mathbf{c}_{(\mathrm{B}_i, \mathrm{B}_{i+1})} - T\|, \qquad i = 0, \ldots, m \\
& \mathbf{c}_{(\mathrm{B}_i, \mathrm{B}_{i+1})} = \sum_{j=n_i+1}^{n_{i+1}} \frac{d_j}{D_i} \cdot \mathbf{p}_j, \text{ where } D_i = \sum_{k=n_i+1}^{n_{i+1}} d_k, \qquad i = 0, \ldots, m, \\
\text{and} \quad & \mathrm{B}_i < \mathrm{B}_{i+1}, \qquad i = 0, \ldots, m.
\end{aligned}$$

A naive and viable approach, for small problems, is to enumerate all possible sets of boundaries and to evaluate each objective function value numerically in order to find the objective-minimizing boundaries. Assuming that the set of keys that are of interest, **T**, is finite and that any piece of music is of finite length with n event onsets, the computational complexity of this approach is $O(n^m)$.

The practical application of this method typically yields far better performance than that implied by the $O(n^m)$ computational time. The number of key changes and associated boundaries is typically small in relation to the number of event onsets ($m \ll n$). In smaller scale compositions such as the two Bach examples from *A Little Notebook for Anna Magdalena* discussed in the next section, the pieces can typically be segmented into three key areas.

Using music knowledge and a little common sense, we can add a few constraints to further reduce the search space. Adjacent key areas should be distinct from each other, that is to say, $T_i \neq T_{i+1} (i = 1 \ldots m)$. If the passage to be analyzed is a complete piece, the first and last key areas could be constrained to be the same, with the corresponding constraint being $T_1 = T_{m+1}$. For example, in pieces with three key areas, the first and last segments are most often in the same key as that of the composition, while the middle portion is in a different key.

For longer pieces, dynamic programming can be invoked to compute the best distance minimizing boundaries.

6.3 Two Examples

We now proceed to apply the solution framework to two examples from J. S. Bach's *A Little Notebook for Anna Magdalena*. The two examples chosen each contain one departure and one return to the primary key area. The task at hand is to find the two boundaries, B_1 and B_2 separating the primary and the secondary key areas.

In these two problem instances, the three segments are required to satisfy the constraint that parts 1 and 3 must be in the same key ($T_1 = T_3$), and that this key must be distinct from that of part 2 ($T_1 \neq T_2$). These additional constraints greatly reduce the solution space. The reduced and modified problem instance is as follows:

$$
\begin{aligned}
\min \quad & d^{\min}_{(\mathrm{B}_0,\mathrm{B}_1)} + d^{\min}_{(\mathrm{B}_1,\mathrm{B}_2)} + d^{\min}_{(\mathrm{B}_2,\mathrm{B}_3)} \\
\text{s.t.} \quad & d^{\min}_{(\mathrm{B}_i,\mathrm{B}_{i+1})} = \|\mathbf{c}_{(\mathrm{B}_i,\mathrm{B}_{i+1})} - T_i\|, \qquad i = 0, 1, 2, \\
& T_i = \arg\min_{T \in \mathbf{T}} \|\mathbf{c}_{(\mathrm{B}_i,\mathrm{B}_{i+1})} - T\|, \qquad i = 0, 1, 2, \\
& \mathbf{c}_{(\mathrm{B}_i,\mathrm{B}_{i+1})} = \sum_{j=n_i+1}^{n_{i+1}} \frac{d_j}{D_i} \cdot \mathbf{p}_j \text{ where } D_i = \sum_{k=n_i+1}^{n_{i+1}} d_k, \qquad i = 0, 1, 2, \\
& T_1 = T_3 \neq T_2, \\
\text{and} \quad & 0 = \mathrm{B}_0 < \mathrm{B}_1 < \mathrm{B}_2 < \mathrm{B}_3 = \mathrm{L}.
\end{aligned}
$$

In both instances, the BSA method identified the correct keys, $T_1(= T_3)$ and $T_2(\neq T_1)$, and determined the boundaries B_1 and B_2. These boundaries are compared to a human expert's choices. The BSA's boundaries (A1, A2) and the expert's choices (EC1, EC2) are close but not identical in both cases. Sections 6.3.1 and 6.3.2 compare and discuss the rationale behind, and the relative merits of, these solutions.

6.3.1 *Example 1:* Minuet in G *by Bach*

The first example is Bach's Minuet in G from *A Little Notebook for Anna Magdalena*. The Minuet is in the key of G, with a middle section in D major. Figure 6.4 shows a segment of the piece that contains the transitions away from and back to the intended key G. In the figure are markings denoting a human expert's choice of boundaries (indicated by EC1 and EC2) between the keys of G and D. Also marked are the BSA's choices for these same boundaries (indicated by A1 and A2).

The difference between A1 and EC1 is one between a synthetic and analytic choice. The human expert chose EC1, a boundary that occurs before the D major key is established conclusively in bar 20. In light of the D major key context established in bar 20, bar 19 is heard retroactively as being in the key of D major rather than G. In D major, the right-hand melody in bar 19 traverses the scale degrees $\{\hat{4}, \hat{2}, \hat{3}, \hat{4}, \hat{1}\}$. It is not an uncommon practice for a human listener to retroactively frame the past in the light of present knowledge. In addition, EC1 also prioritizes phrase symmetry by delineating a 2-bar + 2-bar phrase structure.

Fig. 6.4 Modulation boundaries for Bach's Minuet in G (EC = human expert; A = BSA algorithm)

The Boundary Search Algorithm described in Sect. 6.2 was not designed to have backward and forward analysis capabilities. It selects the boundaries that produce pitch event groupings in such a way that each grouping maps to point clusters on the Spiral Array that aggregate to points closest to the respective key representations. Based on this selection criteria, the BSA chooses the beginning of bar 20 to mark the beginning of the D major section. This choice, in fact, agrees with the textbook definition of modulation.

The textbook boundary for the next key change falls squarely between EC2 and A2. The human expert's decision was influenced by the 4-bar phrasing established from the beginning of the piece. Giving preference to boundaries that lock into this 4-bar period, the human listener waited for the latest phrase to end (at the end of the bar) before placing EC2 at the barline. With this choice of boundary, the D and C♮ on beats two and three in bar 24 are heard as an ending of the middle section. A performer, if attempting to highlight this fact, would play it quite differently than if these two notes were the beginning of the next section.

Now consider A2. The BSA chose to group the notes in beats two and three of bar 24 with the final G major section, a choice that is closer in spirit to the textbook solution. In this case, the D on beat two of bar 24 is heard as the fifth degree $\hat{5}$ in G major, and not the first degree $\hat{1}$ in D major. Strictly speaking, the C♮ is the first note that belongs to G major and not D. Thus, the textbook boundary would start the G major area on beat three of bar 24.

Both the algorithm's and the expert's choices for modulation boundaries are valid for different reasons. The human expert's choices involved retroactive decisions that framed past information in the light of present knowledge. The human's choices were also influenced by phrase structure, preferring to reinforce the 4-bar phrase boundaries. The BSA considered only the pitch events, taking the pitches literally and choosing the boundaries accordingly.

6.3.2 *Example 2:* March in D *by Bach*

The second example is Bach's March in D, also from *A Little Notebook for Anna Magdalena*. This piece was chosen because it is a little more complex than the previous Minuet example. In the transition from A major back to D major, the piece hints at a number of different keys before returning unequivocally to D major. The A major middle section shifts to D major in the second beat of bar 12, portending the return to D major. The D acts as a V to the G major tonality at the end of bar 13; a G♯ in the bass at the end of bar 14 acts as a leading note to A major; there is a momentary hint of B minor between bars 15 and 16 , followed by a root movement by fifths–E minor, A major, D major, [G major]–with the key of D major established by bar 18. Clearly, this is a less than perfect example with which to test an algorithm that looks for boundaries between only three parts. The salient portion of the piece is shown in Fig. 6.5, with the same labeling convention as before.

The first set of boundaries, EC1 and A1, were almost identical. The BSA breaks up the phrase structure in bar 6 to include the last beat in the new A major section, while the human expert waited for the completion of the phrase to place the boundary EC1. The slight discrepancy resulted because the computer algorithm was not designed to recognize the composer's phrasings. Otherwise, the two choices are not very different. However, it is important to note that G♯ occurs on the first beat of bar 6, an indication of the new key area, A major; so, a textbook boundary should begin the A major section at the G♯.

For the second set of boundaries, marked EC2 and A2 respectively in Fig. 6.5, the choices differed quite a bit. The human expert took into account the sequential pattern starting from the middle of bar 13 and ending at the climax in bar 16, and chose to think of the final D major section as beginning in beat one of bar 18. The BSA, on the other hand, did not account for figural groupings and sequential patterns in the notes. It chose, instead, to begin the last D major section as soon as the note material of the third part agreed with the pitch collection for D major.

Fig. 6.5 Modulation boundaries for Bach's March in D (EC = human expert; A = BSA algorithm)

6.4 Conclusions

The Boundary Search Algorithm is designed to segment a stream of music data into contiguous sets by key. The algorithm gave accurate assessments of the goodness of fit between each collection of pitch events and the key candidate to suggest optimal placements of key boundaries. For the purposes of this application, the CE defined in Sect. 6.2.1 summarizes only pitch and duration information. The method does not account for pitch order within each data set, nor does it incorporate beat information. With only pitch and duration information, the BSA was demonstrated to perform well in the problem of determining modulation boundaries. Given only pitch and duration information, a human would probably perform in a comparable manner. Rough temporal information is embodied in the sequence of the key segments.

The BSA does not explicitly use chord functions to determine the key areas. Chord membership and functional relations to each key area are incorporated into the Spiral Array model by design. With some additional work, a functional analysis of a musical passage can be extracted from mappings of the data to the Spiral Array model. The problem of harmonic analysis can perhaps be addressed in a future paper.

In addition, the implementation of the BSA as described in this chapter is designed to analyze a piece of music in its entirety. In order for the algorithm to mimic human musical perception as information is revealed over time, the ideas outlined in this chapter can be adapted to detect key boundaries in real time. For example, in a real time system, analyses can be performed at each time increment and the number of boundaries, m, allowed to increase by at most one.

Finally, the human mind in apprehending music information performs more than sequential data processing. Pitch and duration data are but a fraction of the total information. The mind also organizes music data into figural groupings and phrase structures, and it can sometimes revise prior assessments based on new information. The analysis of the difference in the human and algorithmic solutions indicates that a dynamic interplay exists between memory and present knowledge so that the chances of finding meaningful patterns in the time series data is maximized.

Acknowledgments I thank Jeanne Bamberger for her guidance and cogent advice that made this research possible; and Martin Brody for his insightful comments on interpreting the results in an early version of this document.

References

1. Auhagen, W., Vos, P.G.: Experimental methods in tonality induction research: a review. Music Percept. **17**(4), 417–436 (2000)
2. Bamberger, J.S.: Developing Musical Intuition. Oxford University Press, New York (2000)
3. Brown, R.: Tonal implications of the diatonic set. In Theory Only **5**, 3–21 (1981)
4. Chew, E.: Towards a mathematical model of tonality. Ph.D. Dissertation, MIT (2000)
5. Chew, E.: Modeling tonality: applications to music cognition. In: Proceedings of the 23rd Annual Meeting of the Cognitive Science Society (2001)

6. Cohn, R.: Neo-Riemannian operations, parsimonious trichords, and their tonnetz representations. J. Music Theory **41**(1), 1–66 (1997)
7. Cohn, R.: Introduction to Neo-Riemannian theory: a survey and a historical perspective. J. Music Theory **42**(2), 167–180 (1998)
8. Gjerdingen, R. (ed.): Music Percept. **17**, 4 (2000) Note: Special issue on tonality induction.
9. Krumhansl, C.L.: Cognitive Foundations of Musical Pitch. Oxford University Press, New York (1990)
10. Lewin, D.: Generalized Musical Intervals and Transformations. Yale University Press, CT (1987)
11. Longuet-Higgins, H. C., Steedman, M. J.: On interpreting Bach. B. Meltzer and D. Michie (eds.). Machine Intelligence, vol. 6, p. 221. Edinburgh U Press, Edinburgh (1971)
12. Longuet-Higgins, H.C.: Mental Processes. MIT Press, MA (1987)
13. Rowe, R.: Key induction in the context of interactive performance. Music Percept. **17**(4), 511–530 (2000)
14. Shmulevich, I., Yli-Harja, O.: Localized key-finding: algorithms and applications. Music Percept. **17**(4), 531–544 (2000)
15. Steedman, M.: The well-tempered computer. Phil. Trans. R. Soc. Lond. **349**, 115–131 (1994)
16. Temperley, D.: The perception of harmony and tonality: an algorithmic approach. Ph.D. Dissertation, Columbia University (1996)
17. Temperley, D.: What's key for key? The Krumhansl-Schmuckler key-finding algorithm reconsidered. Music Percept. **17**(1), 65–100 (1999)
18. Vos, P.G., Van Geenen, E.W.: A parallel-processing key-finding model. Music Percept. **14**(2), 185–223 (1996)

Chapter 7
Argus Segmentation Method

Abstract This chapter describes an $O(n)$ algorithm for segmenting music automatically by pitch context using the Spiral Array, a mathematical model for tonality, and applies it to the segmentation of post-tonal music, namely, Olivier Messiaen's Regards IV and XVI from his *Vingt Regards sur l'Enfant Jésus*. Using the idea of the *center of effect* (CE), a summary point in the interior of the Spiral Array, segmentation boundaries map to peaks in the distances between the CEs of adjacent segments of music. The best-case computed boundaries are, on average, within 0.94 % (for Regard IV) and 0.11 % (for Regard XVI) of their targets.

7.1 Introduction

Segmentation by context is a necessary part of music processing both by humans and by machines. Efficient and accurate algorithms for performing this task are critical to computer analysis of music, to the analysis and rendering of musical performances, and to the indexing and retrieval of music using large-scale datasets. Computational modeling of the segmentation process can also lead to insights into human cognition of music. This chapter will focus on the problem of determining boundaries that segment a piece of music into contextually similar sections according to pitch content. In particular, the algorithm will be applied to Olivier Messiaen's (1908–1992) Regard de la Vierge and Regard des prophètes, des bergers et des Mages, the fourth and sixteenth pieces in his *Vingt Regards sur l'Enfant Jésus* [9].

The $O(n)$ segmentation method uses the Spiral Array model [1], a mathematical model that arranges musical objects in three-dimensional space so that inter-object distances mirror their perceived closeness. The Spiral Array represents tonal objects

This is a minor revision of the work, "Regards on Two Regards by Messiaen: Post-tonal Music Segmentation Using Pitch Context Distances in the Spiral Array" by Elaine Chew, published in the Journal of New Music Research **34**(4), 341–354 (2005).

E. Chew, *Mathematical and Computational Modeling of Tonality*, International Series in Operations Research & Management Science 204,
DOI: 10.1007/978-1-4614-9475-1_7,

at all hierarchical levels in the same space and uses spatial points in the model's interior to summarize and represent segments of music. The array of pitch representations in the Spiral Array is akin to Longuet-Higgins' Harmonic Network [8] and the tonnetz of neo-Riemannian music theory [5]. The key spirals, generated by mathematical aggregation, bear structural similarity to Krumhansl's network of key relations [6], constructed using multidimensional scaling of experimental data, and key representations in Lerdahls tonal pitch space [7], formulated from a metric based on a hierarchical (chromatic, diatonic, and triadic) arrangement of pitch relations.

The present segmentation algorithm, named Argus, is the latest in a series of computational analysis techniques utilizing the Spiral Array. The Spiral Array model has been used in the design of algorithms for key-finding [2], pitch spelling [4], and offline determining of key boundaries [3]. The previous model for determining key boundaries is the one most relevant to this chapter. This earlier algorithm requires knowledge of the entire piece and the total number of segments, and does not compute in real-time. The present algorithm scans the piece once from beginning to end, and thus can be realized as a real-time method for computing segmentation boundaries. So far, the Spiral Array has only been used in the analysis of tonal music. This chapter extends the Spiral Array model's applications to the analysis of post-tonal music. The current Argus method is the first that explicitly uses distances in the Spiral Array as a measure of pitch context difference or separation.

The Spiral Array consists of an array of nested spirals each representing a kind of tonal entity at some hierarchical level. The Argus algorithm for automatic segmentation uses only the outermost pitch spiral in the Spiral Array model and the interior space to compute a distance between the local contexts (captured by a pair of sliding windows) immediately before and after each point in time. The distance measure peaks at a segmentation boundary and the peaks can be used to identify such boundaries. Since the segmentation algorithm detects boundaries between sections employing distinct pitch collections, the procedure does not depend on key context and can be applied in general to both tonal and atonal music.

The Argus method is highly efficient, computing in O(n) time, and requires only one left-to-right scan of the piece. The algorithm computes in real-time; however, due to the formulation of the problem, there will necessarily be a lag in the assignment of a boundary that is equivalent to the length of the forward window. The algorithm is tested on Messiaen's Regards IV and XVI and the results presented for various window sizes. The computational results are compared to manual segmentations of the piece.

The related work includes that on finding local tonal context, for example, Temperley's dynamic programming approach to determining local key context [11], Shmulevich and Yli-Harja's median filter approach to local key-finding [10], and Toiviainen and Krumhansl's self-organizing map approach to determining and visualizing varying key strengths over time [12]. These methods center around key-finding, which applies only to tonal music; and the focus of these methods is on the determining of the local key context rather than on finding segmentation boundaries. Segmentation in general is an extremely broad topic, and touches upon motivic patterns, rhythmic structure, metric groupings, phrase groupings, harmonic sequences, and timbral quality, amongst a host of other features. For every recognizable aspect of music for which computational algorithms can be devised, there can be a corresponding study on segmentation by that feature. As a result, the literature review of segmentation in general is beyond the scope of this chapter.

The remainder of the chapter presents a concise overview of the Spiral Array model followed by a description of the segmentation algorithm. Then, in Sect. 7.3, a descriptive analysis of Regard IV is followed by a detailed analysis of the computational results and comparisons between the computed and manually assigned boundaries, and geometric explanations of the effectiveness of the proposed method. A similar treatment of Regard XVI is presented in Sect. 7.4, followed by discussions in Sect. 7.5, and conclusions in Sect. 7.6.

7.2 The Spiral Array Model

This section provides an overview of the structure of, and the underlying concept (namely, the *center of effect*) behind, the Spiral Array model [1]. The description of the proposed segmentation algorithm follows the introduction to the Spiral Array.

7.2.1 The Model's Structure

The Spiral Array model represents pitches on a spiral so that spatially close pitch representations form familiar higher level tonal structures such as triads and keys. The model represents each higher level object as the convex combination of its lower level components. These weighted sums of representations of the components result in spatial points in the interior of the pitch spiral. For example, Fig. 7.1 shows the hierarchical construction of major key representations, from pitches to triads to keys.

Figure 7.1a shows the pitch spiral. Adjacent pitches along the spiral are related by intervals of a perfect fifth; each turn of the spiral contains four pitch representations, and as a result, vertical neighbors are a major third apart. Pitch representations can be generated by the following equation:

$$\mathbf{P}(\mathrm{k}) = \begin{bmatrix} x_\mathrm{k} \\ y_\mathrm{k} \\ z_\mathrm{k} \end{bmatrix} = \begin{bmatrix} \mathrm{r}\sin\frac{\mathrm{k}\pi}{2} \\ \mathrm{r}\cos\frac{\mathrm{k}\pi}{2} \\ \mathrm{kh} \end{bmatrix}, \tag{7.1}$$

where r is the radius of the spiral and h is the vertical ascent per quarter turn. Each pitch representation is indexed by its number of perfect fifths from a reference pitch. For example, C is arbitrarily mapped to the index 0, and represented by $\mathbf{P}(0)$. G, a perfect fifth above C, is mapped to the index 1, and represented by $\mathbf{P}(1)$. The model

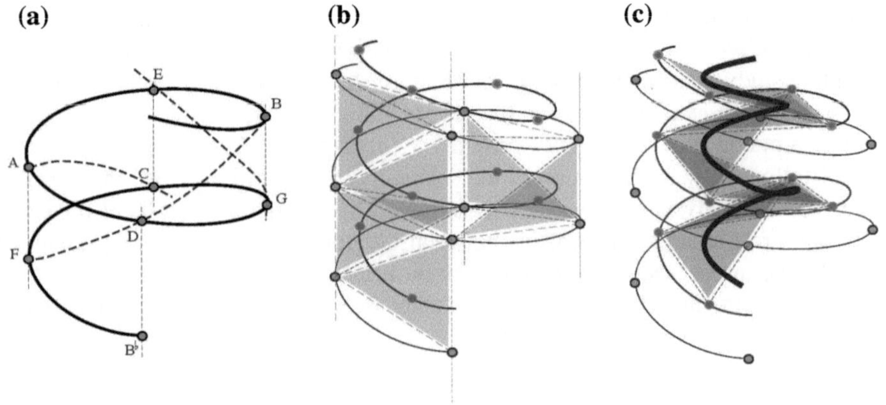

Fig. 7.1 The Spiral Array model. **a** Pitch spiral, **b** major triad representations, **c** major key representations

assumes octave equivalence so that all pitches with the same letter name map to the same spatial point on the pitch spiral.

Based on the arrangement of pitches on the pitch spiral, pitches that define a triad form compact clusters. They also form the vertices of a triangle. Each triad is represented by the convex combination of its component pitches, that is to say, a point in the interior of the triangle. For example, the major triad is defined as

$$\mathbf{C}_{\mathrm{M}}(\mathrm{k}) = w_1 \cdot \mathbf{P}(\mathrm{k}) + w_2 \cdot \mathbf{P}(k+1) + w_3 \cdot \mathbf{P}(\mathrm{k}+4), \tag{7.2}$$

$$\text{where} \quad w_1 \geq w_2 \geq w_3 > 0 \text{ and } \sum_{i=1}^{3} w_i = 1.$$

The sequence of major triad representations also forms a spiral, as shown by the inner spiral in Fig. 7.1b.

Three adjacent major triads uniquely define the pitch collection for a major key. They also form the IV, I, and V chords of the key. Hence, major keys are defined as

$$\mathbf{T}_{\mathrm{M}}(\mathrm{k}) = \omega_1 \cdot \mathbf{C}_{\mathrm{M}}(\mathrm{k}) + \omega_2 \cdot \mathbf{C}_{\mathrm{M}}(k+1) + \omega_3 \cdot \mathbf{C}_{\mathrm{M}}(\mathrm{k}-1), \tag{7.3}$$

$$\text{where} \quad \omega_1 \geq \omega_2 \geq \omega_3 > 0 \text{ and } \sum_{i=1}^{3} \omega_i = 1.$$

Again, the sequence of major key representations forms a spiral. This major key spiral is shown as the innermost spiral in the illustration in Fig. 7.1c. Corresponding definitions exist for the minor triad and key representations—see Chap. 3 for further details.

7.2.2 Parameter Selection

The Spiral Array model is calibrated using mathematical constraints to reflect perceived closeness among the different entities. For example, in the definition of the major triad, the weights are constrained so that the weight on the root is no less than the weight on the fifth, which is no less than the weight on the third. Since the segmentation algorithm uses only the pitch representations, we shall describe here only the parameter selection for the pitch spiral, namely, the choice of r and h.

The pitch spiral is uniquely defined by its aspect ratio h/r. The goal is to constrain the parameters so that the distance between any two pitch representations correspond to their perceived closeness. Suppose the desired rank order of the interval distances is as follows: (P5/P4), (M3/m6), (m3/M6), (M2/m7), (m2/M7), (d5/A4), where P denotes a perfect interval, M a major interval, m a minor interval, d a diminished interval, and A an augmented interval. Figure 7.2 shows examples of these intervals inside the Spiral Array. Admittedly, this is an approximation of reality as it assumes equivalence within interval classes, and of ascending and descending intervals of the same size.

Algebraic manipulation shows that the mathematical constraints on the aspect ratio, $\sqrt{2/15} \leq h/r \leq \sqrt{2/7}$, produce the desired distance-based ranking of interval relations. In the remainder of the paper, h/r is always set to $\sqrt{2/15}$, the case in which major triad pitches form equilateral triangles. Appendix A gives the details of the calibration criteria for the Spiral Array's model parameters.

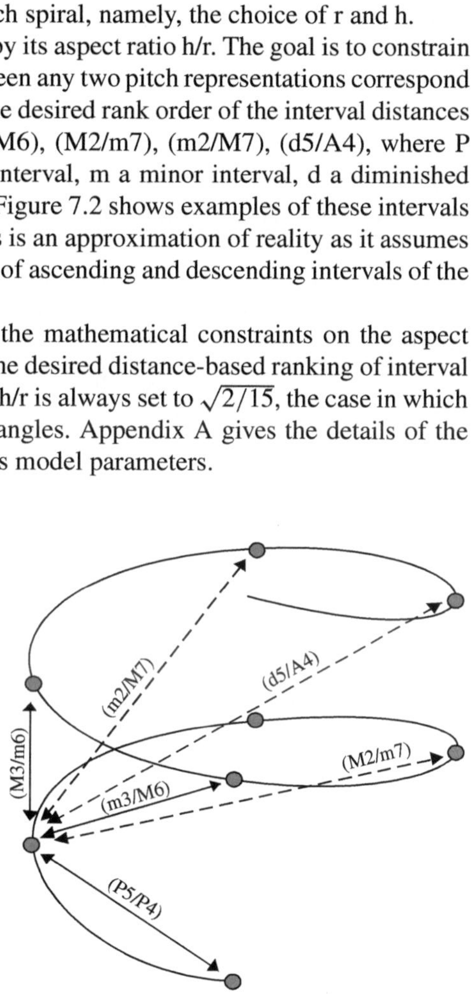

Fig. 7.2 Intervals in the Spiral Array

7.2.3 The Center of Effect

One of the main ideas behind the Spiral Array model is the use of points in the interior of the spiral to represent collections of pitches. More specifically, any pitch collection can generate a *center of effect* (CE), a weighted sum of the pitch representations that is a point in the model's interior. This concept was demonstrated in the definition of the major triad and major key in the previous section. Generalizing the idea, any segment of music can also generate a center of effect.

For example, each pitch in a musical segment can be mapped to the pitch spiral and each pitch position can be weighted by the pitch's proportional duration in the segment to generate the CE. Suppose

n_j = number of active notes in the jth time period, and
$\mathbf{p}_{i,j}$ = the Spiral Array position of the ith note's pitch class in the jth time period,

then, the formal definition of the duration-based CE of the musical segment (duration being accounted for through summation across the time slices) in the time interval $[a, b]$ is:

$$c_{a,b} = \sum_{j=a}^{b} \sum_{i=1}^{n_j} \frac{\mathbf{p}_{i,j}}{N_{a,b}}, \quad \text{where} \quad N_{a,b} = \sum_{j=a}^{b} n_j. \tag{7.4}$$

Other definitions can be devised to emphasize the metric weight of the note. The naïve duration-based definition is the one used in the current implementation of the segmentation algorithm. This is not an unreasonable choice in the case of Messiaen, given the frequent metric changes in his *Vingt Regards*.

7.2.4 The Argus Algorithm

The intuition behind the segmentation algorithm is that at a boundary point, the distance between the CEs of the local section immediately preceding and succeeding the boundary is at a maximum. Consider the degenerate case when two adjacent segments have identical pitch sets in the same proportional durations. In this case, the distance between the CEs of these two segments of music is zero. If the pitch sets are the same, but their proportional durations differ slightly between the two sets, as may be the case for two segments operating in the same tonal context, this CE distance is small. As the difference between the two segments increases, the distance between their CEs will also increase. Note that this distance is measured inside the Spiral Array space, and corresponds to tonal context change across the boundary between the two segments. At the point of transition between two pitch contexts, the CE distance reaches a local maximum as, on each side of the boundary, adjacent segments will necessarily share more pitch material than adjacent segments at this boundary.

The idea behind the segmentation algorithm is shown in Fig. 7.3. Suppose the piece consists of three sections that employ distinct pitch sets, with boundaries at B1 and B2 (B0 and B3 represent the beginning and end of the piece) as shown in Fig. 7.3. At B1, the distance between the CEs of the preceding light gray section and the succeeding dark gray section (represented by disks of the corresponding colors) is maximized.

Sliding the double window across the entire piece produces a plot of d over time that contains local maxima at the boundaries as depicted schematically in Fig. 7.4, where the scissors indicate the actual boundaries of the piece. As can be seen in the diagram, the algorithm requires only one left-to-right scan of the piece and computes in $O(n)$ time.

When to Take a Peak Seriously?

There exist many local maxima in the graph of the distance, d, between the CEs of the segment pairs over time. One question remains: How can one separate the significant peaks in the distance values from the insignificant ones? Statistical process control theory offers a solution. One objective measure is to only select maxima that are above a given threshold, say, one standard deviation, σ_d, from the mean d value, μ_d. Assuming that the distances can be suitably approximated by a normal distribution, $N(\mu_d, \sigma_d)$, values more than two standard deviations above the mean have a statistical significance of approximately 97.5 %. For a real-time realization of the algorithm, this threshold needs to be predetermined. For the purposes of evaluating the algorithm in this chapter, the threshold is assigned after computing all the distances. An advantage of obtaining the distribution for d from the data in this fashion is that the parameters so derived are specific to the piece to be analyzed.

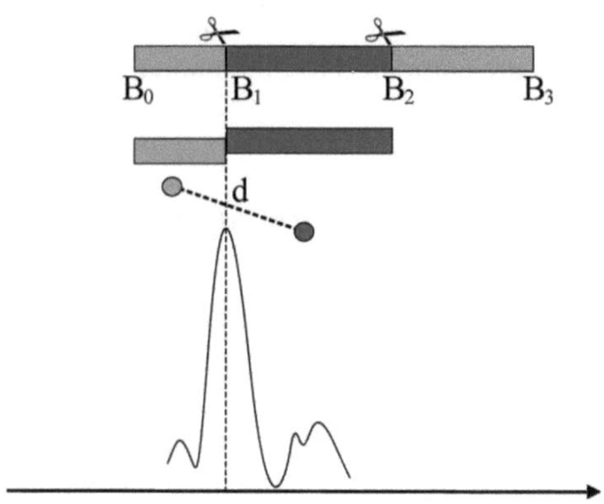

Fig. 7.3 Idea behind the segmentation algorithm

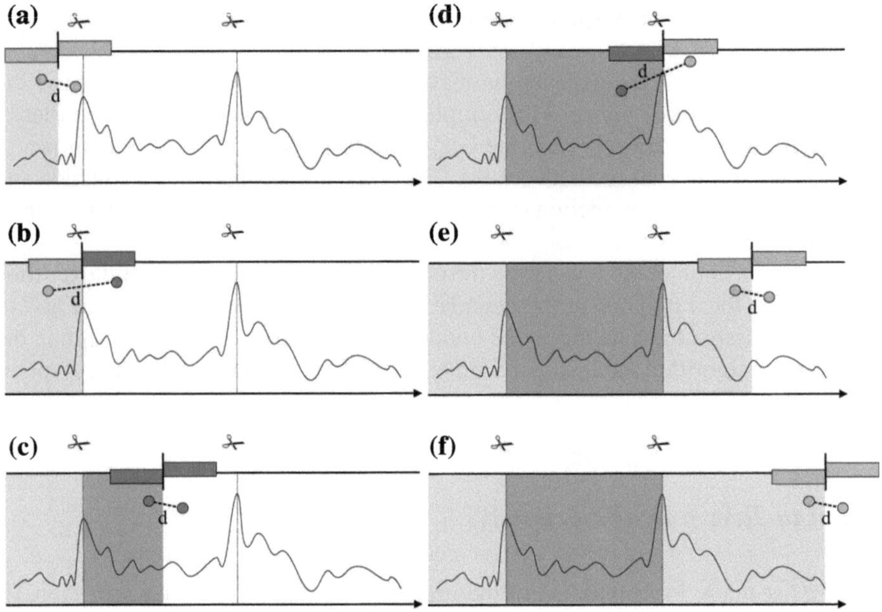

Fig. 7.4 Progression of segmentation algorithm. **a** Before the first boundary, **b** maximum at first boundary, **c** between the first and second boundaries, **d** maximum at second boundary, **e** beyond the second boundary, **f** the end

7.3 Messiaen's Regard IV

We first apply the segmentation algorithm described in the earlier part of this chapter to Regard de la Vierge, the fourth piece in Olivier Messiaen's (1908–1992) *Vingt Regards sur l'Enfant Jésus* [9]. Any discussion on the computational results first requires some groundtruth to which to compare the algorithm's outcomes. In Sect. 7.3.1, a manual analysis of the piece is presented, showing the different sections. Then, the computational results using various forward and backward window sizes are compared to this manual segmentation in Sect. 7.3.2.

7.3.1 Manual Segmentation

A manual analysis of Messiaen's Regard de la Vierge yields three main sections with defining motifs as shown in Fig. 7.5. The first, call it Section A, presents phrases built on an asymmetric 6+7 sixteenth-note-long motif shown in Fig. 7.5. Section B contains the fluid triplet motif and a second more angular and wide-ranging six-chord sequence. Section C is the most energetic of the three and contains a rhythmic motif whose bass line turns into a second motif with stacked octaves. The square patches

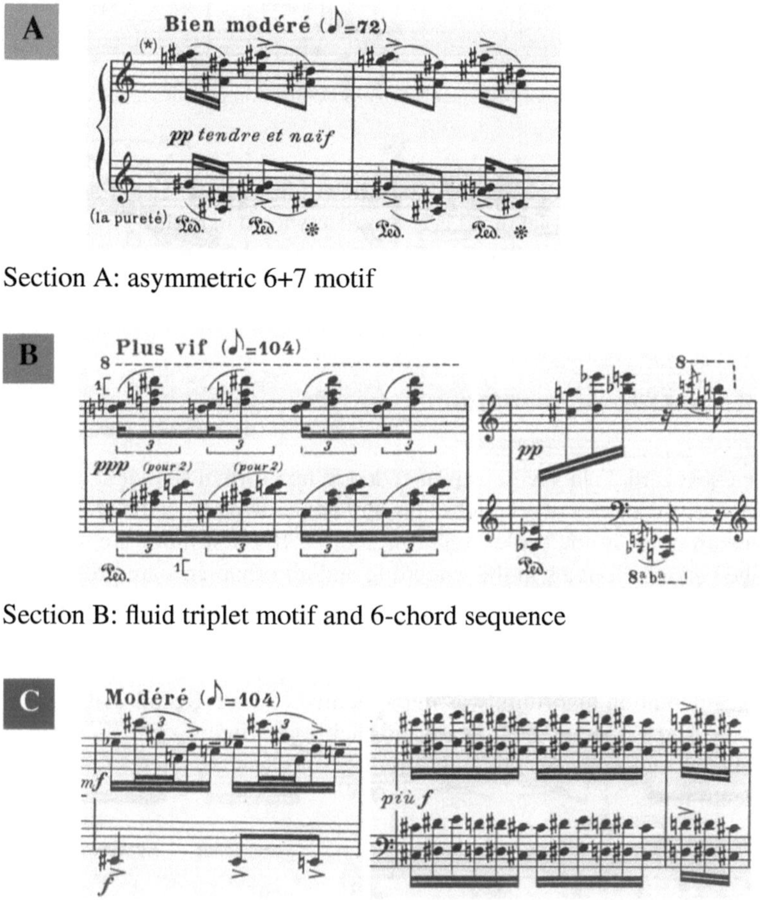

Section A: asymmetric 6+7 motif

Section B: fluid triplet motif and 6-chord sequence

Section C: rhythmic motif and stacked octaves

Fig. 7.5 Sections in the fourth Regard and their defining motifs

on the left show the grayscale tone used to represent that section. The demarcation between sections is clear from the score. In addition to the different motivic material, the composer also writes into the score tempo changes at the beginning of each of the identified sections.

The sequence of sections in the piece, ABACABAC, can be visualized using the bar in Fig. 7.6. The numbers at the boundaries correspond to the number of sixteenth notes from the beginning of the piece. In this particular counting scheme, the sixteenth notes in the triplet motif in Section B are counted as full-valued sixteenth notes and all ornaments are grouped with the pitch set that they embellish.

Some variations occur in the return of each section in the second half of the piece. For example, the A sections in the second half of the piece combine material from

Fig. 7.6 Manual segmentation of Messiaen's Regard IV into component sections

Section A (main motif) and Section C (baseline of first motif) See Figure 7.7a. In the return of Section B, the triplet motif is reversed as shown in Figure 7.7b. However, the note material is still primarily the same in each case, and the sections are recognizable as being similar to their earlier counterparts.

7.3.2 Automatic Segmentation

Messiaen's Regard de la Vierge was encoded in text format so that at each sixteenth note instance, the names of all pitches present are known. As in the previous section, the sixteenth notes in the triplet figure in Section B's first motif are assumed to be full-valued sixteenth notes in this encoding and all ornaments are grouped with the notes that they embellish. Note that an encoding that respects the tempo changes (one that would be closer to the performed timing) would be slightly different, but would be expected to produce similar results.

The segmentation algorithm was implemented in Java. Recall that the algorithm requires the user to specify the forward and backward window sizes shown in Fig. 7.8. In this chapter, we consider the algorithm's results when $f = b = 60, 40, 20$, and 10 sixteenth notes.

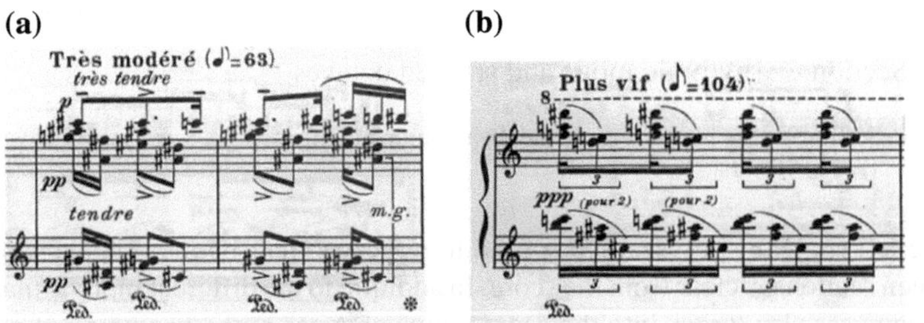

Fig. 7.7 Variations in the return of the sections. **a** combination of motivic material in A sections in second half of piece, **b** retrograde motif in return of Section B

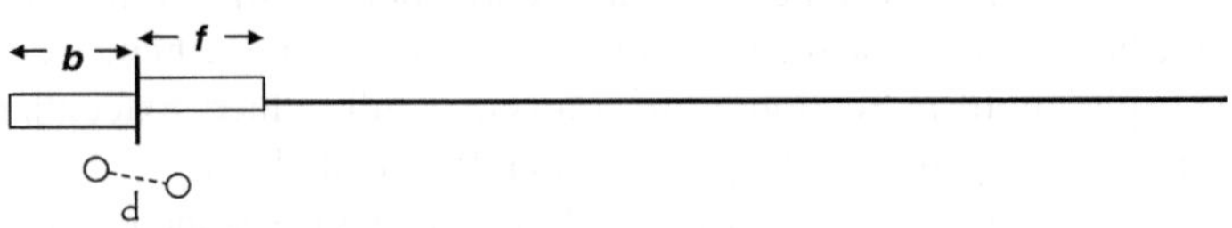

Fig. 7.8 Parameters in the algorithm: the forward and backward window sizes, f and b

Results When $f = b = 60$ Sixteenth Notes

The segmentation boundaries are computed with forward and backward windows of size 60 (sixteenth notes). The resulting distance values are plotted over time and shown in Fig. 7.9. There is a horizontal solid line cutting across the plot of d at around 0.75 on the y-axis. This marks the average d value, μ_d. The dotted horizontal lines correspond to one and two standard deviations above the mean, $\mu_d + \sigma_d$ and $\mu_d + 2\sigma_d$, respectively. Maxima above one standard deviation are marked by vertical lines in the plot that extend into the section analysis bar. Overlaid on the top part of the chart is the manual segmentation of the piece for comparison with the computed segmentation boundaries.

The average absolute discrepancy between the computed and actual boundaries is 8.43 sixteenth notes. Comparing this number with the number of sixteenth notes (893 in total), the boundaries are, on average, off by 0.94 %.

Most of the discrepancies are caused by bridge material that either contains pitch collections of the upcoming sections or motivic material borrowed from other sections. Two examples are given in Fig. 7.10, where the outlined arrows indicate the computed boundaries and the solid arrows indicate the actual boundaries. Figure 7.10a explains the discrepancy between the computed boundary at 196 and the actual boundary at 208. In this case, the final chords in Section B are re-spelt enharmonically (with sharps instead of flats) to prepare for the return of Section A. Figure 7.10b explains the largest discrepancy value (corresponding to the peak at 719), which occurs where the Section C motif is appended to the end of Section B.

Results When $f = b = 40$ Sixteenth Notes

Next, the algorithm is applied with forward and backward context windows that are 40 sixteenth notes wide. The results are documented in Fig. 7.11. At $f = b = 40$,

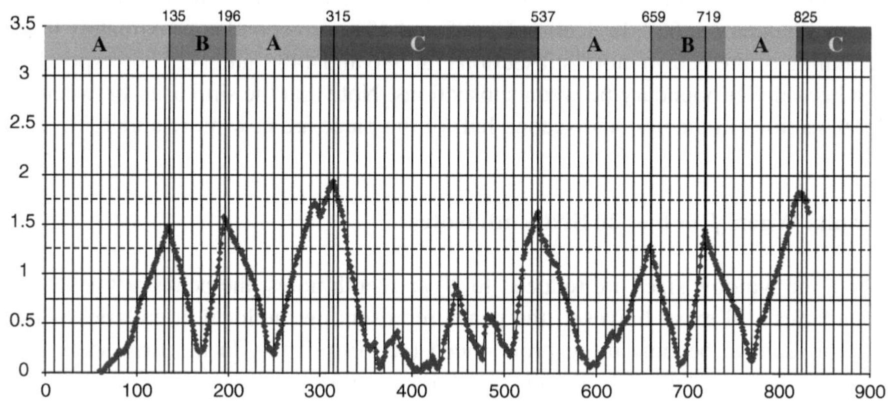

Fig. 7.9 Plot of d over time (in sixteenth notes), $f = b = 60$ ($\mu_d = 0.7418$, $\sigma_d = 0.5083$)

Fig. 7.10 Analysis of computed-actual boundary differences: computed (*white arrows*) versus actual (*black arrows*). **a** Enharmonic spelling of chord pitches in B in preparation for return of A, **b** largest discrepancy caused by motivic material from Section C appended to the end of Section B before the return of Section A

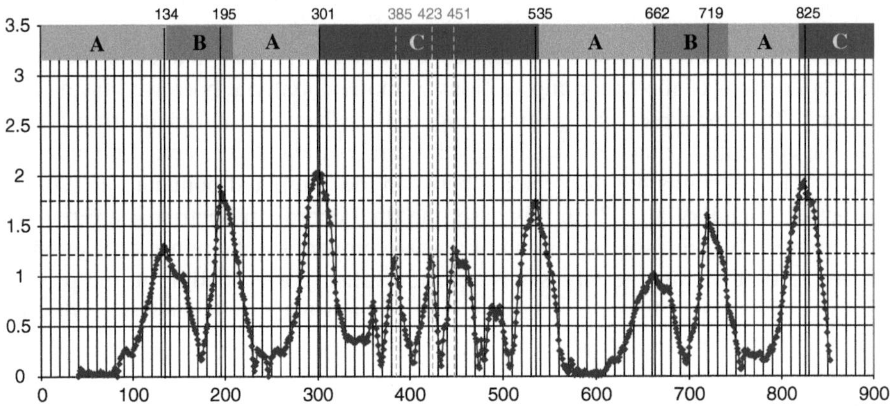

Fig. 7.11 Plot of d over time (in sixteenth notes), $f = b = 40$ ($\mu_d = 0.6821, \sigma_d = 0.5343$)

the plot of d over time shows some ambiguity in determining the significant peaks. The three peaks inside Section C at times 385, 423, and 451 are higher than the one for the boundary at 662. In Section C, 385 and 451 correspond approximately to the beginning of the octave motif subsections (at 386 and 448); and, 423 corresponds to another smaller boundary between subsections. Note that the similarity between the ABA sections in the first and second half of the piece are now more apparent.

Results When $f = b = 20$ Sixteenth Notes

When the window sizes are reduced to 20 sixteenth notes, the peaks at the section boundaries still exist; however, the peaks representing the local boundaries within Section C have become more pronounced (see Fig. 7.12). In the first Section C, the highest peaks (at 385 and 447) mark the beginnings of the octave motif (exact positions in the score are at 386 and 448). The beginning of the second motif in

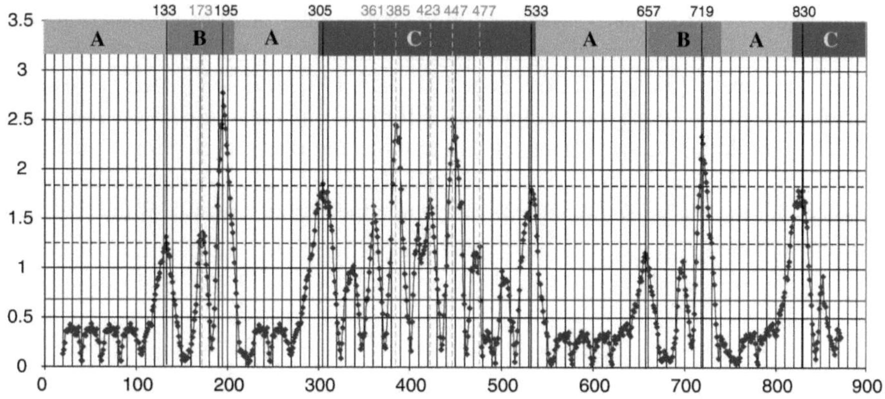

Fig. 7.12 Plot of d over time (in sixteenth notes), $f = b = 20$ ($\mu_d = 0.6832$, $\sigma_d = 0.5753$)

Section B is marked by a peak at 173, and a corresponding but smaller peak can be seen at 699. In addition to the more pronounced similarity between the two ABA sections, the repeated patterns in the more homogeneous Section A are now visible as sequences of low-lying humps.

Results When $f = b = 10$ Sixteenth Notes

When the window sizes are reduced to 10 sixteenth notes, the local peaks within the large sections (especially C) dominate the picture (see Fig. 7.13). The low-lying humps signifying repeated pitch patterns in Section A at window size 20 have now transformed into more defined patterns that indicate the number of repetitions of the main motif phrase. At window size 10, the beginning of the second motif in Section

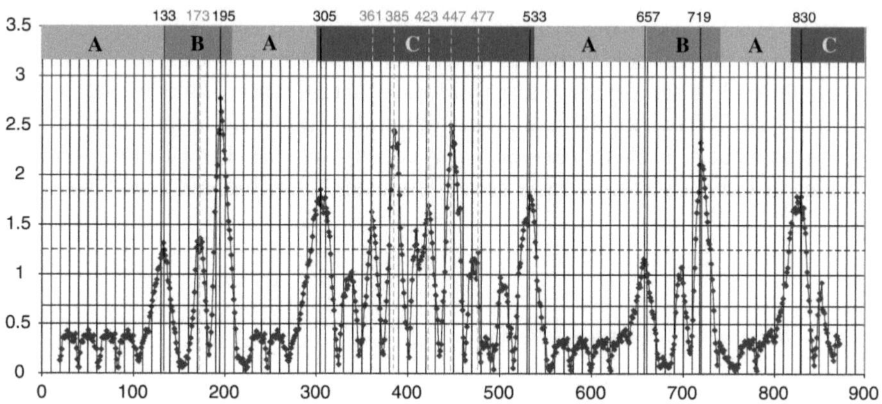

Fig. 7.13 Plot of d over time (in sixteenth notes), $f = b = 10$ ($\mu_d = 0.5380$, $\sigma_d = 0.4930$)

B is marked by a peak at 169, and a similar but smaller peak can be seen at 698. Section C is visibly the most segmented section of the three. In the first Section C, the other peaks {314, 331, 361, 385, 404, 423, 447, 477, 487} each correspond approximately to the appearance of a motif different from the immediately preceding one. The two dotted vertical lines (at 834 and 851) in the final Section C mark the boundaries of the tremolo bar (in the score at 836 and 855). As can be expected, local details rather than large-scale patterns determine the distance profile of the piece for smaller window sizes.

7.4 Messiaen's Regard XVI

To further support the argument for the Argus segmentation algorithm, this section applies the algorithm to another example by Messiaen: the Regard des prophètes, des bergers et des Mages, the sixteenth piece in the *Vingt Regards*. Section 7.4.1 presents a manual analysis of the piece, and Sect. 7.4.2 the computational analyses.

7.4.1 Manual Segmentation

Unlike Regard IV, the composer did not label all section boundaries in Regard XVI. He gives a clue to its structure in the note beneath the title: *Tam-tams et hautbois, concert énorme et nasillard* ... The piece can be subdivided into the tam-tams (T), hautbois (H), and nasal concert (C) sections. The defining motifs of these sections are shown in Fig. 7.14, and the sections (obtained by manual analysis) are shown in Fig. 7.15. The numbers correspond to the thirty-second note count from the beginning of the piece.

The presentation of the tam-tams, then the hautbois sections are followed by the enormous nasal concert, made up of the Section C motif shown in Fig. 7.14, and joined later by the hautbois theme. Bridge material borders the nasal concert section. Nearing the end, the tam-tams return, followed by a reiteration of the H and C motifs superimposed. The two motifs share many of the same pitches, and are superimposed twice in the piece. The clear boxes with gray edges group the H and C motif sections. Together with the T sections, these boxes delineate the pitch-distinct sections in the piece.

7.4.2 Automatic Segmentation

The text encoding of the piece represents note clusters at the thirty-second note level. All ornaments are grouped with the first thirty-second note slice of the notes

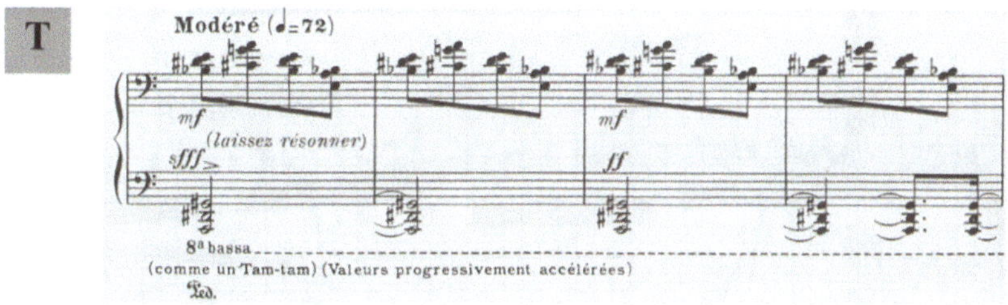

Section T: tam-tams

Section H: hautbois

Section C: concert énorme et nassilard …

Fig. 7.14 Sections in the sixteenth Regard and their defining motifs

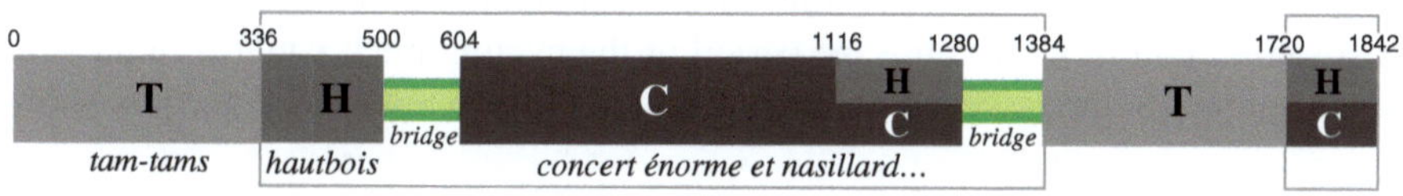

Fig. 7.15 Manual segmentation of Messiaen's Regard XVI into component sections

they embellish, and the two triplets in bars 52 and 54 are snapped to neighboring thirty-second note grids according to the proportion 3:2:3.

Results When $f = b = 256$ (32 Quarter Notes)

When forward and backward context windows are set at 256 thirty-second notes (that is to say, 32 quarter notes), the algorithm found the two major segmentation boundaries within its search range. As shown by the chart in Fig. 7.16, the boundaries at 336 and 1384 were approximated by the two major peaks at 336 and 1380. Considering that the entire piece is 1842 thirty-second notes long, the boundary estimates are, on average, within 0.11 % of the correct boundaries.

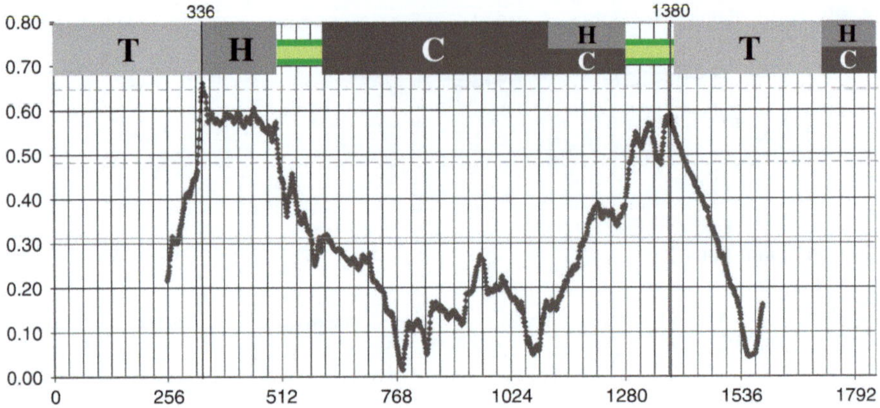

Fig. 7.16 Plot of d over time, $f = b = 256$ ($\mu_d = 0.3122$, $\sigma_d = 0.3122$)

Results When $f = b = 128$ (16 Quarter Notes)

Observe in Fig. 7.17 that, when the window sizes are reduced to 128 thirty-second notes (16 quarter notes), the main peaks above the two standard deviation line are 394, 1,380, and 1,714, approximating the three boundaries at 336, 1,384, and 1,720. Note that 1714 is at the rightmost edge of the search range and the actual peak could exist to the right of it. Because the period of the cycling pitch patterns in the tam-tams section (16 thirty-second notes) is a divisor of the window size, the plot shows

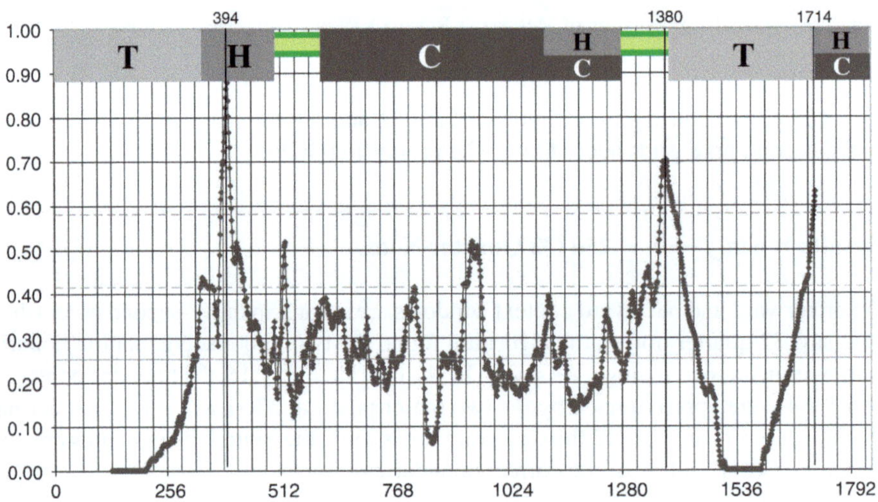

Fig. 7.17 Plot of d over time, $f = b = 128$ ($\mu_d = 0.2534$, $\sigma_d = 0.1634$)

a flat line at zero in the central part of the tam-tams section. The span of this flat line increases as the window size gets reduced to, say, 64 or 16.

7.5 Discussion: Why Does It Work?

It is reasonable to wonder why a geometric model for tonality such as the Spiral Array that is based on close positioning of pitch classes related by perfect fifth and major third intervals could be used to segment a post-tonal work, and why the segmentation algorithm worked as well as it did. This section contains reflections on the geometric interpretations of pitch context within the Spiral Array.

The Spiral Array model clusters entities that are perceptually close, which means that pitches in the same key form neat compact clusters in the Spiral Array space. The CE generated by a segment of music in a given key resides inside the convex hull of the pitches of that key, that is to say, the smallest convex volume that contains all spatial representations of pitches in that key. When the context changes, for example when the key modulates, this convex hull shifts either to a neighboring, or a distant, shape. These shapes are well-defined in the tonal context.

What happens when one uses the same geometry for a post-tonal work such as Messiaen's *Vingt Regards*? Are there other, possibly more viable, structures for representing post-tonal pitch collections? Sections 7.5.1 and 7.5.2 will address these two questions.

7.5.1 Geometric Shapes Corresponding to Pitch Sets in Messiaen's Examples

In the case of Messiaen's two Regards, the pitch contexts do not necessarily form neat compact shapes. However, they generate distinctive volumes such as those shown in Figs. 7.18 and 7.19. The segmentation algorithm will work as long as the CEs generated by the notes in each context as represented by these convex hulls are distinguishable (sufficiently distant) from each other. For example, the shapes for the three examples from Regard IV (shown in Fig. 7.18) are distinct enough from each other such that CEs generated within each of the three contexts (i.e., inside each of the three volumes) are readily separable one from the other, as evidenced by the automatic segmentation results.

The separation is not as obvious in the samples from Sections H and C of Regard XVI (shown in Fig. 7.19b and c.) In this case, the two convex volumes overlap a fair bit, and the CEs of the two sets are not as distinguishable. This observation is further confirmed by the fact that the composer superimposes the H and C motives on two occasions in the piece.

The position of the CE inside each shape is also influenced by the rhythmic structure of the sample. Two samples are distinguishable not only if the pitch sets

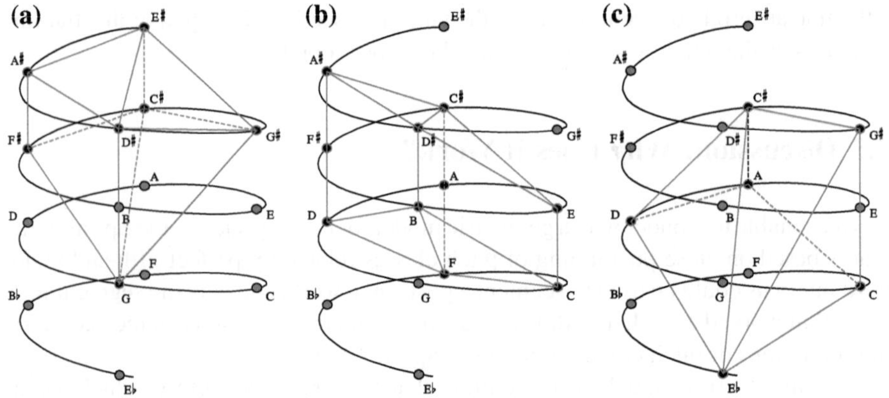

Fig. 7.18 Convex hulls of pitch sets in sections of Regard IV shown in Fig. 7.5. **a** Section A, **b** Section B: triplet motif, **c** Section C: rhythmic motif

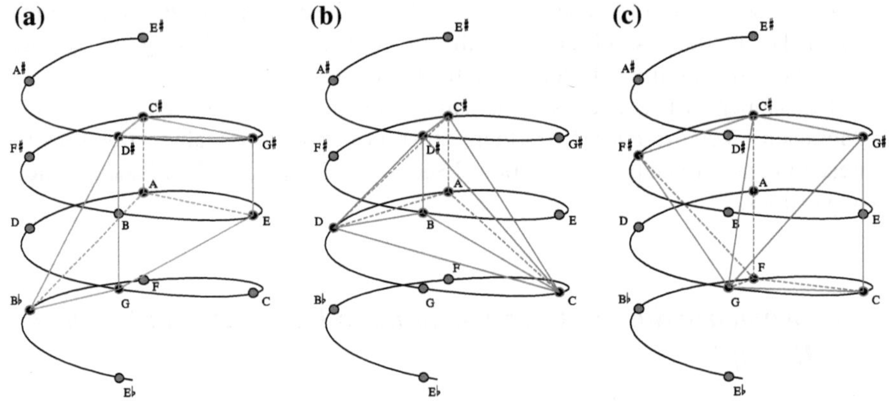

Fig. 7.19 Convex hulls of pitch sets in sections of Regard XVI shown in Fig. 7.14. **a** Section T, **b** Section H, **c** Section C

are different, but also if the distributions of the pitch set are different. The convex hull of the pitch sets appears to be the dominating factor affecting the success of the segmentation algorithm for these examples.

7.5.2 *What are the Structural Alternatives?*

The question arises as to whether the Spiral Array space is the best one to create the separate CEs. Conceivably, one could construct a model with the best pitch set separation for segmenting a specific piece of music. It is unlikely that such a model tailored to one piece would generalize to many others. Since the Spiral Array exhibits strong correspondence to models rooted in theory and psychology, one might argue that it is a reasonable one to use to model perceived distances between pitch clusters designed to sound distinct one from another. However, readers may still wonder, as one reviewer did, if and what other structural representations for musical pitch might be more appropriate for segmenting the Messiaen examples by pitch content. I shall first paraphrase the reviewer's questions, then expand on some answers and explanations.

According to the one reviewer, one might argue that other, non-tonal models may be more appropriate from a perceptual point of view when dealing with non-tonal music. From a practical computational point of view, the question remains open. What happens if one replaces the fifth-related pitches of the pitch spiral in the Spiral Array, for instance, (a) by pitches separated by semitones, (b) by pitches organized in thirds, (c) by random pitch name classes, and so on, and run the algorithm on the same pieces for the same parameter settings?

In answer to part (c) of the question, one can almost always find a more appropriate positioning of pitch classes to maximize the separation between pitch sets in different sections of a piece. Let us consider the extreme and simplistic case shown in Fig. 7.20. In a piece that can be segmented into sections described by two pitch class sets, one strategy is to simply cluster pitch classes common to both section types (shown as black dots) in the center of a space, and to place nonoverlapping pitch class sets (shown as the gray set and the clear set) as far as possible in the given space. This structure would be more effective than the Spiral Array, and optimized for this particular example.

Next, we consider generalizable (positions fixed for all pieces) and regular pitch class representations, specifically, helical representations that exhibit a fixed interval between consecutive pitch class representations. The line of fifths is a one-dimensional structure that fits this description. Another constraint inherent in the Spiral Array that may not be immediately obvious is that, although the model assumes octave equivalence, enharmonic pitches are not grouped into the same class, i.e., A♭ = G♯, hence the helix is not folded back on itself into a torus. Pitch spelling is preserved so as to distinguish among diminished, minor, major, and augmented intervals and chords, and voice leading implications.

Let us now constrain the discussion to helical structures that cover all pitch class names (the line of fifths can be considered a special class of these structures, when the aspect ratio, height-increase-per-step/radius, goes to infinity). Then, assuming enharmonic equivalence, replacing the fifth-related pitches of the pitch spiral by pitches separated by intervals other than minor seconds or perfect fifths would cover only a subset of all pitch classes, assuming enharmonic equivalence.

Without enharmonic equivalence, stepping by perfect fifths is a necessary requirement to cover all pitch classes. As an illustration using the Spiral Array as the reference structure, stepping by major seconds and by major and minor thirds are shown in the diagram sequence in Fig. 7.21. Assuming that an only thirds (or only seconds) representation covers all the pitch classes in the piece, the result would not differ from that using the Spiral Array. This answers questions (a) and (b).

There remains one more question that could be asked. Why not assume enharmonic equivalence and use the corresponding toroid structures? A reasonable question,

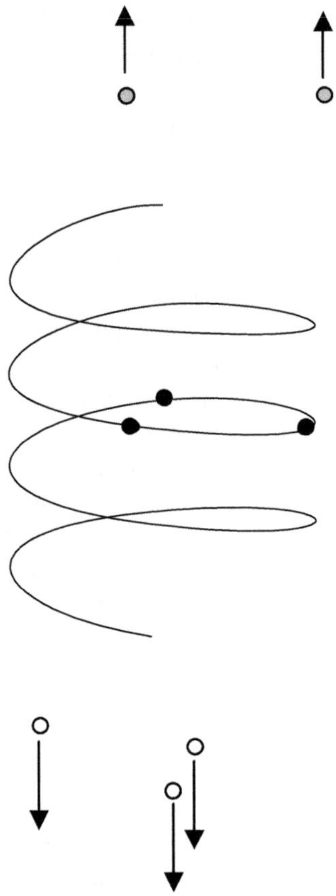

Fig. 7.20 Best separation of pitch class sets from two distinct sections

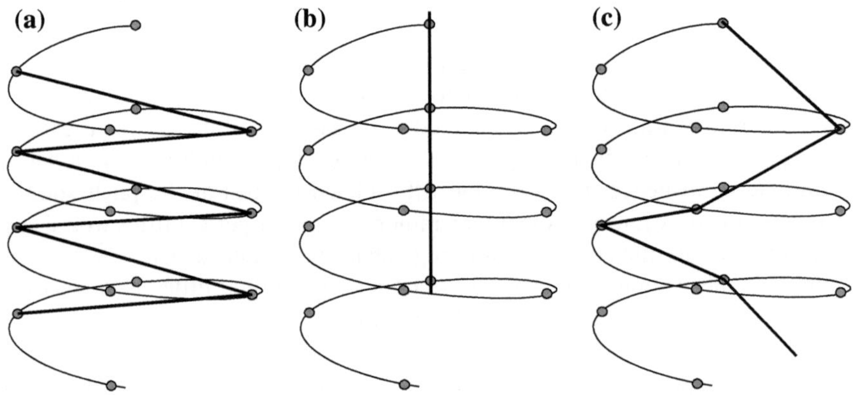

Fig. 7.21 Illustrations of pitch class sequences based on other intervals. **a** seq based on M2 intervals, **b** seq based on M3 intervals, **c** seq based on m3 intervals

especially with reference to post-tonal music. The argument can be further supported by Messiaen's use of his modes of limited transposition, scales that map back to the same pitch classes after only a small number of transpositions; the most limited being the whole tone and the octatonic scales, which repeat after only two transpositions. The extent to which Messiaen's pitch spelling is meaningful then comes into question, a topic for another paper. For now, I have assumed that the composer meant the spelling that he chose, and that each notated pitch is not lightly interchangeable with its enharmonic equivalent. Without further experimentation, this assumption seems reasonable given that the composer, with few exceptions (and for these, only for a very limited time), uses the same pitch spellings whenever the same motivic material recurs.

7.6 Conclusions

In this chapter, I have presented an algorithm for segmenting music according to pitch context based on a technique for tracking and assigning a distance measure to pitch context differences between consecutive segments of music, and applied it to the automated analysis of Messiaen's Regards IV and XVI. The computational results have shown that large search windows identify section boundaries and small search windows find local section and pattern boundaries. Some future work remains to identify the instantaneous optimal window size, if one exists, and to determine the best statistical model for the distance time series, and a robust method for setting the threshold for categorization of a peak as being significant. The use of dynamically varying window sizes may produce better results; however, studying the distance time series for each window size can also offer insights into the pitch context variations at multiple timescales. The method is also amenable to a dynamic programming formulation and implementation. However, an algorithm with a parsimonious description is preferred, and is the one described in this chapter.

The strength of the algorithm lies in its use of the interior space in a 3D model to summarize pitch information, making the method highly efficient, and easy to visualize and explain. By using an interior point to summarize pitch context, one gains computational efficiency but loses much information, such as the convex hull of the pitch set, and the actual pitches themselves and the sequence in which they are sounded. In the case when a section is transposed, which did not occur in the Messiaen, the local patterns are preserved, and the relative distances within the section remain unchanged. In the case where the same thematic material is repeated in sequence, the second time some interval higher than the first, the Argus method could still detect a pitch context change, even though the pitch patterns remain unchanged. The Argus algorithm is designed to separate a piece by pitch context, and not by thematic material. In the Messiaen examples, the thematic material corresponded to distinct pitch contexts. Hence, the automatic segmentation not only separated sections with distinct pitch contexts, but also different thematic material.

The algorithm works well on the two Messiaen examples because the pitch collection in each section generates a CE that is sufficiently distant from that of its neighboring sections. This is aided in part by the fact that pitch classes with the same letter name but differing by an accidental, although close in frequency, are far apart and have a distinctive span in the Spiral Array model. In general, the greater the separation of the CEs of the distinct pitch sets, the better the accuracy of the Argus algorithm. Future investigations will include the comparison of models assuming enharmonic equivalence with the current one to determine the effect, or lack thereof, of pitch spelling and other arrangements of pitch representations (when assuming enharmonic equivalence) on the relative positions of the CEs and the resulting segmentation boundaries.

Acknowledgments This chapter grew out of a talk given at the MaMuX seminar organized by Moreno Andreatta on March 21–22, 2004, at IRCAM, whereby all participants were encouraged to concentrate on a few selected pieces from Messiaen's *Vingt Regards*. I thank Moreno for sparking my interest in analyzing selections from Messiaen's *Vingt Regards*, and the attendees of the MaMuX seminar for their valuable feedback at that initial analysis of Regard IV. I also thank Gérard Assayag for his encouragement to submit the paper to the First International Conference on Sound and Music Computing, and the opportunity to expand on the paper. The reviewers' detailed comments that have helped shape the current version of the paper are much appreciated. This material is based upon work supported by the National Science Foundation under Grant No. 0347988. The work made use of the Integrated Media Systems Center Shared Facilities supported by the National Science Foundation under Cooperative Agreement No. EEC-9529152. Any opinions, findings, conclusions, or recommendations expressed in this material are those of the authors, and do not necessarily reflect the views of the National Science Foundation.

References

1. Chew, E.: Towards a mathematical model of tonality. Ph.D. thesis, Massachusetts Institute of Technology, Cambridge, MA (2000)
2. Chew, E.: Modeling tonality: applications to music cognition. In: Moore, J.D., Stenning, K. (eds.) Proceedings of the 23rd Annual Meeting of the Cognitive Science Soc., pp. 206–211. Lawrence Erlbaum Assoc. Pub., Edinburgh (2001)
3. Chew, E.: The Spiral Array: an algorithm for determining key boundaries. In: Anagnostopoulou, C., Ferrand, M., Smaill, A. (eds.) Music and Artificial Intelligence Proceedings of the Second International Conference on Music and Artificial Intelligence. LNCS/LNAI, vol. 2445, pp. 206–211. Springer, Heidelberg (2002)
4. Chew, E., Chen, Y.-C.: Real-time pitch spelling using the Spiral Array. Comput. Music J. **29**(2), 61–76 (2005)
5. Cohn, R.: Introduction to neo-Riemannian theory: a survey and a historical perspective. J. Music Theory **42**, 167–180 (1998)
6. Krumhansl, C.L.: Cognitive Foundations of Musical Pitch. Oxford University Press, Oxford (1990)
7. Lerdahl, F.: Tonal Pitch Space. Oxford University Press, Oxford (2000)
8. Longuet-Higgins, H.C., Steedman, M.J.: On interpreting Bach. Mach. Intell. **6**, 221–241 (1971)
9. Messiaen, O.: Vingt Regards sur l'Enfant Jesus. Durand Editions Musicales, Paris (1944)
10. Shmulevich, I., Yli-Harja, O.: Localized key-finding: algorithms and applications. Music Percept. **17**(4), 531–544 (2000)
11. Temperley, D.: What's key for key? The Krumhansl-Schmuckler key-finding algorithm reconsidered. Music Percept. **17**(1), 65–100 (1999)
12. Toiviainen, P., Krumhansl, C.L.: Measuring and modelling real-time responses to music: the dynamics of tonality induction. Perception **32**(6), 741–766 (2003)

Part V
Pitch Spelling

Chapter 8
Real-Time Pitch Spelling

Abstract This chapter describes and presents a real-time bootstrapping algorithm for pitch spelling based on the Spiral Array model [3]. Pitch spelling is the process of assigning appropriate pitch names that are consistent with the key context to numeric representations of pitch, such as MIDI or pitch class numbers. The Spiral Array model is a spatial model for representing pitch relations in the tonal system. It has been shown to be an effective tool for tracking evolving key contexts (see [4, 5]). Our pitch-spelling method derives primarily from a two-part process consisting of the determining of context-defining windows and pitch-name assignment using the Spiral Array. The method assigns the appropriate pitch names without having to first ascertain the key. The Spiral Array model clusters closely related pitches and summarizes note content by spatial points in the interior of the structure. These interior points, called *centers of effect* (CEs), approximate and track the key context for the purpose of pitch spelling. The appropriate letter name is assigned to each pitch through a nearest-neighbor search in the Spiral Array space. The algorithms utilize windows of varying sizes for determining local and long-term tonal contexts using the Spiral Array model.

8.1 Why Spell?

The problem of pitch spelling is an artifact of equal temperament tuning in Western tonal music whereby several pitches are approximated by the same frequency so as to ensure that music in different keys can be played using a reasonably sized pitch set. Pitches of the same frequency but denoted by different pitch names are said to be enharmonically equivalent. In MIDI format and many other numeric music rep-

---

 This is a minor revision of the work, "Real-Time Pitch Spelling Using the Spiral Array" by Elaine Chew and Yun-Ching Chen, published in the Computer Music Journal, **29**(2), 61–76 (2005).

E. Chew, *Mathematical and Computational Modeling of Tonality*, International Series in Operations Research & Management Science 204, DOI: 10.1007/978-1-4614-9475-1_8, © Springer Science+Business Media New York 2014

resentations, enharmonically equivalent pitches are represented by the same number indicating the note's ostensible frequency and not its letter name. Hence, each number can map to more than one letter name. For example, MIDI number 56 can be spelt as A♭ or G♯. The spelling of a given pitch number is primarily dependent on the key context, and to a lesser extent, the voice leading tendencies in the music.

One might argue that enharmonic spellings are part of an artificial system that may not entirely be necessary for the representation and analysis of music. However, as noted by Cambouropoulos ([2], p. 412), enharmonic spelling "allows higher level musical knowledge to be represented and manipulated in a more precise and parsimonious manner and thus may facilitate other musical tasks such as harmonic analysis, melodic pattern matching, and motivic analysis." Meredith ([15], p. 174) adds that "a correctly notated score ... is a structural description of the passage that is supposed to represent certain aspects of the way that the passage is intended to be interpreted by an expert listener... [and] serves a similar function to, say, a Schenkerian analysis." We proceed to make a case for the necessity for better spelling algorithms.

Pitch spelling is the prelude to all automated transcription systems. The spelling of each pitch directly affects its notation. For example, the choice of A♭ or G♯ for MIDI number 56 affects its notation in a score. There is much room for improvement in spelling technology as demonstrated by the performance of commercial notation software in transcribing MIDI to score. Figure 8.1 shows examples of pitch-spelling errors in an excerpt from the beginning of Beethoven's Piano Sonata Op.109. The notation shown is the correct spelling, and the circles indicate places in the music where a misspelling (not shown) occurred. The spelling errors made by two leading notation software programs are circled in the two examples. In each case, the default settings were used. The process typically involved the reading of the key from the file followed by pitch-name assignments based on the given key. The majority of errors in these two cases were caused by an inability to detect the changes in the local tonal context. On the other hand, the arrows indicate the spelling errors made by our algorithm.

Pitch spelling is also critical for building accurate computer systems for music analysis. Accurate spelling will result in more robust key-finding algorithms. For example, G♯ is an important pitch (the leading tone) in the key of A major, whereas A♭ is not a member of the same key and would hardly be used at all except in passing. Pitch spellings also affect chord-recognition algorithms and the interpretation of chord function. Consider the problem of determining the chord label and function of the MIDI numbers { 60, 64, 68 }. The pitches { C, E, G♯ } form an augmented triad with C as its root, whereas the pitches { A♭, C, E } form an augmented triad with A♭ as its root with very different implications for the ensuing harmonic motion.

The local key context determines the pitch spelling; in turn, the name of the pitch determines the notation and serves as a clue to the key context. When creating and notating a new piece, the composer's choice of musical notation, such as pitch spelling, stems from his or her concept of the local context (key) and voice leading. The notated score then serves as a guide for the expert listener or interpreter to the composer's intended musical structure (contextual delineations) for the piece. This interdependence between notation (such as pitch spelling) and structure (for example,

Fig. 8.1 Spelling errors by two leading notation software programs (locations circled, misspellings not shown) compared to those made by our proposed algorithm (*arrowed*): (**a**) Finale; (**b**) Sibelius

context) creates a potentially problematic cycle in the design of an accurate algorithm for pitch spelling, the classic "chicken and egg" problem.

We de-couple the problems of pitch spelling and key-finding by creating an algorithm that does not require explicit knowledge of the key context to determine the appropriate spelling. Using the Spiral Array, any musical fragment that has been correctly spelled will generate a center of effect (CE), a summary point in the model's interior, that is closest to—and hence can act as a proxy for—the key. Using this evolving CE, the algorithm maps each numeric representation to its plausible pitch names on the Spiral Array and selects the best match through a nearest-neighbor search. Our algorithm computes in real time, requiring only present and past information, thus facilitating its integration into real-time music systems.

In this chapter, we test the algorithms using MIDI rather than audio data to isolate the pitch-spelling process for more accurate assessment. A system that takes audio as input, even audio generated from MIDI data, would introduce additional error through transcription. By starting with MIDI information, we directly assess the accuracy of the proposed pitch-spelling algorithms. Note that the algorithms apply to any numeric input that is related to frequency and not just MIDI data. Examples of such representations include, and are not limited to, audio input that has been transcribed to frequency numbers, pitch class sets, and piano rolls.

We present and analyze the computational results of three variants of our real-time bootstrapping pitch-spelling algorithm when applied to examples from two different cultures. The first set consists of selected movements from Beethoven's Piano Sonatas: the third movement of Op.79 and the first movement of Op.109. The second is a set of variations for piano and violin on a popular Taiwanese folksong by composer Youdi Huang with a primarily pentatonic melody. The late Beethoven sonata (Op. 109) is a complex example, a mature work by the composer containing many key changes, both sudden and subtle. By careful analysis of the pitch spelling results using a piece of such complexity, we gain insight into the challenges in pitch spelling so as to guide the algorithm design and parameter selection. We found few examples of music of harmonic complexity comparable to the late Beethoven sonata with pitch-spelling information in the public domain. In our tests, all musical examples were optically scanned and manually checked for errors, then converted to MIDI from score as input for the program. It is important to point out that the aim of this chapter is not to provide a comprehensive or comparative evaluation of the algorithm; see [15, 16] for examples of such studies.

8.2 Related Work

Pitch spelling has been a concern in computer music research since the early work of Longuet-Higgins [13], where the goal of automated transcription was achieved by a combination of meter induction, pitch spelling, and key induction. Although in his earlier work with Steedman ([12], p.237), Longuet-Higgins touches on the problem of pitch spelling, that work focussed on key-finding, and the authors noted near the end of the article that "after the key has been established accidentals can be met with equanimity." In [13], appropriate pitch spellings were selected such that the distance between each note and the tonic of the given key was less than six steps in the line of fifths, thus favoring diatonic intervals over chromatic ones. Longuet-Higgins defined diatonic intervals to be the distance between pitches separated by less than six steps on the line of fifths (P5/P4, M2/m7, M6/m3, M3/m6, M7/m2, where P = perfect, M = major, m = minor, d = diminished and A = augmented); the "diabolic" tritone interval has a separation of exactly six steps on the line of fifths; and any interval separated by more than six steps is considered chromatic. The tonic is derived by assuming that the first note of the melody is either the tonic or the dominant. As the melody unfolds, the algorithm revisits the tonic choice by evaluating a small local set of notes to see if an alternative interpretation of the local key would result in diatonic intervals among the pitches. The program was applied to the cor anglais solo from the prelude to Act III of Wagner's *Tristan und Isolde*.

More recent pitch-spelling algorithms include Temperley's line-of-fifths approach [20], Cambouropoulos' interval-optimization approach [1, 2] and Meredith's ps13 algorithm ([14, 15]). These pitch-spelling algorithms decouple the problem of key-finding and pitch spelling; they produce output that can be used as input for key-finding programs. Admittedly, they all incorporate implicit knowledge of key structure in their design, for example, through the representations used, their interval preferences, and/or the use of harmonic analysis.

Temperley's algorithm [20], like that of Longuet-Higgins, uses the line-of-fifths pitch representation. Temperley's method maps pitches to the line of fifths in a way such that the spellings map to close clusters on the line. Each collection of spelt notes generates a center of gravity on the line of fifths. The best spelling is determined by seeking the nearest neighbor to the current center. In addition, the algorithm avoids chromatic semitones adjacent to each other through voice-leading rules and prefers good harmonic representations through the harmonic feedback rule. The accuracy of the pitch-spelling algorithm depends on the reliability of the metrical and harmonic analysis. The method was implemented in the Melisma Harmonic Analyzer by Sleator and Temperley [19] and tested on the Kostka-Payne corpus. Temperley reports that the algorithm made 103 spelling errors out of a total of 8,747 notes, giving a correct percentage of 98.82 % ([20], p. 136).

Cambouropoulos' method [1, 2] focuses on interval relations, and selects pitch spellings based on intervals more likely to occur within the tonal scales. The fundamental principles of this algorithm are notational parsimony (minimizing the use of accidentals) and interval optimization (avoiding diminished and augmented intervals that are rare or do not appear at all in the tonal major and minor scales). By limiting the interval choice, Cambouropoulos essentially constrains the pitch-spelling search to a local region along the line of fifths. Cambouropoulos argues that the line-of-fifths approach can be viewed as a special case of the interval-optimization algorithm. The algorithm was tested on ten complete sonatas by Mozart (K279–284, K330–K333) and on three waltzes by Chopin (Op.64 Nos.1–3). Because Chopin's harmonic language was more complex than Mozart's, the algorithm performed better on the Mozart examples. Cambouropoulos reports an overall correct spelling rate of 98.59 %.

Table 8.1 Percentage of intervals and notes spelt correctly for selected composers based on [15] comparison of pitch-spelling algorithms

Author(s)	Method	Percentage correct			
		Vivaldi 1678–1741	Bach 1685–1750	Mozart 1756–1791	Beethoven 1770–1827
		223,678 notes	627,083 notes	172,097 notes	48,659 notes
Meredith	ps13	99.23	99.37	98.49	98.54
Cambouropoulos	Interval Opt	99.13	97.92	98.58	98.63
Longett-Higgins	Line-of-fifths	98.48	98.49	96.26	97.46
Temperley	Line-of-fifths	98.69	99.74	93.40	92.34
Average		**98.88**	**98.88**	**96.68**	**96.74**

Meredith's ps13 algorithm [14] can be broken down into two stages. The first stage votes for the most likely spelling while the second stage scans and corrects for local voice-leading errors. In stage 1, the algorithm counts the number of times each pitch occurs in a context surrounding the note to be spelt; then, these numbers, added over all plausible harmonic chromatic scale tonics that would result in a given letter name, return the votes for the given note having that letter name. The letter name with the highest number of votes wins.

Meredith compares the ps13 algorithm's performance with that of his implementations of the algorithms by Longuet-Higgins, Temperley, and Cambouropoulos. In the 2004 pilot study, he tested the algorithms on all Preludes and Fugues in Book 1 of J. S. Bach's *Well-Tempered Clavier* (BWV 846–869) and reported the following correct percentage rates: ps13 (99.81 %), Cambouropoulos (93.74 %), Longuet-Higgins (99.36 %) and Temperley (99.71 %). In an updated study using a large corpus of tonal music containing a total of 1.73 million notes, Meredith reports the following results: ps13 (99.33 %), Cambouropoulos (98.71 %), Longuet-Higgins (97.65 %) and Temperley (97.67 %). The corpus consists of music by nine composers ranging from Corelli to Beethoven. The span of musical styles is limited: 80 % of the notes are contributed by Baroque music, and there are only ten movements (less than 3 % of the corpus, note-wise) by Beethoven, the most recent composer. A selection of the results is shown in Table 8.1. In these excerpted evaluations, the Vivaldi-Bach selections generally scored much higher than the Mozart–Beethoven examples, with the Beethoven garnering the most errors in three out of the four methods tested.

Table 8.2 summarizes the results from the different approaches to pitch spelling as reported by the respective authors, including the one presented in this chapter. It is important to stress that, given the varied test data, it is difficult to compare the algorithms. As can be seen from Meredith's studies [14, 15], and later from our reports on two Beethoven pieces, the results of the pitch-spelling algorithms depend highly on the test corpus. It is not the purpose of this chapter to draw numerical comparisons among the various pitch-spelling algorithms. Our goal here is to present a computational approach to pitch spelling utilizing the Spiral Array and its development

Table 8.2 Result summary for recent pitch-spelling algorithms as reported by the authors

Author(s)	Algorithm	Test set	No. of notes	% correct
Chew and Chen (present study)	Spiral Array	Beethoven Sonata Op.79 (mvt 3)	1,375	99.93
		Beethoven Sonata Op.109 (mvt 1)	1,516	98.21
		Huang Song of Ali-Shan	1,571	100.00
		Overall	**4,462**	**99.37**
Cambouropoulos	Interval optimization	Mozart Piano Sonata. (K279–284, K330–333)	54,418	98.80
		Chopin Waltzes (Op.64 Nos.1–3)	4,876	95.80
		Overall	**59,294**	**98.59**
Meredith	ps13	Bach WTC Book 1 (BWV 846–869)	41,544	99.81
		Corelli through Beethoven	**1.73M**	**98.50**
Temperley	Line of fifths	Kostka-Payne corpus	**8,747**	**98.82**

through computational testing on a challenging work that reveals some core requirements for robust pitch spelling. We highlight the similarities and differences between our Spiral Array approach and the various algorithms of other authors.

8.3 The Algorithm

We propose a real-time computational algorithm for pitch spelling using the Spiral Array model [3]. Any approach to pitch spelling must address the problem of determining the local context, for example, by determining a center of gravity on the line of fifths [13, 20], by calculating the likelihood of a given context [14, 15] or by optimizing intervals over a sliding window [2]. Our approach consists of two phases, not necessarily in chronological order, namely: (1) a method for assessing the tonal context of a section and assigning the appropriate pitch names; and, (2) strategies for selecting the window of events that define the tonal context.

Our method for assigning appropriate pitch names uses the Spiral Array model. The Spiral Array is a geometric representation for tonal entities that spatially clusters pitches in the same key and closely related keys. The Spiral Array model uses aggregate points inside the structure called *centers of effect* (CEs) to summarize tonal contexts. The CE approach is similar to Temperley's line-of-fifths approach in that a center of gravity is generated by the data. The depth added by going from one to three dimensions allows the modeling of more complex hierarchical relations in the Spiral Array. Given a CE that represents the local context, each pitch name is assigned through a nearest-neighbor search in the Spiral Array space.

The Spiral Array Model consists of spatial representation for pitch classes, chords, and keys. Higher level entities are generated as weighted sums of their lower level components. Each type of tonal entity forms a spiral, resulting in an array of nested spirals. The same idea of spatial aggregation extends to the concept of the CE. Any segment of music can be mapped to the Spiral Array structure and can generate a CE, the sum of the pitch class representations weighted by their durations. This CE has been shown to be an effective tool for determining the key of the section through a nearest-neighbor search for the closest key representation [4]. For this reason, we use the CE as a proxy for the tonal context in the pitch-spelling algorithm.

Tracking an evolving tonal context is a difficult problem. Since the Spiral Array space clusters closely related tonal entities, we do not need to track precisely the occurrence of shifts to nearby keys for the purpose of pitch spelling. It suffices to be able to catch abrupt changes to more distant tonalities in the context of the overall key of the piece. We achieve this goal by using a combination of long-term (cumulative windows) and short-term (sliding windows) information to determine the local context.

8.3.1 The Spiral Array Model

The main difference between our approach and other recent algorithms is the use of the Spiral Array representation for tonal entities. The model is described in Chap. 3. The part of the Spiral Array model relevant to the pitch-spelling algorithm is the pitch class spiral, presented in Sect. 3.1 and portrayed in Fig. 8.2. In the Spiral Array, pitches map to spatial positions at each quarter-turn of an ascending spiral, and

Fig. 8.2 A section of the pitch class spiral in the Spiral Array

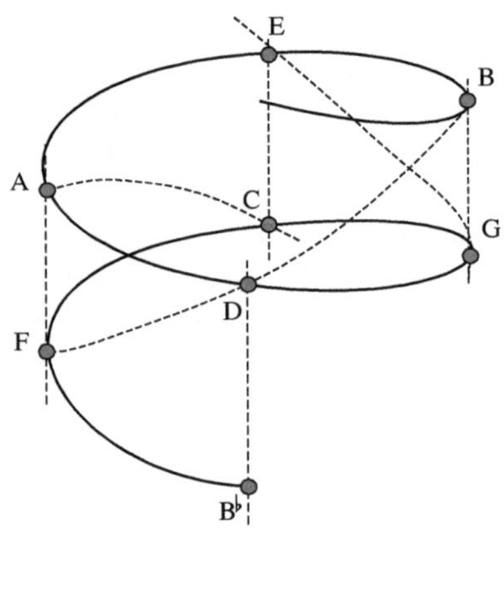

neighboring pitches are four scale steps (that is to say, the distance of a perfect fifth or seven semitones) apart, which results in vertically aligned pitches being two major scale steps (or four semi-tones) apart:

$$\mathbf{P}(\mathrm{k}) \stackrel{\mathrm{def}}{=} \begin{bmatrix} x_{\mathrm{k}} \\ y_{\mathrm{k}} \\ z_{\mathrm{k}} \end{bmatrix} = \begin{bmatrix} \mathrm{r}\sin\frac{\mathrm{k}\pi}{2} \\ \mathrm{r}\cos\frac{\mathrm{k}\pi}{2} \\ \mathrm{kh} \end{bmatrix},$$

where r is the radius of the spiral and h is the vertical ascent per quarter turn. Each pitch is indexed by its distance, according to the number of perfect fifths, from an arbitrarily chosen reference pitch, C. Note that the pitch class representations is a spiral configuration of the line of fifths, and of the *tonnetz* attributed to the mathematician Euler (see [9]) or the Harmonic Network used by Longuet-Higgins [10, 11]. The pitch class representations in the Spiral Array are shown in Fig. 8.2. Higher level objects, such as triads and keys are generated as convex combinations of their defining pitches and triads respectively [3].

Someinterval preference is implicitly incorporated into the geometric model. As part of its design, the Spiral Array is calibrated so that spatial proximity corresponds to perceived relations among the represented entities. The calibration criteria that pertain directly to the pitch spiral correspond to perceived closeness between pairs of pitches. The aspect ratio, r/h, is set as $\sqrt{2/15}$, in which case, intervals in a major triad (P5/P4, M3/m6) map to the shortest inter-pitch distances, other intervals in the diatonic major scale (M6/m3, M2/m7, M7/m2) map to farther distances, and the tritone (d5/A4) maps to an even farther distance.

In the Spiral Array model, any collection of notes generates a CE. For a sequence of music time series data, the CE is a point in the interior of the Spiral Array that is the convex combination of the pitch positions weighted by their respective durations. Formally, if

$$\begin{aligned} n_j &= \text{number of active pitches in chunk } j \\ \mathbf{p}_{i,j} &= \text{the Spiral Array position of the } i\text{th pitch in chunk } j, \text{ and} \\ d_{i,j} &= \text{the duration of the } i\text{th pitch in chunk } j, \end{aligned}$$

then we divide music information in equal time slices that we call chunks and define $\mathbf{c}_{a,b}$ to be the CE of all pitch events in the window from chunks a through b:

$$\mathbf{c}_{a,b} = \sum_{j=a}^{b} \sum_{i=1}^{n_j} \frac{d_{i,j}}{D_{a,b}} \cdot \mathbf{p}_{i,j}, \quad \text{where} \quad D_{a,b} = \sum_{j=1}^{b} \sum_{i=1}^{n_j} d_{i,j}.$$

The CE method has been shown to be an effective way to determine the tonal context of a section through a nearest-neighbor search for the closest key representation in the Spiral Array model [4]. In [4] the CE method using the Spiral Array was shown to determine the key in fewer steps than the shape-matching algorithm of Longuet-Higgins and Steedman [12] using the Harmonic Network or the probe-tone profile method of Krumhansl and Schmuckler [9] when tested on the canonical *Well-Tempered Clavier* Book 1 fugue subjects. Consequently, we use the CE as a proxy for the key context without explicitly having to determine the actual key. In the Spiral Array, pitches that are in the same key form compact clusters in the three-dimensional space. The fact that the Spiral Array serves as an effective model for key-finding and clusters pitches in a given key leads to the hypothesis that it would perform well in tracking the key by proxy and assigning pitches by a nearest-neighbor search.

8.3.2 Pitch-Name Assignment

Using the CE method for generating a center of effect inside the Spiral Array, we use the Spiral Array model to determine spellings for unassigned pitches. Pitches in a given key occupy a compact space in the Spiral Array model. Thus, the problem of finding the best pitch spelling corresponds to finding the pitch representation that is nearest to the current key context. Each plausible spelling of an unassigned pitch is measured against the current CE, and the pitch that satisfies the nearest-neighbor criteria is selected to be the appropriate pitch name. See Fig. 8.3 for a graphical depiction of the pitch-name assignment process.

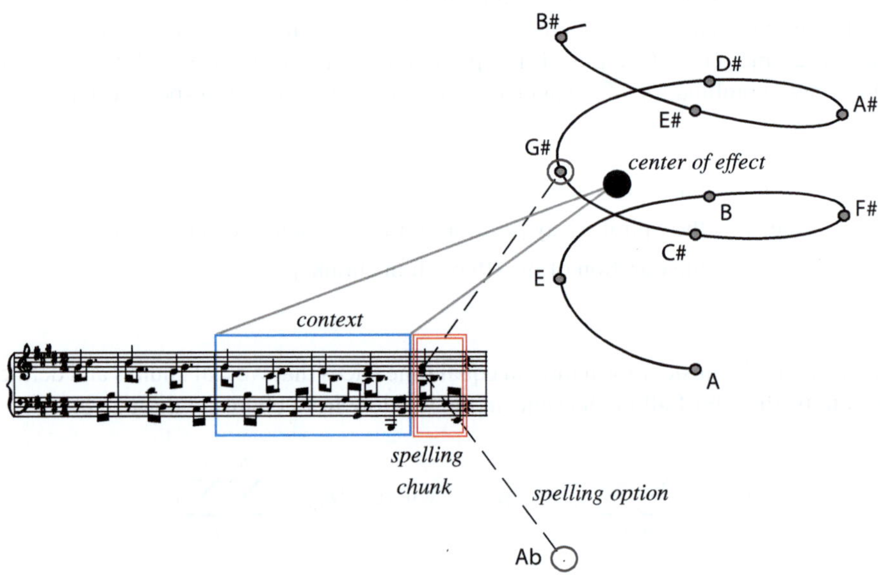

Fig. 8.3 Pitch-name assignment using the Spiral Array

Table 8.3 Table of equivalent spellings

Option 1	(SA index)	Option 2	(SA index)	Option 3	(SA index)
B♯	(12)	C	(0)	D♭♭	(−12)
C♯	(7)	D♭	(−5)	B♯♯	(19)
C♯♯	(14)	D	(2)	E♭♭	(−10)
D♯	(9)	E♭	(−3)	F ♭♭	(−15)
D♯♯	(16)	E	(4)	F♭	(−8)
E♯	(11)	F	(−1)	G♭♭	(−13)
E♯♯	(18)	F ♯	(6)	G♭	(−6)
F♯♯	(13)	G	(1)	A♭♭	(−11)
G♯	(8)	A♭	(−4)		
G♯♯	(15)	A	(3)	B♭♭	(−9)
A♯	(10)	B♭	(−2)	C♭♭	(−14)
A♯♯	(17)	B	(5)	C♭	(−7)

More concretely, each pitch read from the MIDI file will correspond to two or three most probable letter names as shown in Table 8.3 (the numbers in the brackets indicate the index of the pitch in the Spiral Array).

Note that the index triplets are of the form <index−12, index, index+12 >. These indices correspond to pitch positions separated vertically by three cycles of the spiral. The pitch spiral extends to indices of larger magnitude than shown in the table, for example, index −16 or 20 for the (G♯, A♭) option. In Table 8.3, we have shown only the most parsimonious (and common) spellings, that is, spellings up to and including two sharps or two flats.

Having determined the candidates for spelling, the algorithm then selects the spelling that is closest to the current CE. Suppose we are currently assigning names to the pitches in chunk j, and the probable pitch names of the ith pitch in this chunk are $< I_{i,j} - 12, I_{i,j}, I_{i,j} + 12 >$. Assume that $\mathbf{c}_j$ represents the CE that defines the context of chunk j. The index that is consistent with the key context is given by

$$I^*_{i,j}(\mathbf{c}_j) = \arg\min\{\|\mathbf{P}(I_{i,j} - 12) - \mathbf{c}_j\|, \|\mathbf{P}(I_{i,j}) - \mathbf{c}_j\|, \|\mathbf{P}(I_{i,j} + 12) - \mathbf{c}_j\|\}.$$

(The standard mathematical construction "arg min" means the argument that minimizes the function. In this case, "arg min" returns the index, that minimizes the function from the choices $I_{i,j} - 12$, $I_{i,j}$, and $I_{i,j} + 12$. The optimal choice is called $I^*_{i,j}$.

The concept is illustrated graphically in Fig. 8.3. The single-lined box denotes the contextual window. The notes in this window generate a CE inside the Spiral Array. The double-lined box denotes the next chunk to be spelt. The first note in this spelling chunk can be spelt as G♯ or A♭ (note that this is the only case where the number of options is two, see Table 8.3). The spelling option closest to the CE is selected–in this case, G♯.

8.3.3 Comparison to Other Pitch-Spelling Models

Note that the current pitch-name assignment mechanism bears some similarities to that of previous models. Like Temperley's center of gravity approach, the CE serves as a kind of center of gravity in the Spiral Array space. The difference lies in the representation (the Spiral Array vs. the line of fifths) and the segmentation scheme present in our approach. At this point, it would be useful to compare the interval distance rankings in the different models, including Cambouropoulos' interval optimization model, as shown in Table 8.4. Distinct from other models, the Spiral Array prioritizes the major third relation above the seconds and sevenths, and ranks the tritone (A4/d5) much lower than the other models.

In Longuet-Higgins' approach, all "unspelled" pitches are compared to a key choice. Although we do not explicitly determine the key context, the CE acts as a proxy for the key, and key-to-pitch distances also exist in the Spiral Array. The key representation resides on an interior spiral and serves as a prototypical CE in a given context. Table 8.5 shows the key-to-pitch distance rankings for the Spiral Array model and the line-of-fifths approach used by Longuet-Higgins'. In Longuet-Higgins' method, the key is mapped to the same line of fifths as the pitches, whereas in the Spiral Array, the added dimensions allow the key representation to be defined in the interior of the structure, thus resulting in asymmetry in the interval relations. For example, key C to pitch G (a P5 with C as lower pitch) is not the same as key C to pitch F (a P5 with F as lower pitch or a P4 with C as lower pitch). For the Spiral Array, the intervals are given with the key as the lower pitch. Note that a P4 is ranked lower than P5, M3, M2 and M6, and A1 (for example, C♯ in the key of C) is ranked higher than m2 (D♭ in the key of C).

Apart from the pitch representation, an important distinction between our method and that of others (with the possible exception of Longuet-Higgins' method), is that our method processes pitch names in real time using only present and past information

Table 8.4 Spiral Array pitch-to-pitch distances and interval preference comparison

Order of preference	Spiral Array	Line of fifths	Interval optimization
Closest (most preferred)	P4/P5	P4/P5	P4/P5
	M3/m6	M2/m7	m2/M7
	m3/M6	m3/M6	M2/m7
	M2/m7	M3/m6	m3/M6
	m2/M7	m2/M7	M3/m6
	d4/A5	A4/d5	A2/d7
:	d1/A8	A1/d1	d3/A6
	A1/d8	d4/A5	d4/A5
	A4/d5	A2/d7	A4/d5
	A2/d7	D3/A6	d1/A1
	d3/A6	A3/d6	A3/d6
Farthest (least preferred)	d6/A3	d2/A7	d2/A7

Table 8.5 Pitch-to-key distances in Spiral Array versus line of fifths

Order of preference	Spiral Array	Line of fifths
Closest (most preferred)	P5	P4/P5
	M3	M2/m7
	M2	m3/M6
	M6	M3/m6
	P4	m2/M7
	M7	A4/d5
:	m3	A1/d1
	m7	d4/A5
	m6	A2/d7
	A4	D3/A6
	A1	A3/d6
Farthest (least preferred)	m2	d2/A7

as will be evident in the upcoming section. This feature allows the algorithm to be integrated easily into real-time systems for music processing.

8.3.4 Segmentation Strategy

Another important aspect of our algorithm is the determining of context-defining windows for generating the contextual CE. The context of any segment of music is the result of the interaction between the local changes and the larger-scale key of the piece. We propose a two-phase bootstrapping approach to capture the effects of both the local changes as well as the global context in real-time.

We segment the MIDI data into chunks for analysis, so as to batch process the pitch-name assignments. Each iteration of the algorithm consists of two phases, and two iterations of the algorithm are shown in Fig. 8.4. In the first phase, we assign the pitch names to the current chunk (indicated by the double-lined box) using the last w_s chunks as the contextual window for generating a CE. In the second phase, the context is given by a convex combination of a smaller local window (the smaller box of size w_r that overlaps both the current chunk and the cumulative window) and the cumulative window. The current chunk is then re-spelt using the hybrid CE.

Formally, if chunk j contains the pitches to be assigned letter names, the assignments are made using the following CEs:

Phase I: pitch names are assigned using the CE: $\mathbf{c}_j = \mathbf{c}_{j-w_s,j-1}$.
Phase II: pitch names are assigned using a hybrid CE:
$\mathbf{c}_j = f \cdot \mathbf{c}_{j-w_r+1,j} + (1-f) \cdot \mathbf{c}_{1,j-1}$, where $0 \leq f \leq 1$.

Qualitatively, f determines the balance between the local and global contexts. When $f = 1$, the current chunk is re-spelt using only the local context; when $f = 0$, the chunk is re-spelt using only the global context. To initialize the process, the notes in the first chunk are assigned pitch names with indices closest to 2, thus biasing the

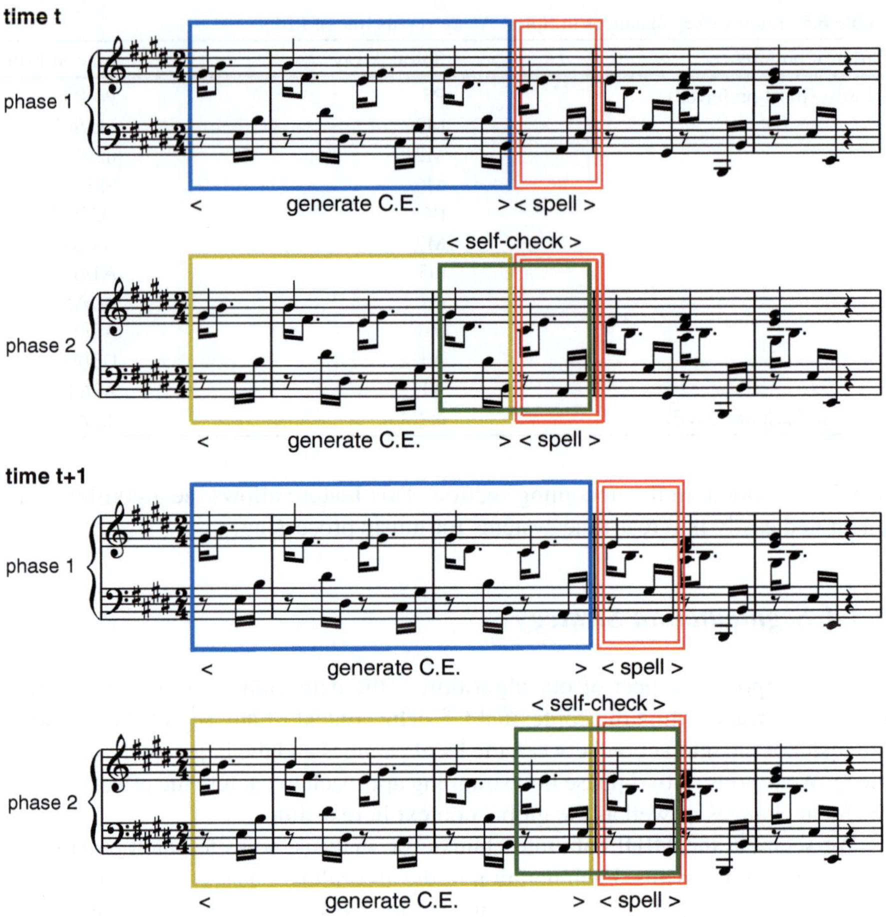

Fig. 8.4 Two-phase assignment method ($w_s = 4$, $w_r = 2$)

notation toward fewer sharps and flats. A CE is generated based on this preliminary assignment, and the original assignments are revisited to make them consistent with the chosen key context. The second pass on the spelling ensures that the first batch of name assignments are self-consistent.

Note that this bootstrapping algorithm also serves as a general framework for other techniques for capturing tonal contexts, such as the cumulative-window approach [6] and the sliding-window approach [7]. The cumulative-window approach uses only the global context to assign pitch spellings: this is achieved by setting $w_s = 0$ and $f = 0$ (see Fig. 8.5a). The sliding-window approach uses only the recent window of events to assign pitch spellings: this is achieved by setting w_s to the desired window size and by setting $w_r = 0$ and $f = 1$ (see Fig. 8.5b). When $b < a$ in $\mathbf{c}_{a,b}$, a CE is not generated and the algorithm defaults to a do-nothing step.

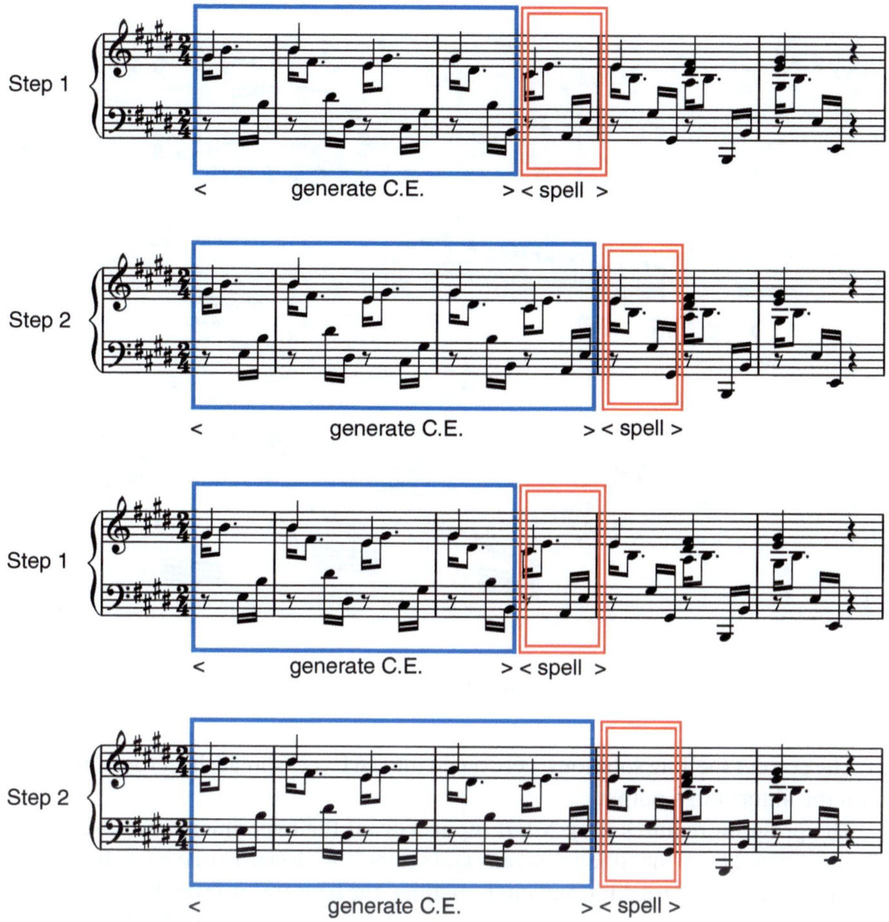

Fig. 8.5 Special cases of the bootstrapping algorithm: **a** cumulative CE method: $w_s = 0$ and $f = 0$; **b** sliding window method: $w_s = 4$, $w_r = 0$ and $f = 1$

8.4 Computational Results

The pitch-spelling algorithm was implemented in Java as part of the Music on the Spiral Array (MuSA) package, an implementation of the Spiral Array and its associated algorithms. The algorithm assigns letter names to MIDI numbers in real-time as the music unfolds. We compare these algorithm-assigned pitch names with those chosen by the composer as notated in the score to evaluate the algorithms' efficacy. We created our data sets by manually scanning the scores using the Sibelius music notation program, checking the digital scores visually and aurally for optical recognition errors, and generating spelling solution tables from the score by hand to ensure accuracy.

Fig. 8.6 Spelling challenges: intermediate keys

The correct spelling solutions were typed into an Excel spreadsheet for comparison with the algorithm's output.

The program requires only MIDI input. For all the experiments, we used MIDI files generated from the Sibelius scores and set the chunk size to one beat. To separate the problems of beat tracking and pitch spelling, the beat size was read from the MIDI files. Note that the algorithms generalize to real-time spelling of MIDI data captured from live performance, in which case, the chunk size could be some unit of time.

The real-time bootstrapping algorithm was tested on two movements from Beethoven's (1770–1827) Piano Sonatas and on You-Di Huang's (b.1912) "Song of Ali-Shan", a set of variations for voice and piano. Beethoven's 32 piano sonatas are a staple in the piano literature; the two movements that we have chosen are the third movement of Sonata No. 25 in G major, Op.79 (composed 1809) and the first movement of Sonata No. 30 in E major, Op.109 (composed 1820). The late sonata, Op.109 is a particularly challenging example because it is harmonically complex and contains numerous subtle and sudden key changes. This is the data set that inspired the bootstrapping algorithm for pitch spelling. Figure 8.6 shows the local keys present in each part of the selection shown in Fig. 8.1, one of the explanations for the challenges posed by this test set.

You-Di Huang is a celebrated Chinese-born composer and music educator who currently resides in Taiwan. The "Song of Ali-Shan" (1967), A028 in d minor, composed for violin/chorus and piano, are based on a popular Taiwanese song extolling

Fig. 8.7 Spelling challenges in the "Song of Ali-Shan": **a** theme; **b** transition to the parallel major key with change from triple to duple meter

the beauty of the maidens of Ali-Shan (Mount Ali). This data set presents some different challenges for the bootstrapping algorithm. The theme (shown in Fig. 8.7a), and much of the piece, is pentatonic with a distinct Chinese flavor. The first variation is pentatonic in style and in duple (2/4) meter. The second variation switches to a triple meter. The third variation returns to a duple meter but transitions to a seven-pitch scale (see Fig. 8.7b), using chromatic pitches to elaborate on the original pentatonic melody. The meter change tests the robustness of the choice of window size under different pitch-time organizations. The first half of the piece is predominantly in a D minor mode, whereas the second half proceeds in the key of D major. Even though D minor and D major share the same tonic, they differ by two important pitches (F / F♯ and B♭/B).

Table 8.6 Table of results for cumulative window experiments

Piece	Number of notes	Number of errors	% correct
Beethoven Op.79 (3rd mvt)	1375	1	99.93
Beethoven Op.109 (1st mvt)	1516	73	95.18
Total	2891	74	97.44

8.4.1 Brief Review of Prior Experiments

We briefly summarize the results of an earlier pitch-spelling experiment using the cumulative window algorithm, leading to the motivation for the bootstrapping approach. Note that a cumulative window approach is similar in spirit to Longuet-Higgins' [12] algorithm in that a global context is used to assign pitch names. In the Beethoven Op.79 example, the cumulative window algorithm achieved a correct spelling rate of 99.93 % (one error out of 1,375 notes; overlapping and tied notes were counted only once). In the Beethoven Op.109 example, the same algorithm achieved a correct spelling rate of 95.18 % (73 errors out of 1,516 notes). The overall correct rate was 97.44 % as summarized in Table 8.6.

The errors in both examples can be categorized into two types: those based on voice leading conventions and those due to unexpected local key changes. Figure 8.8a shows an example of an error (circled) due to ignoring spelling conventions for stepwise motion. Errors of the second type can be caused by insensitivity to either subtle key changes or rapid adoption of a new (and distant) key context. Figure 8.8b shows an example of errors (circled) owing to an inability to adapt quickly enough to a new key context (bookended by the double bar lines). As before (in Fig. 8.1), the notation shown is the correct spelling, and the circles indicate places in the music where a misspelling (not shown) occurred.

8.4.2 Approaching the Bootstrapping Algorithm

A natural solution to the lack of sensitivity in local contextual changes in the cumulative-window method is to use a sliding-window approach. Ensuing experiments using a sliding-window approach produced the results shown in Table 8.7. The tests were carried out using only the more challenging Beethoven Op.109 example so as to better detect spelling improvements. As expected, the sliding window allowed the contextual CE used in the pitch-spelling assignments to be more sensitive to local key changes and the number of spelling errors was reduced from the original 73 for the cumulative window algorithm (shown in Table 8.6) to 31 for $w_s = 4$ in the sliding window algorithm (see Table 8.7).

Although the sliding-window approach eliminated many of the errors due to local key changes, it remained insensitive to sudden key changes, unable to adapt quickly

Fig. 8.8 Sources of error in the cumulative window approach (notation shown is correct spelling, and *circles* indicate locations of misspellings, not shown): **a** linear motion in Op. 109 resulting in spelling error in bar 10 (*circled*); **b** sudden key change prior to bar 61 in Op. 109 resulting in spelling errors (*circled*)

Table 8.7 Table of results for sliding-window experiments using the Beethoven Op.109 example

Parameters			Number of errors	% correct
w_s	w_r	f	(total: 1516 notes)	
4	0	1	**31**	**98.00**
8	0	1	47	96.90
16	0	1	40	97.36

enough to a switch to an unexpected new key, which occurs on several occasions in the Beethoven Op. 109 example. The sliding window algorithm's inability to adapt to sudden key changes resulted in the design of the bootstrapping algorithm described in this chapter.

Table 8.8 Table of results for bootstrapping approach for Beethoven Op.109 example

Parameters			Number of errors	% correct
w_s	w_r	f	(total: 1516 notes)	
4	2	0.6	28	98.15
4	3	0.9	32	97.89
4	3	0.8	**27**	**98.22**
4	3	0.7	**27**	**98.22**
8	2	0.9	31	97.96
8	2	0.8	40	97.36
8	2	0.7	40	97.36
8	4	0.9	40	97.36
8	4	0.8	40	97.36
8	4	0.7	43	97.16
8	6	0.9	**27**	**98.22**
8	6	0.8	**30**	**98.02**
8	6	0.7	47	96.90
16	4	0.8	40	97.36
16	4	0.7	40	97.36
16	6	0.9	31	97.96
16	6	0.8	**30**	**98.02**
16	8	0.9	37	97.56
16	8	0.7	41	97.30

8.4.3 Bootstrapping Algorithm Experiments

We next applied the bootstrapping algorithm to the Beethoven Op.109 example and the computational results are shown in Table 8.8. The algorithm was tested using various parameter values for the phase I window size, w_s, the phase II window size, w_r, and the relative weight of the local context versus the cumulative context, f. The best result had 27 spelling errors (out of a total of 1,516 notes) and was accomplished by the following parameter values: $(w_s, w_r, f) = (4,3,0.8)$, (4,3,0.7) and (8,6,0.9). The next best result had 30 errors using the parameters (8,6,0.8) and (16,6,0.8).

Since the best results were achieved with high values for f, we can deduce that the local context is more important than the global context in pitch spelling. Consequently, a purely sliding window method should perform better than a purely cumulative window method, as exhibited by the results shown in Tables 8.6 and 8.7. However, the better results achieved by the combination of both a local and cumulative effect in the bootstrap algorithm implies that pitch spelling, like tonal perception, is a function of both local and global tonal contexts.

8.4.4 Tests on the "Song of Ali-Shan"

Having shown the pitch-spelling algorithm to be effective in assigning pitch names for classical music by Beethoven, we next tested it on a piece by Taiwanese composer You-Di Huang. The pitch-spelling experiment was performed using some of the best parameters from Table 8.8, all the parameters from Table 8.7 and the cumulative window parameters. The list of experiments is documented in Table 8.9. The bootstrap parameters are (4,3,0.8), (8,6,0.9) and (16,6,0.8); and, the sliding window parameters are (4,0,1), (8,0,1) and (16,0,1). Despite the shifting meters and pitch contexts in the piece, all experiments returned 100 % perfect spellings (1,571 notes out of 1,571 notes).

It is important to note that the results, although seemingly uninformative due to its perfection, is not a trivial one. In the piece, not only does the key signature change, the meter (time structure) is also subject to variation while the window sizes, w_s and w_r, remained constant. As a point of comparison, the default Java MIDI-to-note name translator, part of the Java Development Kit, returned results that had 156 errors (a 10 % error rate). Recall that the Beethoven Op.79 example also performed well (only 1 error in 1,375) using just the cumulative window. This suggests that the pitch-name assignment technique is reasonably robust for pieces of intermediate complexity, and the bootstrapping only improves the results in cases where the key changes tend to the extreme.

Compiling the results from the three sets of test data, we assembled the overall pitch spelling results for parameters (4,3,0.8) and (8,6,0.9) in Table 8.10. A composite

Table 8.9 Table of results for experiments using "The Song of Ali-Shan"

Parameters			Number of errors	% correct
w_s	w_r	f	(total: 1571 notes)	
0	0	0	0	100
4	0	1	0	100
8	0	1	0	100
16	0	1	0	100
4	3	0.8	0	100
8	6	0.9	0	100
16	6	0.8	0	100

Table 8.10 Overall pitch spelling results

Piece	Number of notes	Number of errors	% correct
Beethoven Op.79 (3rd mvt)	1375	1	99.93
Beethoven Op.109 (1st mvt)	1516	27	98.22
You-Di Huangs "Song of Ali-Shan"	1571	0	100.00
Overall result	4462	28	99.37

result using these parameters for the Op.79 and Ali-Shan examples yields a total of 1 error out of 2,496 notes (that is to say, 99.97 % correct). To better estimate the performance of the algorithm on a test set of more varied difficulty, we include the Op.109 results to obtain the overall performance estimate. Overall, there were 28 spelling errors out of a total of 4,462 notes, giving an estimated percentage correct rate of 99.37 %. The errors remaining are primarily due to voice-leading conventions for chromatic motion, which result in pitch spellings that are not consistent with the local key context or its most closely related keys. For example, in Fig. 8.1a, b, the F double sharp is the result of linear upward motion, and not the local key context, C♯ minor. A simple rule-based system for detecting linear chromatic motion may produce more errors than it corrects. To effectively eliminate voice-leading errors, one approach is to separate out the various voices in a polyphonic piece, a topic for another chapter.

8.5 Conclusions

This chapter has presented the problem of assigning pitch names to MIDI note numbers. Pitch spelling is an essential component in automated transcription and computer analysis of music. Proper naming of pitch numbers results in more accurate key tracking chord recognition algorithms. Better algorithms for transcription and analysis improve the state of the art in music retrieval and other interactive music applications.

A bootstrapping framework for real-time pitch name assignments using the Spiral Array model was described which used a variety of data windows to summarize both local and global key contexts. The bootstrapping framework captures both the cumulative-window as well as the sliding-window variations on the algorithm. The proposed algorithms provide accurate spellings in real time, are computationally efficient, and scale well to large data sets. The algorithms were tested on three different test sets: movement 3 of Beethoven's Piano Sonata No.25 in G major, Op.79, movement 1 of Beethoven's Piano Sonata No.30 in E major, Op.109, and a set of variations on the "Song of Ali-Shan". The challenging Beethoven Op.109 example inspired the bootstrapping two-step assignment algorithm described in this chapter. The first and third test sets posed little problems for all variants of the bootstrapping algorithm, including the sliding-window and the cumulative-window algorithms. The best overall correct assignment rate was 99.37 %.

The two main features of our algorithm are its use of the Spiral Array and the bootstrapping algorithm. Future research that separates the two for independent testing will help reveal the relative importance of each component to accurate pitch spelling and lead to the development of better algorithms. Such tests can incorporate aspects of other researchers' models with one of the two components identified, for example, the Spiral Array with Cambouropoulos' windowing approach, or the bootstrapping algorithm with the line-of-fifths representation.

As part of future work, more urgency will be placed on the testing of our pitch-spelling algorithm on databases such as Meredith's for comparison of our method to other spelling algorithms. We have shown in this chapter the algorithm's behavior over a range of parameter values and reported an overall performance estimate of 99.37 %. Included in this estimate was the result from the data used in development in order to capture the performance of the algorithm over a more varied test set in terms of tonal complexity. This reuse of development data for the performance estimate can be avoided in the future by conducting quantitative and comparative tests using a larger test set. A large corpus will also result in more statistical significance regarding the performance of the proposed algorithm.

Based on our experience with a limited test set, it appears that the numerical differences between the performance scores of the different algorithms would be larger if the input data were tonal but more complex. For example, late Beethoven is likely to reveal the differences between the pitch-spelling algorithms more distinctly than early Beethoven. This leads us to suggest that future testing should also be performed on a larger corpus of complex tonal music. It would also be an interesting exercise to test the pitch-spelling algorithm on contemporary compositions, many of which are tonal in nature, or atonal but adhering to the conventions of tonal spelling.

Addendum

Since the publication of the bootstrapping algorithm, Meredith and Wiggins [16] have conducted an extensive evaluation of pitch-spelling algorithms, including the one in this chapter, using a classical and baroque test corpus containing 195,972 notes. For this corpus, they showed that the bootstrapping method performed as well as the optimal version of Cambouropoulos' algorithm (99.15 %), better than the Longuet-Higgins (95.11 and 98.21 %, respectively), Temperley (99.04 %), and Temperley-Sleator (ranging from 97.25 to 97.79 %) methods, and slightly worse than Meredith's ps13 algorithms (99.31 and 99.43 %, respectively). The performance of the bootstrapping method depended primarily on the window sizes, and on the relative weighting between the local versus global CEs. The results also showed the bootstrapping algorithm to be the least dependent on style, having the lowest standard deviation in algorithm performance across the eight composers.

Meredith further optimized the Chew-Chen pitch-spelling algorithm [17], and demonstrated that the pitch class spiral, as used in the bootstrapping method, was equivalent to the line of fifths [18]; and, indeed, the pitch class spiral is the line of fifths coiled around a cylinder, with a pitch class represented at every quarter turn. While the pitch class spiral can be replaced by the line of fifths with impunity in the bootstrapping method, there is merit in using the helical configuration of the line of fifths (i.e., the pitch class spiral) in a number of situations. The line of fifths representation is no longer an equivalent (or sufficient) substitute for the pitch class spiral when multiple interrelated levels of representations in the Spiral Array are invoked for music analysis, as in the case of key-finding or chord recognition. When pitch spelling is employed in tandem with such applications, representational parsimony can be achieved by using the pitch class spiral for spelling [9]. For example, MuSA.RT (see Chap. 9) uses the Spiral Array for pitch spelling, key-finding, chord recognition, and visualization.

Acknowledgments We thank the referees and editors who have provided many detailed comments and helpful suggestions for improving this manuscript. The research has been funded by, and made use of shared facilities at, USCs Integrated Media Systems Center, an NSF ERC, Cooperative Agreement No. EEC-9529152. Any opinions, findings and conclusions or recommendations expressed in this material are those of the authors and do not necessarily reflect those of NSF.

References

1. Cambouropoulos, E.: Automatic pitch spelling: from numbers to sharps and flats. In: Proceedings of the VIII Brazilian Symposium on Computer Music. Fortaleza, Brazil (2001)
2. Cambouropoulos, E.: Pitch spelling: a computational mode. Music Percept. **20**, 411–429 (2003)
3. Chew, E.: Towards a mathematical model of tonality. Ph.D. thesis, Massachusetts Institute of Technology, Cambridge, MA (2000)
4. Chew, E.: Modeling Tonality: applications to music cognition. In: Moore, J.D., Stenning, K. (eds.) Proceedings of the 23rd Annual Meeting of the Cognitive Science Society, pp. 206–211. Lawrence Erlbaum Assoc. Pub., Edinburgh (2001)
5. Chew, E.: The Spiral Array: an algorithm for determining key boundaries. In: Anagnostopoulou, C., Ferrand, M., Smaill, A. (eds.) Music and Artificial Intelligence—Proceedings of the Second International Conference on Music and Artificial Intelligence, pp. 18–31. Springer, Heidelberg (2002)
6. Chew, E., Chen, Y.-C.: Mapping MIDI to the Spiral Array: disambiguating pitch spellings. In: Computational Modeling and Problem Solving in the Networked World–Proceedings of the 8th INFORMS Computer Society Conference, pp. 259–275, Kluwer Academic Publishers (2003)
7. Chew, E., Chen, Y.-C.: Determining context-defining windows: pitch spelling using the Spiral Array. In: Proceedings of the 4th International Conference for Music Information Retrieval, pp. 26–30 (2003)
8. Chew, E.: Eliane Chew Responds. In: Letters, Comp. Mus. J. **31**(4), 7–10 (2007)
9. Krumhansl, C.L.: Cognitive Foundations of Musical Pitch. Oxford University Press, Oxford (1990)
10. Longuet-Higgins, H.C.: Letter to a musical friend. Mus. Rev. **23**, 244–248 (1962)
11. Longuet-Higgins, H.C.: Second letter to a musical friend. Mus. Rev. **23**, 271–280 (1962)
12. Longuet-Higgins, H.C., Steedman, M.J.: On interpreting Bach. Mach. Intell. **6**, 221–241 (1971)
13. Longuet-Higgins, H.C.: The perception of melodies. Nature **263**, 646–653 (1976)
14. Meredith, D.: Pitch spelling algorithms. In: Proceedings of the Fifth Triennial ESCOM Conference, pp. 204–207. Hannover, Germany (2003)
15. Meredith, D.: Comparing pitch spelling algorithms on a large corpus of tonal music. In: Wiils, U. K. (ed.) Computer Music Modeling and Retrieval 2004, pp. 173–192. LNCS 3310, Springer, Heidelberg (2004)
16. Meredith, D., Wiggins, G.: Comparing pitch spelling algorithms. In: Proceedings of the International Conference on Music Information Retrieval, pp. 280–287, London (2006)
17. Meredith, D.: Optimizing Chew and Chen's pitch-spelling algorithm. Comp. Mus. J. **31**(2), 54–72 (2007)
18. Meredith, D.: Proof of the equivalence of the Spiral Array and the line of fifths in Chew and Chen's pitch-spelling algorithm. In: Letters, Comp. Mus. J. **31**(4), 5–7 (2007)
19. Sleator, D., Temperley, D.: The Melisma music analyzer. http://www.link.cs.cmu.edu/music-analysis (2001)
20. Temperley, D.: The Cognition of Basic Musical Structures. MIT Press, Cambridge (2001)

Part VI
Visualization

Chapter 9
MuSA.RT

Abstract The purpose of this chapter is to describe MuSA.RT Opus 2, an interactive system for tonal visualization of music at multiple scales, and to present examples of the types of musical features and attributes that can be abstracted and visualized by the system. MuSA.RT aims to create an environment by which musical performances can be mapped in real-time to a concrete and visual metaphor for tonal space, wherein we can see the establishment and evolution of the tonal context. In this environment, expert musicians will be able to see the tonal structures of what they play, initiated listeners will be able to visually follow the structures that they hear, and novices can learn to hear the structures that they see. MuSA.RT is both an interactive art installation that can convert musical performances into mathematically elegant graphics and a scientific tool for visualizing the inner workings of tonal induction and tracking algorithms. In this chapter we describe the mapping strategies for transforming a MIDI stream into tonal structures in 3D space, and our solution for overcoming the challenge of real-time concurrent processing of data streams; we will also give examples and present case studies of visual mappings of music by Pachelbel, Bach, and Barber. The reader can download the latest version of the MuSA.RT software, MuSA_RT, from the Mac App Store (https://itunes.apple.com/ca/app/musa-rt/id506866959?mt=12, cited 30 August 2013), follow the examples posted at https://musa-rt.blogspot.com, or try out new ones themselves.

9.1 Introduction

This chapter presents MuSA.RT, an interactive multi-scale music visualization system that tracks and displays the trajectory of the tonal content and context of music in real time. Tonal music, which includes almost all the music that we hear, is perceived

---

 This is a minor revision of the work, "Interactive multi-scale visualizations of tonal evolution in MuSA.RT Opus 2" by Elaine Chew and Alexandre R. J. Francois, published in Computers in Entertainment (CIE)—Theoretical and Practical Computer Applications in Entertainment, **3**(4), 1–16 (2005) http://doi.acm.org/10.1145/1095534.1095545

E. Chew, *Mathematical and Computational Modeling of Tonality*, International Series in Operations Research & Management Science 204, DOI: 10.1007/978-1-4614-9475-1_9,

as operating within a tonal context that evolves over time. The context organizes the musical information with reference to a tonal center. Although the ability to perceive and to know this tonal context is ingrained in our hearing of music, tonality remains a concept that eludes easy explanation. The purpose of tonal visualization is to reveal high-level pitch structures in the music, including the pitch classes present, the chords, and the keys. This goal is distinct from music visualizations that map low-level signal attributes to visual patterns. We describe the MuSA.RT Opus 2 system and its application to three popular classical pieces by Johann Pachelbel, Johann Sebastian Bach, and Samuel Barber.

As a companion to the book, the MuSA.RT blog, https://musa-rt.blogspot.com, contains further examples of visualizations generated by MuSA.RT. A newer version of MuSA.RT, named MuSA_RT [11], is available for download from the Mac App Store.

MuSA.RT stands for Music on the Spiral Array, Real-Time. MuSA.RT Opus 2, shown in Fig. 9.1, is a real-time, interactive music system that displays and tracks the tonal context of music in the Spiral Array model [1], using state-of-the-art algorithms for tonal induction. The spiral array is a geometric model for tonality that represents tonal elements in three-dimensional space so that perceptually close entities are near one to another. The Spiral Array serves as both the medium for visualizing the evolution of tonal structures over time as well as the mathematical model on which the algorithms are based [2]. Hence, the visual allegory for contextual evolution serves concurrently as a manifestation of the algorithms' performance. The visual results of the algorithms will allow musicians to see the tonal structures of the pieces

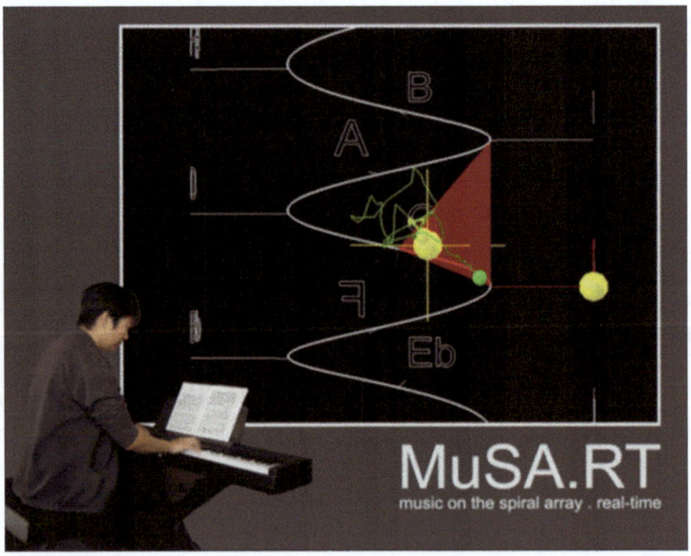

Fig. 9.1 MuSA.RT Opus 2

they play, and listener–viewers will be able to follow the tonal structures in each piece as they unfold.

In the Spiral Array model, spatial points called centers of effect (CEs) represent pitch clusters, at both the micro and macro levels, in the interior of the helix. The visual display shows the CEs, with traces marking their recent history. These CE trails create distinctive paths inside the Spiral Array that characterize each piece of music; patterns they trace range from tight squiggles to angular lines to aimless meanderings. The display also shows the pitches present, the closest chord, and the evolving tonal context as labeled spheres in the three-dimensional Spiral Array space; unique triangles mark each major and minor triad.

To design the real-time interactive system, we use the software architecture for immersipresence [10], a software architectural style for specifying modular dynamic systems that perform asynchronous parallel processing of generic data streams. We implement our system using a corresponding open source architectural middleware, available on SourceForge (François 2001). An earlier version of MuSA.RT (Opus 1) was described and demonstrated at the 2003 ACM Multimedia conference [7]. Opus 2 incorporates new features such as the dual-level CE tracking for chords and keys, and improved highlighting of recognized structures.

The input in MuSA.RT Opus 2 is restricted to musical instrument digital interface (MIDI) format. Input data can be acquired through a MIDI controller such as a keyboard (as shown in Fig. 9.1), or from a software MIDI player or sequencer. The tonal structures are computed and mapped to the display instantaneously in response to the MIDI input. The camera view can be manipulated via a gamepad controller, allowing the user to navigate through the three-dimensional space. Alternatively, an autopilot option automatically orients the model so that the viewer looks directly at the center of the action. When in autopilot mode, each note onset produces an impulse that propels the CE in a new direction, and the spiral dances to the rhythm of the music.

The most closely related work is a recent project by Petri Toiviainen and Carol Krumhansl [16] on the measuring and modeling of real-time responses to music. Human listeners are asked to rate the tonal contexts continuously throughout a piece, and the resulting data is visualized as a self-organizing map on a toroid representation of keys that is shown cut open in two dimensions.

Another related project by Craig Sapp [15] produces two-dimensional maps that show the tonal contexts for a piece of music at various hierarchical levels of time. The tonal patterns at each hierarchical level is computed offline and shown as a line of colored points where each color corresponds to a different key. The stacking of these lines results in a multi-(time) scale map of the tonal contexts of a piece of music.

The remainder of this chapter is divided into three main parts. We first give an overview of the Spiral Array model and the relevant computational algorithms for tonal induction. Next, we describe the system design using SAI and the implementation using modular flow scheduling middleware (MFSM). Case studies follow using examples from Pachelbel's *Canon in D*, J. S. Bach's *Suite for Cello*, and Barber's *Adagio for Strings*.

9.2 The Spiral Array Model for Tonality

One of the distinguishing features of our approach is the use of the Spiral Array representation for tonal entities [1]. We give an overview of the Spiral Array model and present the pitch-spelling, chord-tracking, and key-finding algorithms employed in MuSA.RT Opus 2.

9.2.1 The Spiral Array Model

The Spiral Array representation, first proposed by Chew in 2000, consists of a collection of nested spirals, each comprising a type of tonal object. In the model, the outermost spiral consists of pitch class representations. Pitch classes consist of pitches that map to the same class by octave equivalence. For example, the pitches A3, A6, and A4 all map to the same pitch class, A. In the Spiral Array, pitch classes map to spatial positions at each quarter turn of an ascending spiral, and neighboring pitches are the distance of a perfect fifth (or seven half-steps) apart, and vertically aligned pitches are a major third (or four semitones) apart. Assume that each pitch is indexed by its distance, k, according to the number of perfect fifths, from an arbitrarily chosen reference pitch, C. Then a pitch class of index k can be described by the following equation:

$$\mathbf{P}(\mathrm{k}) = \begin{bmatrix} x_{\mathrm{k}} \\ y_{\mathrm{k}} \\ z_{\mathrm{k}} \end{bmatrix} = \begin{bmatrix} \mathrm{r}\sin\frac{\mathrm{k}\pi}{2} \\ \mathrm{r}\cos\frac{\mathrm{k}\pi}{2} \\ \mathrm{kh} \end{bmatrix}.$$

where r is the radius of the spiral and h is the vertical ascent per quarter turn. Note that the pitch class representations is a spiral configuration of the line of fifths, and of the tonnetz attributed to the mathematician Euler (see [8]) or the harmonic network proposed by [13, 14]. The pitch class representations in the Spiral Array are shown in Fig. 9.2.

Higher level objects, such as major and minor triads, $\mathbf{C}_{\mathrm{M}}(\mathrm{k})$ and $\mathbf{C}_{\mathrm{m}}(\mathrm{k})$, and major and minor keys, $\mathbf{T}_{\mathrm{M}}(\mathrm{k})$ and $\mathbf{T}_{\mathrm{m}}(\mathrm{k})$ are generated as convex combinations of their defining pitches and triads respectively. For example, the following equations describe the major triad and major key representations respectively:

$$\mathbf{C}_{\mathrm{M}}(\mathrm{k}) = w_1 \cdot \mathbf{P}(\mathrm{k}) + w_2 \cdot \mathbf{P}(\mathrm{k}+1) + w_3 \cdot \mathbf{P}(\mathrm{k}+4),$$
$$\mathbf{T}_{\mathrm{M}}(\mathrm{k}) = \omega_1 \cdot \mathbf{C}_{\mathrm{M}}(\mathrm{k}) + \omega_2 \cdot \mathbf{C}_{\mathrm{M}}(\mathrm{k}+1) + \omega_3 \cdot \mathbf{C}_{\mathrm{M}}(\mathrm{k}-1),$$

where $w_1 \geq w_2 \geq w_3 > 0$, $\omega_1 \geq \omega_2 \geq \omega_3 > 0$, $\sum_{i=1}^{3} w_i = 1$, and $\sum_{i=1}^{3} \omega_i = 1$.

The objects represented in the Spiral Array model will be addressed at greater length in the latter part of this chapter; more detailed descriptions can be found in Chap. 3.

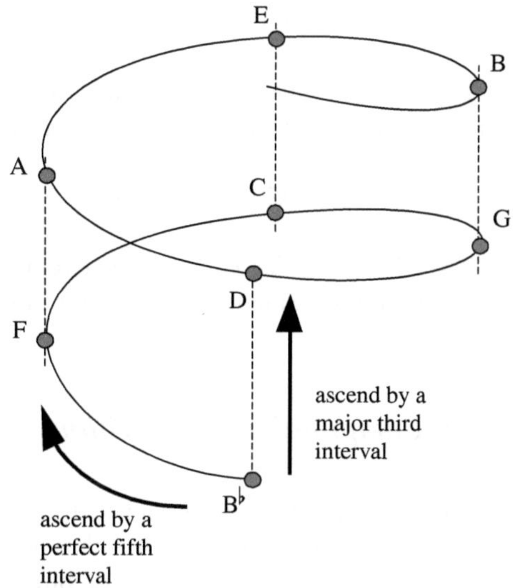

Fig. 9.2 A section of the pitch class spiral in the Spiral Array

The Spiral Array is calibrated so that spatial proximity mirrors perceived closeness among the represented entities. For example, the aspect ratio, h/r, is chosen to be $\sqrt{2/15}$, in which case, intervals in a major triad (P5/P4, M3/m6) map to the shortest inter-pitch distances, other intervals in the diatonic major scale (M6/m3, M2/m7, M7/m2) map to farther distances, and the tritone (d5/A4) maps to even farther distances.

9.2.2 The Center of Effect

Because the Spiral Array model resides in three-dimensional space, it possesses an interior, and any collection of notes can be mapped to their respective pitch classes and represented by their *center of effect* (CE), the convex combination of the pitch class positions. For example, if a piece of music is segmented by time into short time slices, n_j is the number of active pitches in time chunk j, and $\mathbf{p}_{\mathrm{i,j}}$ is the spiral array position of the ith pitch in time chunk j, then the evenly weighted CE of all pitch events in the window $(a, b]$, then

$$\mathbf{c}_{a,b} = \frac{1}{b-a} \sum_{j=a+1}^{b} \frac{1}{n_j} \sum_{i=1}^{n_j} \mathbf{p}_{i,j}.$$

The CE definition used in MuSA.RT is a time series that is a linear combination of the present and past:

$$\mathrm{CE}_\alpha(t) = \alpha \cdot \mathbf{c}_{t-1,t} + (1-\alpha) \cdot CE_\alpha(t-1).$$

In particular, MuSA.RT maintains two different CEs, one macro-level CE and one micro-level CE. Hence, if $\beta >> \alpha$, then we define a second (more local) CE:

$$\mathrm{CE}_\beta(t) = \beta \cdot \mathbf{c}_{t-1,t} + (1-\beta) \cdot CE_\beta(t-1).$$

The CEs are updated continually over time. Hence, pitches of longer duration and more recent pitches will have a greater influence on each CE.

9.2.3 Computational Algorithms

We use the Spiral Array model to track pitch classes, chords, and keys in real-time. In the Spiral Array, pitches that are in the same key form compact clusters in the 3D space. The clustering of pitches in a triad or key in the Spiral Array allows for the design of algorithms that use the CE and nearest-neighbor searches to perform pitch spelling, triad recognition, and key finding. Computational algorithms for pitch spelling, chord tracking and key-finding using the Spiral Array are described at greater length in [1–7]. We describe here variants of these algorithms that are implemented in MuSA.RT.

Each MIDI pitch event labeled by a pitch number, such as 68, that is assessed by the system needs to be mapped to a pitch class name, such as G♯ or A♭. This process is called pitch spelling. Various algorithms for pitch spelling have been proposed by [4–6] using the Spiral Array. Pitch spelling is important in MuSA.RT because the Spiral Array model does not assume enharmonic equivalence, so pitch names with more sharps are near the top and pitch names with more flats are near the bottom. Hence, we need to perform pitch spelling in order to map MIDI pitch numbers to the model. In MuSA.RT, we use the macro-level CE and nearest-neighbor search to determine the appropriate spelling. For each MIDI pitch event active at time t, amongst all its plausible pitch class name assignments, our strategy is to choose the one that is closest to $\mathrm{CE}_\alpha(t)$, that is to say, we select:

$$\mathrm{k}^* = \arg\min_{\mathrm{k}} ||\mathbf{P}(\mathrm{k}) - \mathrm{CE}_\alpha(t-1)||,$$

where k is the index of an appropriately named pitch class. Once the proper pitch names have been assigned, the MIDI pitch events can be mapped to the Spiral Array to update both the local and long-term CEs.

The key defines the context of the music and can be detected most stably using long-term information. Hence, we use the long-term CE, $\mathrm{CE}_\alpha(t)$, to determine key. The most likely key is given by the key representation closest to the CE under consideration. The likelihood or strength of this key is inversely proportional to the distance between the CE and the key representation. The closer the distance from

CE to key the stronger the key, and the farther the distance the weaker the key. The basic algorithm for, and some benchmarking tests using, the Spiral Array center of effect generator (CEG) key-finding algorithm are described in Chew [1] (2001). In MuSA.RT, we adapt the algorithm to the CE defined in the system:

$$\mathbf{T}^* = \arg\min_{\mathrm{T}} ||\mathrm{T} - \mathrm{CE}_\alpha(t-1)||,$$

where T is any major or minor key representation. The current chord is a local pitch structure. Correspondingly, we use the local CE, $\mathrm{CE}_\beta(t)$, to determine the closest triad:

$$\mathbf{C}^* = \arg\min_{\mathrm{C}} ||\mathrm{C} - \mathrm{CE}_\beta(t-1)||,$$

where C is any major or minor triad representation.

9.3 The MuSA.RT Opus 2 System

This section describes MuSA.RT's system design and implementation. A challenging aspect of the MuSA.RT system is the simultaneous processing of multiple data streams in real-time, such as MIDI, tonal structures computation, graphics, and viewpoint control. Another challenge is the organization of, and interaction between, the software components carrying out these tasks. We designed the system using the software architecture for immersipresence (SAI) framework [10]. The resulting architecture is modular and extensible, which will benefit future development and research.

9.3.1 Software Architecture for Immersipresence

In order to describe the MuSA.RT system, we first introduce the components of the SAI architectural style shown in Fig. 9.3. SAI distinguishes between persistent and

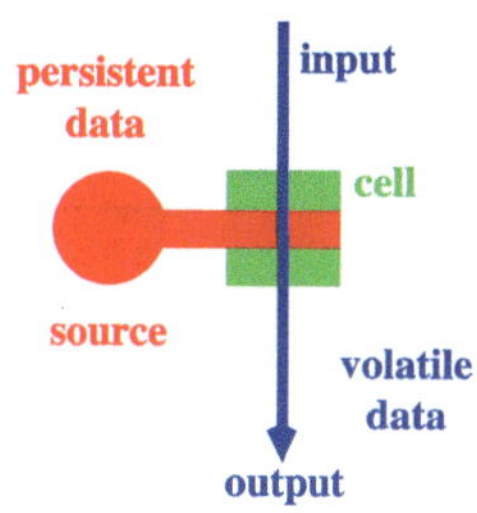

Fig. 9.3 Primitives of SAI

volatile data. Persistent data is held in repositories called sources (red disks), while volatile data flows down streams (thin blue arrows) in the form of pulses. Pulses travel down streams through processing elements called cells (green squares). The ordering of the cells along a stream captures their dependency relations. Each cell is connected to exactly one source, while many cells may be connected to the same source. All the cells connected to a given source have concurrent shared access to the data stored by the source. When a pulse reaches a cell, it is used as input data to a process that may result in the generation of new volatile data that is added to the pulse. The process may also result in the update of some of the persistent data held by the cell's source, also used as input to the cell's process. When the process is completed, the pulse continues its journey downstream. The dynamic evolution of a stream may be thought of as a cascade of processes triggered by pulses. Cells are specialized to implement required functionalities. The pulsar is a special cell type that generates empty pulses at fixed (specified) time intervals.

9.3.2 MuSA.RT System Architecture

Figure 9.4 shows a conceptual-level system diagram for MuSa.RT Opus 2, expressed in the SAI notation. The MuSA.RT system design employs SAI architectural patterns that are typical of interactive systems [9]. A central repository holds persistent data structures that are accessed independently by cells placed on independent streams. These data structures are maintained during the whole life of the application, and

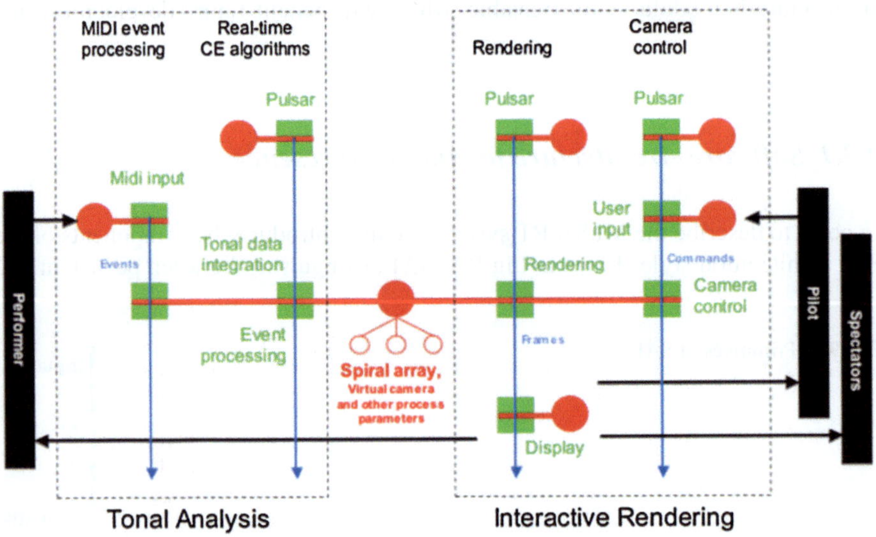

Fig. 9.4 MuSA.RT system diagram

include in particular the Spiral Array and its tonal structures, and the virtual camera used for 3D graphic rendering. Each stream embodies a specific aspect of the system, namely MIDI event processing, real-time tonal context algorithms, rendering, and interactive (virtual) camera control. Each stream is regulated by an individual pulsar (explicitly or not) and thus operates independently of the others. In particular, each stream can operate at a separate frequency adapted to the specific requirements of its function. These streams form two functionally distinct subsystems, (1) real-time music analysis on the Spiral Array and (2) real-time interactive 3D visualization. The details of each subsystem are described in the following sections.

9.3.3 Implementation

Successive MuSA.RT designs have been implemented in C++ using the MFSM [12], an open source architectural middleware that provides code support for SAI's architectural abstractions. MFSM also offers a number of useful open source functional modules. Using some of these modules significantly reduces the coding effort and allows us to focus on MuSA.RT-specific aspects during the development of successive prototypes.

9.3.4 Visual Display

The visual display in MuSA.RT is shown in Fig. 9.5. The display shows the CEs with traces marking their recent history. The local CE is shown as a small green sphere with a green trail, while the long-term CE is shown as a small golden sphere with a corresponding trail. To distinguish these CE trails from the many straight lines in the picture, random noise is added to add a minute flicker to the tails.

Each MIDI note event that is active is mapped to a pitch class representation through a nearest-neighbor search for the contextually most likely pitch spelling, and marked by a silver sphere under its letter name (not shown in Fig. 9.5). The triad closest to the local CE (small green tracker) lights up as a triangle, red for major triads, blue for minor. A sphere of the same color marks the triad and its letter name, also colored the same hue. In general, red denotes an object in the major mode, and a blue one that is in the minor mode.

A throbbing golden sphere inside the spiral, with lines radiating from its center, marks the closest key representation determined by nearest-neighbor search using the long-term CE (small golden tracker). The maximum size of this pulsating object is inversely proportional to its distance from the long-term CE, and corresponds to the strength of the key. In addition, a medium-sized golden sphere marks the letter name of the closest key. The letter name is highlighted in red if the key is in the major mode and blue if it is minor.

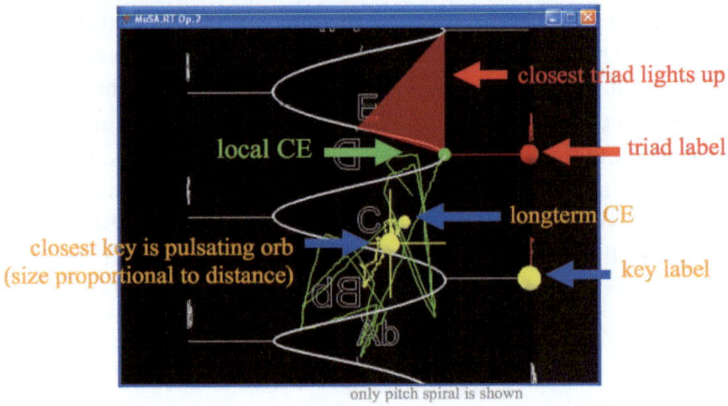

Fig. 9.5 Visual interface in MuSA.RT

9.3.5 3D Rendering

Dynamic visual experience of the virtual environment is achieved by regularly synthesizing a pictorial representation of the current state of the Spiral Array data structure. An example of this synthesized representation is shown in Fig. 9.5. An independent stream (labeled Rendering in Fig. 9.4) handles the production of such images. A pulsar triggers empty pulses at regular intervals. Upon receiving a pulse, the visual rendering cell creates a synthetic composite image that is placed on the stream, and that the image display cell will then present in a window on the screen. This is an established SAI pattern, implemented in an open source white-box module [9] that uses the OpenGL library for 3D rendering. The existing open source module was adapted to the specifics of this project. The pulsing frequency on the stream sets the refresh rate of the display on the screen. We typically run our rendering stream at frequencies of at least 25 Hz so as to provide a smooth visual experience.

9.3.6 View Manipulation

MuSA.RT is equipped with two view manipulation modes: an interactive camera control mode, and an auto-pilot mode.

In the interactive camera-control mode, the pilot can control the camera view by zooming in and out, rotating to the right or left, moving up and down, and tilting the camera angle. To facilitate viewing of the spiral structure, the camera is constrained to movements on a cylinder centered on the same axis as the spiral structure (shown in the left of Fig. 9.6). Right and left directional controls on the gamepad allow the pilot to circle to the right or left, while the view remains centered on the axis of the cylinder. Gamepad buttons allow the pilot to move up and down the cylinder parallel

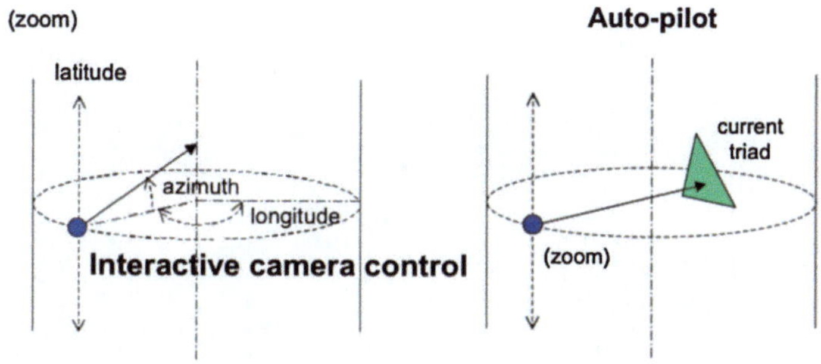

Fig. 9.6 Interactive viewpoint manipulation

to the axis. Up and down directional controls allow the viewer to tilt the camera view along the spiral axis.

In the auto-pilot mode, the camera view is automatically computed by the system. The camera's objective is to face the current triad. It is constrained to move on a cylinder of a fixed radius, and the movements are damped to allow smooth transitions between different views. Each note onset produces an impulse that propels the CE in a new direction, with possibly a change in the closest triad. As a result, both the CE and the camera view in auto-pilot mode move to the rhythm of the music.

9.4 Case Studies

We use MuSA.RT to see the tonal patterns in three pieces: Pachelbel's *Canon in D*, the Prelude from Bach's *Suite for Cello No. 1*, and Barber's *Adagio for Strings* Op.11. In each case, the CE trails create distinctive paths inside the Spiral Array that characterize each piece of music. For each piece, we provide a descriptive analysis of the visual tonal patterns, along with accompanying videos generated by MuSA.RT in auto-pilot mode for each piece. We used the MIDI files encoded by Craig Sapp at kern.humdrum.net as input. As the pieces are played at different tempi, and the tonal patterns evolve at different rates, we manually adjusted the weights, α and β, to provide reasonable multi-scale analyses of the pieces.

9.4.1 Case I. Pachelbel's Canon in D

Pachelbel's *Canon in D* provides visualizations of variations on a ground bass, a bass line that is continually repeated over the course of the piece. A video of the MuSA.RT visual display over the course of the piece is included in the accompanying video, labeled pachelbel-canon.mp4. The MIDI file was played at the default tempo of 60 beats per minute (bpm), and α was set to 0.003 and β to 0.1. Screenshots of the first occurrence of the ground bass figure are shown in Fig. 9.7.

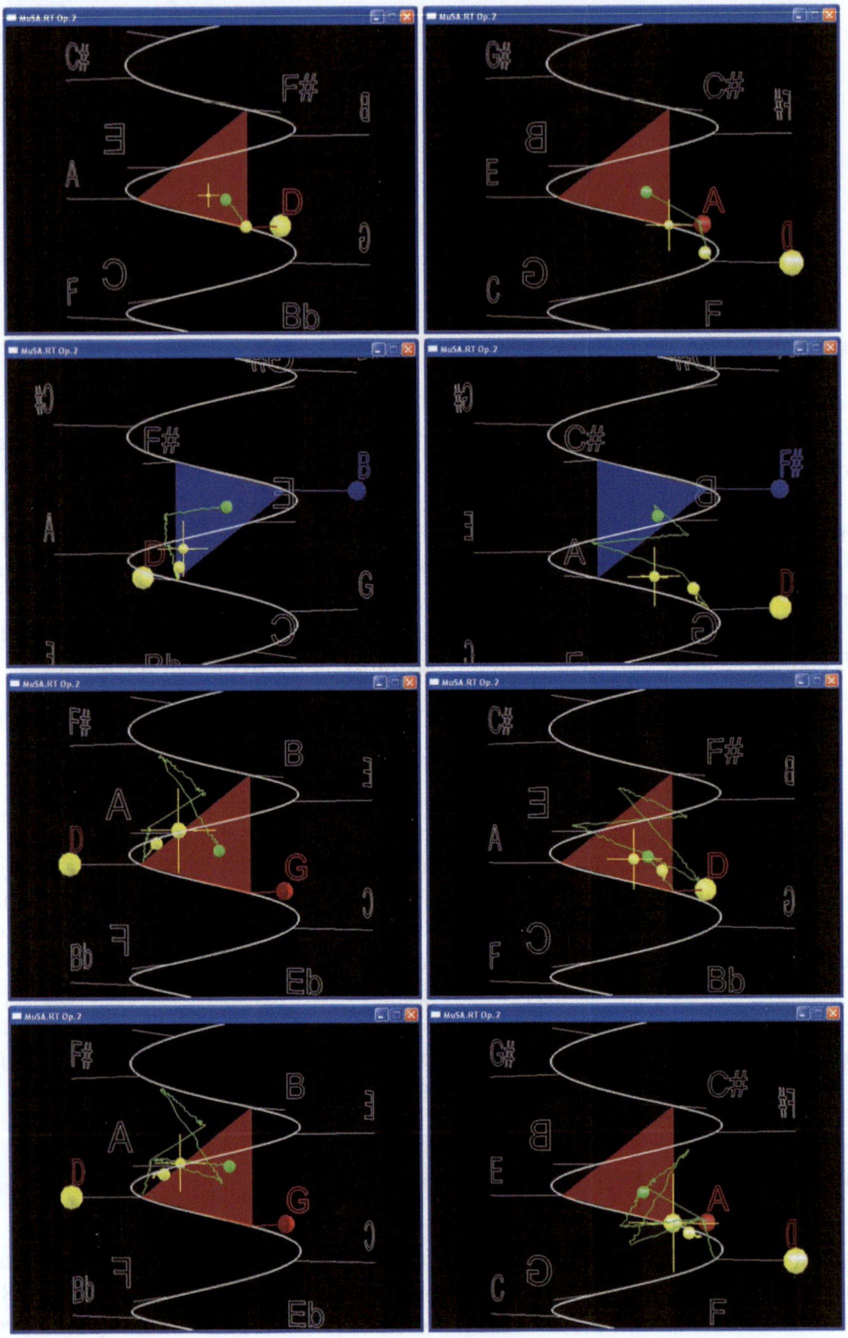

Fig. 9.7 MuSA.RT screenshots of the Pachelbel Canon ground bass pattern (*left-to-right*, *top-to-bottom*): D maj, A maj, B min, F♯ min, G maj, D maj, G maj, A maj

Fig. 9.8 Labeled key and chords in the canon's ground bass figure (bar 7–8)

As a composition on a ground bass, not only the bass line, but also for the most part the harmony as well, is repeated continually throughout the piece. The chords, as exemplified by the harmony between the parts and in the basso continuo accompaniment, change at every beat. The chord sequence, at least for the first several iterations, is {D major, A major, B minor, F♯ minor, G major, D major, G major, A major}. The key stays constant at D major. The chord sequence and the key are reflected in the screenshots in Fig. 9.7 as well as the score shown in Fig. 9.8.

As a set of variations on the ground bass, the Canon evolves by changing the rhythm of the figures elaborated over the ground bass. The original simplicity of the quarter and eighth note figures in the first four iterations of the ground bass (shown in Fig. 9.9a) give way to the more complex parts incorporating first sixteenth note figures (shown in Fig. 9.9b), then some added thirty-second note figures (shown in Fig. 9.9c). As shown in the trajectories in Fig. 9.9, even though the general pattern outlined by the trajectories, the chords, and the keys stays the same, the variations on the ground bass can be distinguished by the types of patterns they trace in the interior of the spiral array. The simple quarter (and eighth) note variation, shown in Fig. 9.9a, traces straight lines from one chord to the next; the variation with sixteenth notes, shown in Fig. 9.9b, takes a more filigreed path, while the more rhythmic variation of Fig. 9.9c, due to the speed of the notes, creates tighter patterns around the chord-to-chord trajectories.

9.4.2 *Case II. Prelude from Bach's Cello Suite No. 1*

The Prelude from Bach's *Cello Suite No.1*, BWV 1007, is a fine example of tonal writing. A video of the MuSA.RT visual display over the course of the piece is

(a)

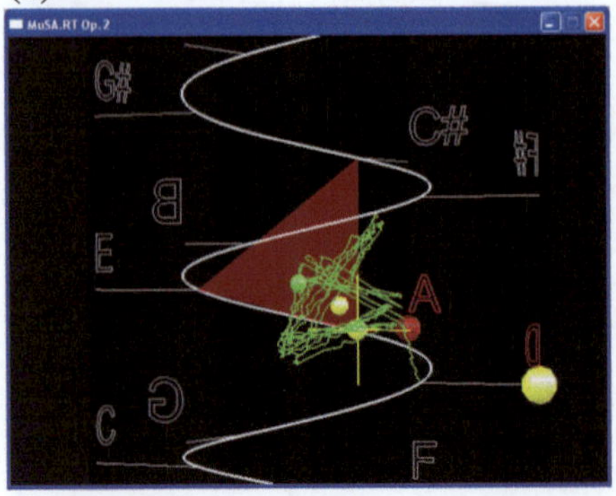

(b)

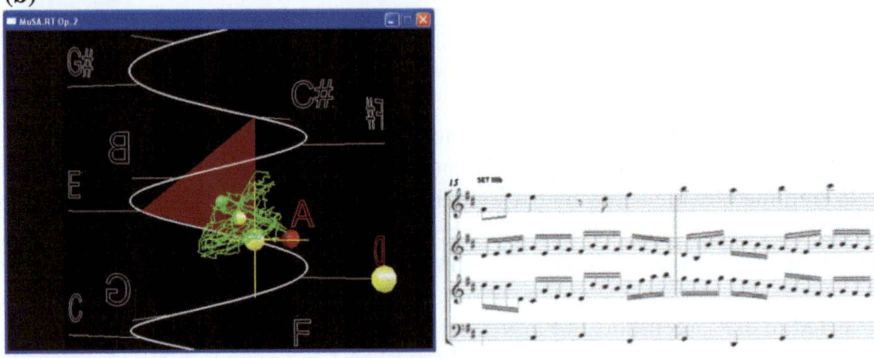

(c)

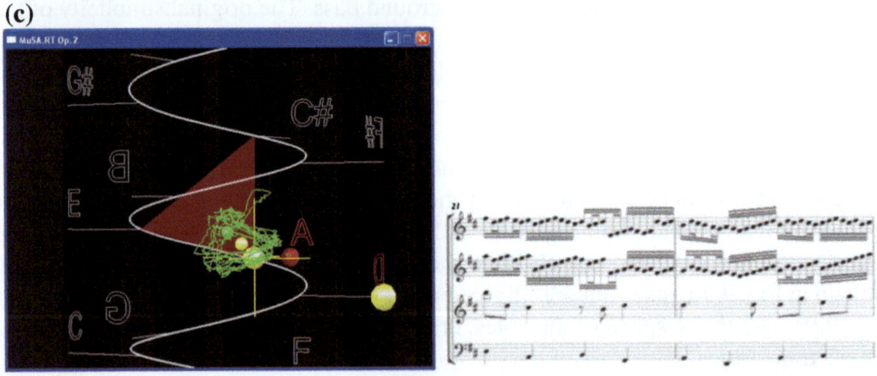

Fig. 9.9 Trajectories traced by the local CE in different parts of Pachelbel's *Canon,* **a** Screenshot showing trajectory of local CE at the end of bar 8 (score shown in Fig. 9.8), **b** Screenshot showing trajectory of local CE at the end of bar 16 (above are bars 15–16), **c** Screenshot showing trajectory of local CE at the end of bar 22 (above are bars 21–22)

included in the accompanying video, labeled bach-bwv 1007.mp4. We set the tempo at 72 bpm, α to 0.003, and β to 0.03. One of the challenges for this piece was the determinination of a β value that would be flexible enough to detect the frequent chord changes, yet sufficiently stable to detect the chord structures.

The resulting visualizations concur with what one might expect in the analysis of the Prelude, with the keys being predominantly G or D major. Figure 9.10 shows some displays from the visual analysis of the piece. The first, Fig. 9.10a, shows the rather stable beginning with trajectories tracing squiggly lines on the plane defined by the G, C, and D major triad representations. Each triad representation is a point inside the triangle outlined by the pitches of the triad and is not shown in visuals. For example, the G major triad representation is a point inside the triangle defined by the pitch representations for G, B, and D.

Figure 9.10b was taken from a tonally more adventurous part of the piece, as evidenced by the span of the local CE trail (green line). Figure 9.10c is a snapshot from the middle of the cadenza-like section on D major, which occupies a larger area than Fig. 9.10a but is centered on the key of D major. Technically, this section beginning in the middle of bar 22 and continuing until the end of bar 38, is not in the key of D, but consists of a very long elaboration of the dominant region in relation to the actual key of G major, which is D major. At one point in the section, the pitch A is repeated so much that the local CE trail shrinks down to a tight ball, as shown in Fig. 9.11.

9.4.3 Case III. Barber's Adagio for Strings Op. 11

Barber's *Adagio for Strings Op.11* is a study in unfulfilled longing, an ambivalent piece that never embraces its key of B♭ minor. The key of B♭ minor is only hinted at, achieved briefly for two bars at the climax in the middle of the piece after 49 bars of music, and never arrived at conclusively, not even in the end.

Figure 9.12 highlights the momentary acknowledgment of the key of B♭ minor in bar 50 (marked by a box), a triumphant arrival on the B♭ minor chord on a strong beat of the measure, only to move on to other chords toward the end of the phrase.

The expectation of closing a phrase conclusively on B♭ minor is never consummated. Even at the end of the piece, where we might logically expect a conclusive ending on B♭ minor, we hear an F major chord, which is spatially close to a B♭ minor chord. Figure 9.13 shows bars 62–69 of the piece, where the final chord in bar 69 is F major, even though the key indicated by the key signature on the left is B♭ minor. This contradiction is reflected in the visual analysis using MuSA.RT. Figure 9.14 shows the final screenshot at the end of the Adagio. The visual display shows that the key is B♭ minor, but the final chord is F major.

A video of the MuSA.RT tonal visualization over the course of the piece is included in the accompanying video labeled barber-adagio.mp4. The MIDI file was played at the default tempo of 104 bpm. To detect the chords, we set β to 0.1; to detect the nebulous key, we chose the highly stable α value of 0.001.

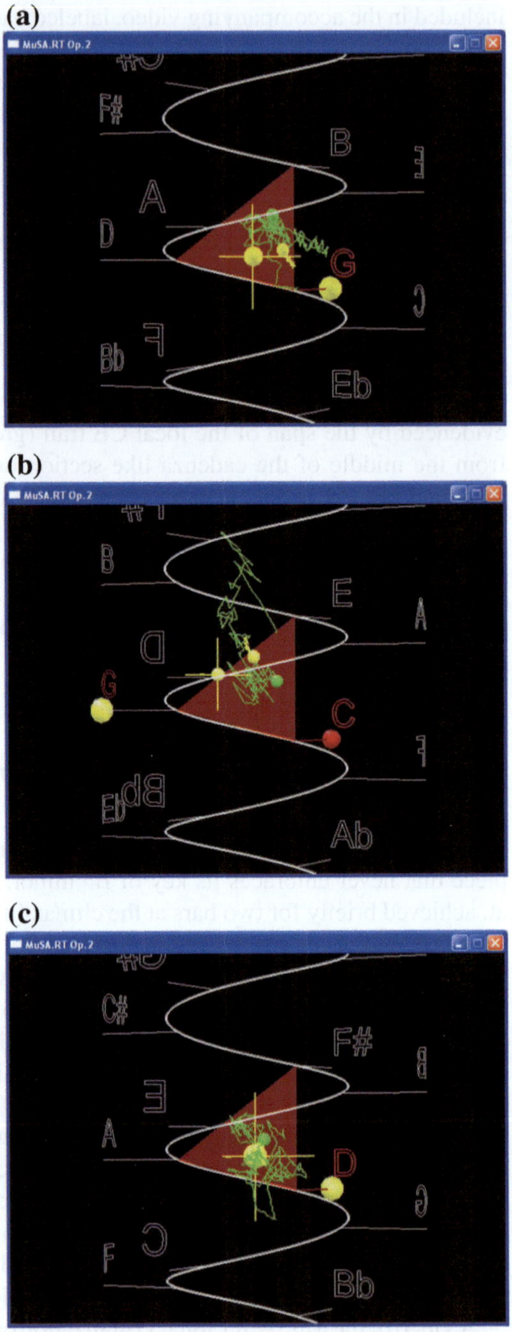

Fig. 9.10 Sample screenshots for the Prelude from Bach's *Cello Suite No. 1* **a** end of bar 3, **b** end of bar 17, **c** beginning of bar 29

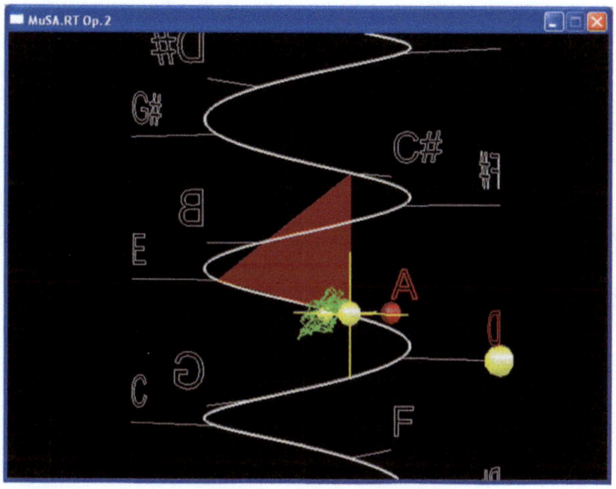

Fig. 9.11 Shrinking down to a tight ball in by beat 4 of bar 33

Fig. 9.12 Brief arrival on B♭ minor in bar 50

Fig. 9.13 Unfulfilled longing for B♭ at end of piece

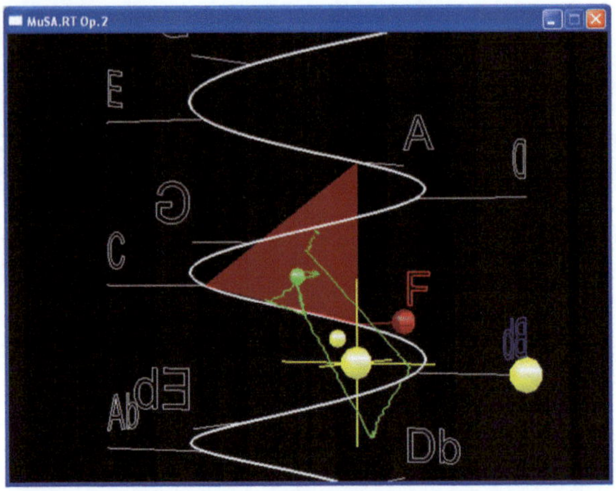

Fig. 9.14 Final screenshot for Barber's *Adagio for Strings*, final chord is F major while key is B♭ minor

An interesting problem arose with respect to pitch spelling in the *Adagio*. The section before the general pause (GP) in bar 53 (indicated in Fig. 9.12 by the letters "GP") ends on the F♭ major triad. A snapshot of the screen just before the GP is shown in Fig. 9.15. If we followed the natural logic of choosing pitch spellings near the long-term CE after the GP, we would end up in a deeply flatted zone low down in the spiral display, activating pitches with numerous flats. To mimic the composer's switch to the enharmonic spellings in the middle of bar 54 (last bar in the excerpt shown in Fig. 9.12) to reduce the number of accidentals, we reset the CEs at the GP (which corresponds to 4 min and 22 s into the video, barber-adagio.mp4) and begin assigning pitch spellings and calculating CEs anew after the GP.

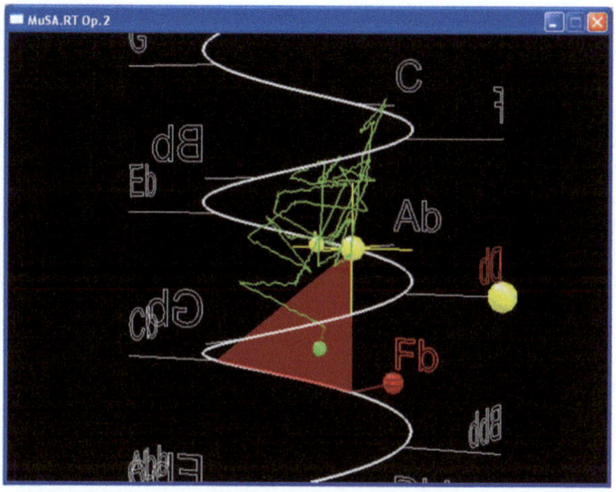

Fig. 9.15 Screenshot of the *Adagio* just before the GP in bar 53

9.5 Discussion and Conclusions

We have presented MuSA.RT Opus 2, an interactive system for multiscale visualization of tonal patterns in music, described the underlying Spiral Array model and associated algorithms, and the design of the system using SAI. We have shown that MuSA.RT is a viable tool, not only for visualizing tonal patterns, but also for understanding tonal function and assessing tonal similarity and difference over time.

A drawback of the current system is the need for human intervention in the setting of the parameters α and β. Future developments for MuSA.RT include automatic detection and dynamic varying of these α and β values. We also plan to improve the visual interface, such as exploring the visualization of keys as volumetric shapes. For now, we have chosen the simplicity of visualizing only two types of chords, major and minor triads, and seventh chords mapped visually to their closest triad. We are considering ways of incorporating more complex chords into the visual display.

Acknowledgments We thank Carol Krumhansl for her comments and suggestions that have helped improve the clarity of this chapter, and Craig Sapp for encoding the MIDI examples.

References

1. Chew, E.: Towards a mathematical model of tonality. Ph.D. Thesis, Massachusetts Institute of Technology, Cambridge (2000)
2. Chew, E.: Slicing it all ways: mathematical models for induction, approximation and segmentation using the Spiral Array. INFORMS J. Computing **18**(3), 305–320 (2006)
3. Chew, E.: Out of the grid and into the spiral: geometric interpretations of and comparisons with the Spiral Array model. Comput. Musicology **15**, 51–72 (2008)

4. Chew, E., Chen, Y.-C.: Determining context-defining windows: pitch spelling using the Spiral Array. In: Proceedings of the 4th International Conference for Music Information Retrieval, pp. 26–30 (2003b)
5. Chew, E., Chen, Y.-C.: Mapping MIDI to the Spiral Array: disambiguating pitch spellings. In: Bhargava, H.K., Ye, N. (eds.) Computational Modeling and Problem Solving in the Networked World, Proceedings of the 8th INFORMS Computer Society Conference, pp. 259–275, Kluwer Academic Publishers (2003a)
6. Chew, E., Chen, Y.-C.: Real-time pitch spelling using the Spiral Array. Comput. Music J. **29**(2), 61–76 (2005)
7. Chew, E., François, A.R.J.: MuSA.RT—Music on the Spiral Array. Real-Time. In: Proceedings of the ACM Multimedia Conference, pp. 448–449 (2003)
8. Cohn, R.: Introduction to Neo-Riemannian Theory: A survey and historical perspective. J. Music Ther. **42**(2), 167–180 (1998)
9. François, A.R.J.: Components for immersion. In: Proceedings of the IEEE International Conference on Multimedia and Expo, pp. 777–780 (2002)
10. François, A.R.J.: A hybrid architectural style for distributed parallel processing of generic data streams. In: Proceedings of the International Conference on Software Engineering, pp. 367–376 (2004)
11. François, A.R.J.: MuSA_RT https://itunes.apple.com/ca/app/musa-rt/id506866959?mt=12 (2012). Accessed 30 August 2013
12. François, A.R.J.: Modular flow scheduling framework (MFSM). http://mfsm.sourceforge.net. Accessed 4 July 2013
13. Longuet-Higgins, H.C.: Letter to a musical friend. The music review, **23**, 244–248 (1962a)
14. Longuet-Higgins, H.C.: Second letter to a musical friend. The music review, **23**, 271–280 (1962b)
15. Sapp, C.S.: Harmonic visualizations of tonal music. In: Proceedings of the International Computer Music Conference, pp. 423–430. http://ccrma.stanford.edu/craig/keyscape (2001). Accessed 4 July 2013
16. Toiviainen, P., Krumhansl, C.L.: Measuring and modeling real-time responses to music: the dynamics of tonality induction. Perception **32**(6), 741–766 (2003)

Chapter 10
Visible Humor

Abstract In this chapter, we explore the use of the Spiral Array spatial visualization of tonal evolution through time for the visual analysis of P. D. Q. Bach's *The Short-Tempered Clavier*. In particular, we analyze situations in which we can see some of the humor devices abstracted in an earlier study by David Huron. We conclude that although we can see a good number of Schickele's humour devices—such as incongruent styles, musically improbable tonality and harmony shifts, and excessive repetition—we do not yet have sufficient information to form a robust computer-based method for detecting musical humor. The reader can download the latest version of the MuSA.RT software, MuSA_RT, from the Mac App Store (http://itunes.apple.com/ca/app/musa-rt/id506866959?mt=12, cited 30 August 2013), follow the examples posted at http://musa-rt.blogspot.com, or try out new ones themselves.

In this chapter, we explore the use of the Spiral Array spatial visualization of pitches, chords, and keys, and the tracking of their dynamic evolution through time for the visual analysis of musical humor. We visually analyze, using the Music on the Spiral Array. Real-Time (MuSA.RT) system [2], pieces from *The Short-Tempered Clavier*—Preludes and Fugues in All the Major and Minor Keys Except the Really Hard Ones (S. 3.14159, easy as), composed by P. D. Q. Bach, a.k.a. Peter Schickele [6]. In particular, we focus on the points of departure from musical expectation (and good taste), and determine if these strategies for violating expectation and thus eliciting laughter can be traced visibly in MuSA.RT.

This chapter is inspired by Huron's work on music-engendered laughter [5], where he systematically identifies and classifies laughter-eliciting musical devices in P. D. Q. Bach's music, each involving expectation violation. P. D. Q. Bach is Peter Schickele, a classically trained contemporary composer who has built a career

This is an expanded version of the work, "Visible Humour—Seeing P. D. Q. Bach's Musical Humour Devices in The Short-Tempered Clavier on the Spiral Array Space" by Elaine Chew and Alexandre R. J. François, published in Klouche, T., Noll, T. (eds.): Mathematics and Computation in Music, CCIS **37**, pp.11-18, Springer (2009) http://link.springer.com/chapter/10.1007/978-3-642-04579-0_2

E. Chew, *Mathematical and Computational Modeling of Tonality*, 179
International Series in Operations Research & Management Science 204,
DOI: 10.1007/978-1-4614-9475-1_10,

on writing laughter-eliciting musical compositions under his pseudonym. Huron's article presents the results of a study of 629 instances of laughter-eliciting music by Schickele. Huron states that, unlike most musical humourists, "beyond the visual gags and language-based humor, most of Schickele's humour devices are to be found in the core musical domains of pitch, time, and timbre." Excluding visual and language-based comedy, he proceeds to identify nine categories of musical devices that elicit laughter: (1, 2) incongruity of instrumental sounds and of musical genres, (3, 4) musically improbable tonal and metric shifts, (5, 6) implausible delays and excessive repetition, (7) incompetence cues, and (8, 9) incongruous quotations and misquotations.

In this chapter, we will show the visual counterparts of some of these humour-evoking devices, particularly the ones based on pitch and time, using examples from *The Short-Tempered Clavier*.

10.1 MuSA.RT and Visualization

The tool we use for our explorations is MuSA.RT (see http://musa-rt.blogspot.com), an interactive music analysis and visualization system, based on Chew's Spiral Array mathematical model for tonality [1] that is capable of tracking and displaying the trajectory of the tonal content and context of music in real-time during a live performance. The Spiral Array is a geometric representation of tonal entities (pitches, chords, and keys) in three-dimensional space.

The chords and keys are determined via the *centers of effect* (CEs) generated by the notes in the piece, which are mapped to their respective pitch class representations in the Spiral Array using a pitch-spelling algorithm. In MuSA.RT, each CE is located at the aggregate position of the notes present and in the recent past, where the effect of notes past are decreasing according to a linear decay function.

MuSA.RT stands for *Music on the Spiral Array. Real-Time*. The system takes as input MIDI from, say, a piano keyboard. The tonal structures are computed and mapped to the display instantaneously in response to the MIDI input. The virtual viewpoint can be manipulated via a game controller or an automatic pilot. To coordinate the concurrent processing of the multiple data streams, the system is implemented in the style of François Software Architecture for Immersipresence [3].

MuSA.RT maps musical performances to distinctive spatiotemporal paths and patterns inside the Spiral Array that are characteristic of the piece, and of the particular interpretation. It provides a means of visualizing not only pitch clusters, and key-finding and chord tracking algorithms and their results, but also the unfolding and history of the patterns over time. When in autopilot, the movement of the spiral is in synchrony with the rhythm of the music.

As a companion to the book, the MuSA.RT blog, http://musa-rt.blogspot.com, contains examples of visualizations generated by MuSA.RT. The latest version of MuSA.RT, named MuSA_RT [4], can be downloaded from the Mac App Store.

10.1.1 Seeing Style Differences

The visualizations in MuSA.RT can reveal pitch structures, and to some extent, time structures as well. However, timbre is presently not a feature that is visualized in the system. We show, by the example of the Prelude No. 5 in D minor from *The Short-Tempered Clavier*, the differences visible when the same theme is repeated in multiple styles (or genres).

In MuSA.RT Opus 2.6, a silver helix shows the outermost spiral on which the pitch classes reside, and the inner and intertwined blue and red helices denote the spirals on which the minor and major keys are represented, respectively. The closest major or minor triad is highlighted in pink or blue, and its name colored and indicated by a sphere of the same color. The closest major or minor key appears as a sphere on the corresponding major or minor key spiral, and a sphere of the appropriate color marks the letter name of the key. When the chord and key are of the same name, the sphere denoting the chord appears inside the sphere for the key. A violet trail traces the history of the key CE, and a purple trail that of the chord CE. Figure 10.1 shows the complete history of Prelude No. 5 in D minor.

Next, we segment the piece into its three stylistic sections, and visualize each part independently. The initial arpeggiated chords that present the theme create relatively smooth trajectories around D minor, as shown in Fig. 10.2. The middle section with its staccato walking bass and interplay of groupings of two against three traces a more jagged version of the initial trails outlined by the first section; see Fig. 10.3. The final section with the boogie bass, in a faster triplet rhythm, again outlines the same tonal regions, but results in a tangled mess, as shown in Fig. 10.4. All score excerpts are adapted from reference [6].

(i) (ii)

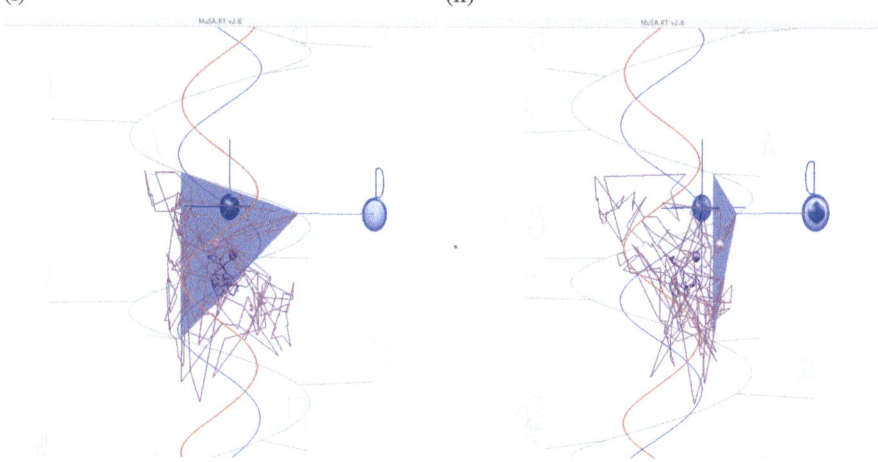

Fig. 10.1 MuSA.RT visualization of P. D. Q. Bach's Prelude No. 5 in D minor. (i) view from center; (ii) view from right

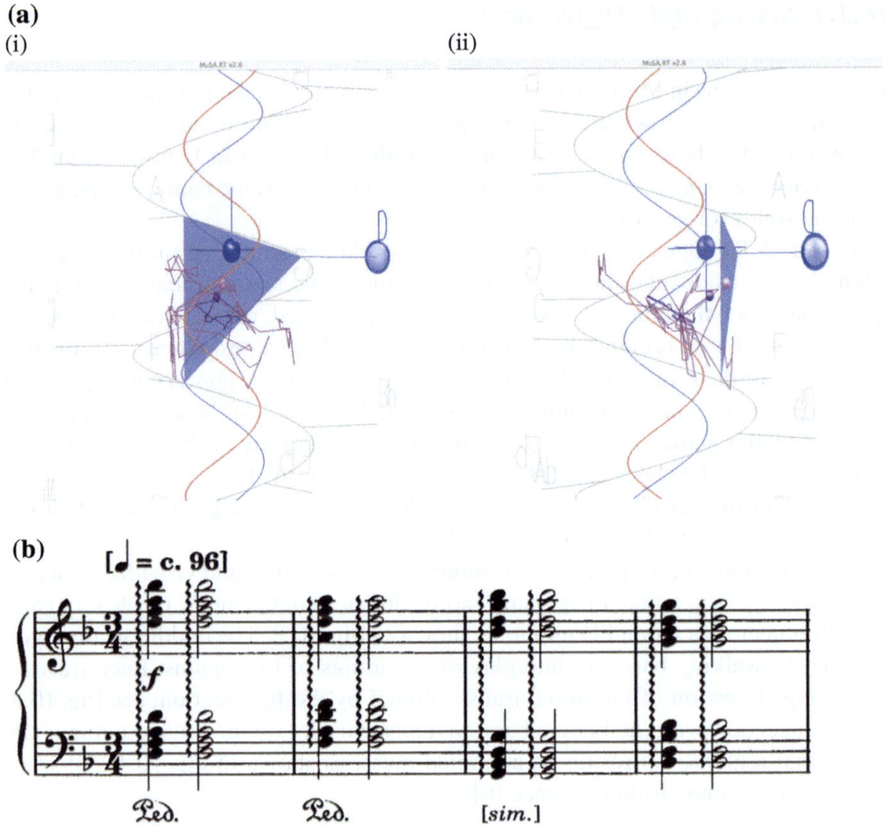

Fig. 10.2 MuSA.RT visualization of P. D. Q. Bach's Prelude No. 5, first part. **a** CE trajectory of first part of piece: bars 1–16. (i) view from center; (ii) view from right. **b** Prelude No. 5 excerpt showing arpeggiated chords: bars 1–4

10.2 Expectations Violated

In this section, we explore the visualizations of incongruous musical genres, musically improbable tonality and harmonic shifts, and excessive repetitions.

10.2.1 The Jazz Ending

The Preludes and Fugues in Schickele's *The Short-Tempered Clavier* by and large begin in a classical or baroque style. In the category of incongruous musical genres, Schickele frequently switches from baroque/classical to jazz sporadically, and particularly at the ends of pieces.

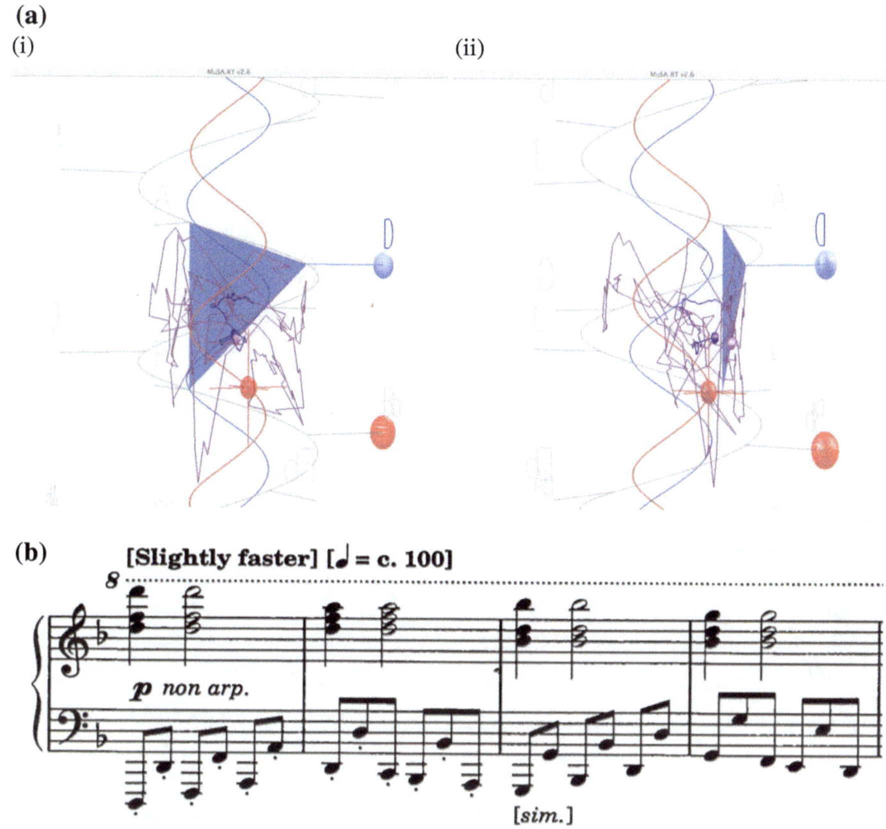

Fig. 10.3 MuSA.RT visualization of P. D. Q. Bach's Prelude No. 5, middle part. **a** CE trajectory of second part of piece: bars 17–32. (i) view from center; (ii) view from right. **b** Prelude No. 5 excerpt showing eighth note walking bass, two against three: bars 17–20

For example, the Prelude No. 1 in C, a parody of J. S. Bach's Prelude No. 1 in C, occasionally slips from the classical triadic treatment to a jazzy minor seventh broken chord, with a definitive next-to-last measure switch to the C minor seventh chord, before ending on the C major triad, evoking a tierce de picarde-like effect. This shift toward the jazz genre toward the end is visible as a southward swing of the trajectory of the piece, a departure from the tightly wound main cluster, shown in Fig. 10.5.

Schickele employs a similar strategy in Prelude No. 8 in G minor, where the entry into the last measure veers sharply from G minor, using the F♯ major triad to step chromatically into the G major ending. Again, the resulting and unexpected tonal shift is visible in the views of the trajectories of the piece, shown in Fig. 10.6; the trajectories travel north sharply to end in the G major triad, giving this tierce de picarde a jazzy twist.

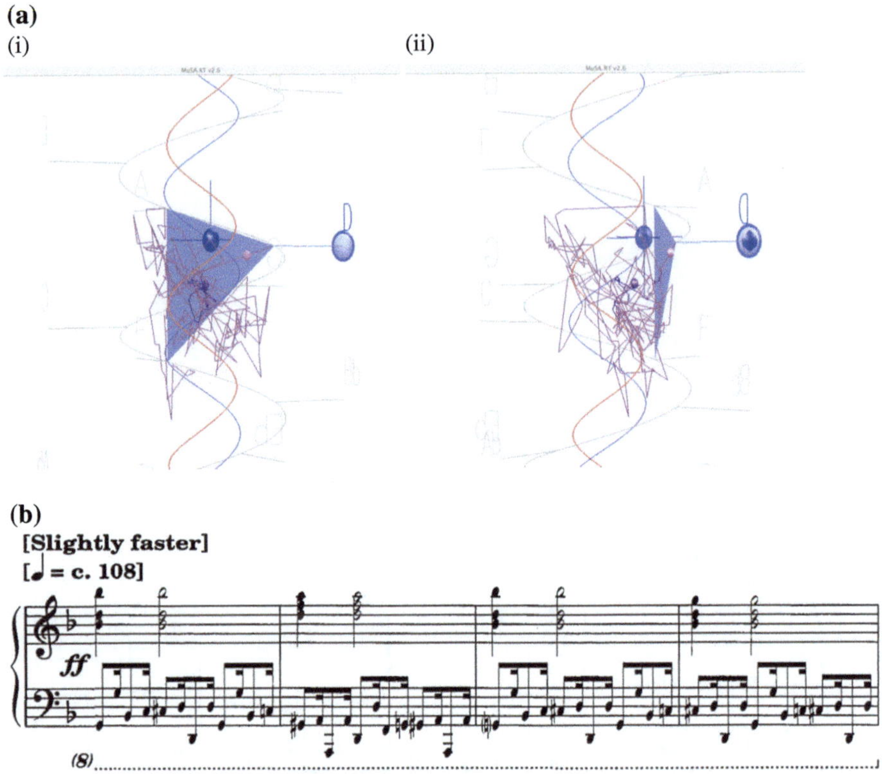

Fig. 10.4 MuSA.RT visualization of P. D. Q. Bach's Prelude No. 5, last part. **a** CE trajectory of third and final part of piece: bars 33–49. (i) view from center; (ii) view from right. **b** Prelude No. 5 excerpt showing boogie bass, two against three: bars 33–36

10.2.2 Improbable Harmonies

Another technique employed by Schickele is the insertion of musically improbable harmonies (chords) and tonality shifts, which typically map to distant locations in the Spiral Array space.

In the Fugue No. 2 in C minor, beginning in measure 26, Schickele begins building toward, and preparing for, a climax in C minor, only to land on a cheerful and triumphant C major in measure 34. This is visible in the CE trails for the piece shown in Fig. 10.7, where the trajectories meander around the B♭ region, only to shoot up toward the E♮ pitch class in a rather acrobatic fashion at measure 34.

In Fugue No. 6 in E♭ major, the composer inserts a jarring chord based on stacked perfect fourths (E♮, A♮, D) in the last four measures of the piece. This departure from expectation is visible as the large upward swoop in the CE trails for the piece shown in Fig. 10.8.

(a)

(i)

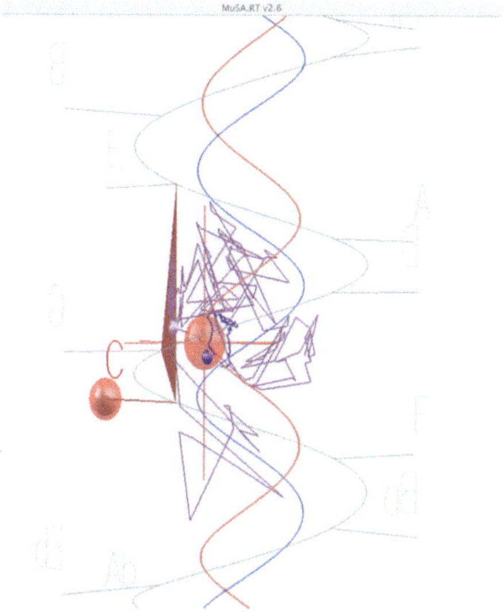

(ii)

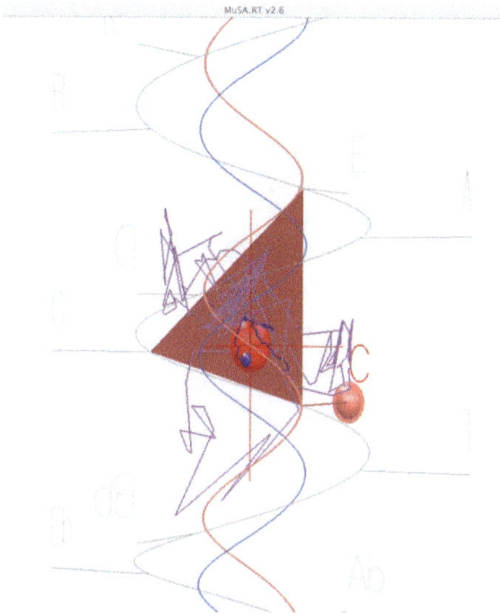

(b)

(i)

(ii)

Fig. 10.5 MuSA.RT visualization of P. D. Q. Bach's Prelude No. 1 in C major. **a** CE trajectory of entire piece. (i) view from center; (ii) view from right. **b** Prelude No. 1 excerpts (i) excerpt showing classical triadic treatment: bars 1–2; (ii) excerpt showing minor seventh (broken) chord causing the CE's southward swing: bars 18–19

(a)

(i) (ii)

(b)

Fig. 10.6 MuSA.RT visualization of P. D. Q. Bach's Prelude No. 8 in G minor. **a** CE trajectory of entire piece. (i) view from center; (ii) view from right. **b** Prelude No. 8 excerpt showing F♯ major triad leading to G major ending: bars 23–24

10.2.3 Excessive Repetition

One of the Schickele's comedic strategies most easy to identify is that of excessive repetition. Typically, a figure is repeated thrice, once to introduce it, the second time as confirmation, and the third to reaffirm the figure. Anything beyond this level of repetition, unless it is an intentional ostinato, can be deemed to be in bad taste. Figure 10.9 shows an example of a ludicrous and awkward amount of repetition in Prelude No. 10 in A major.

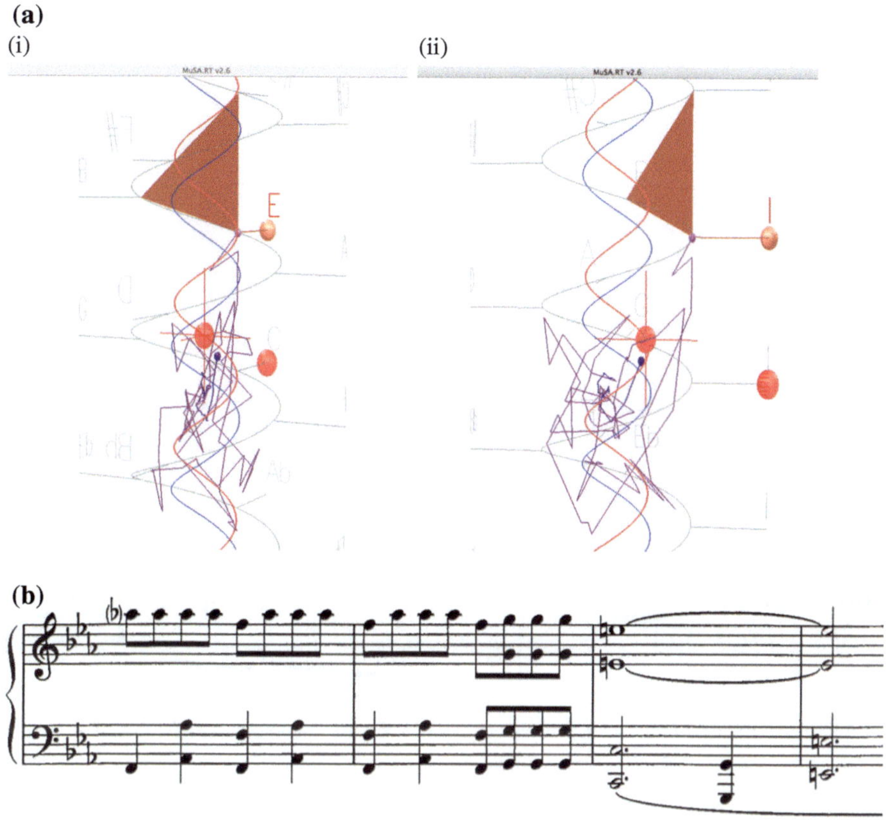

Fig. 10.7 MuSA.RT visualization of P. D. Q. Bach's Fugue No. 2 in C minor. **a** CE trajectory of bars 26–35. (i) view from center; (ii) view from right. **b** Fugue No. 2 excerpt showing landing on E♮ (instead of the expected E♭): bars 33–35

10.3 Remarks on Humor and Tonality

In conclusion, we are able to see stylistic changes, improbable tonal and harmonic shifts, and incessant repeats, some of the laughter-eliciting devices uncovered by Huron in Schickele's music. These strategies are distilled from examples of laughter-eliciting music, thus implying that Schickele's humor derives from subsets of these devices. However, it is unclear that the converse is true; that is to say, whether all such expectation-violating events would result in humorous situations. For example, Pachelbel's *Canon in D*, and Ravel's *Bolero* all exhibit fairly large numbers of repeats, but most listeners do not find either of these pieces funny. In much post-tonal music, tonal (key) and harmonic (chord) progressions that may have been previously deemed to be improbable in earlier genres, are frequently invoked in practice.

(a)

(i) (ii)

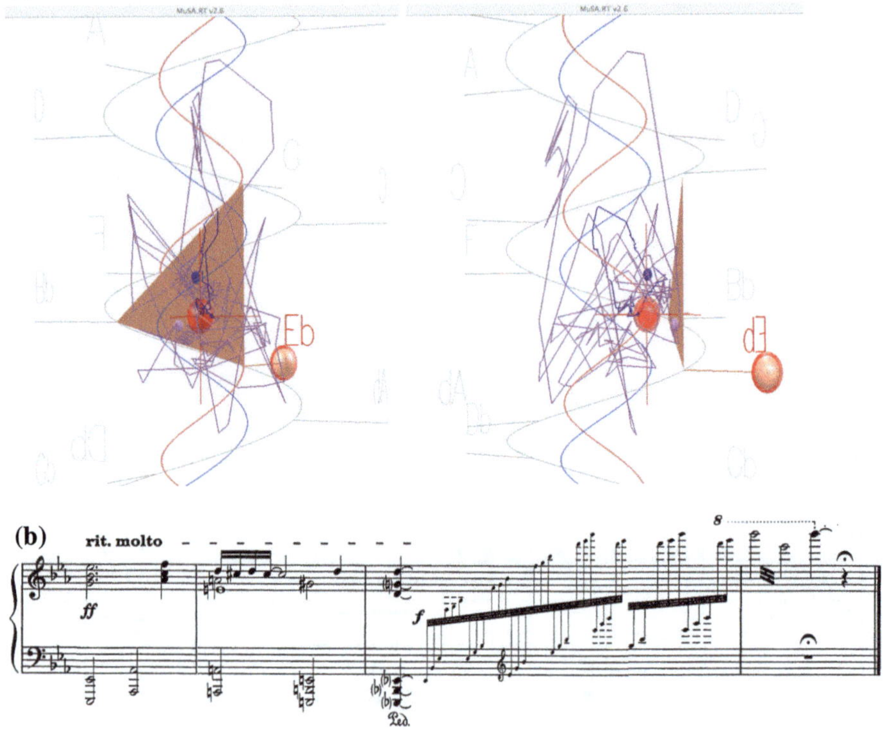

Fig. 10.8 MuSA.RT visualization of P. D. Q. Bach's Fugue No. 6 in E♭ major. **a** CE trajectory starting around 39 to end (bar 55). (i) view from center; (ii) view from right. **b** Fugue No. 6 excerpt showing unusual harmonies causing the CE's upward swoop: bars 52–55

Thus, although we can see a good number of Schickele's humor devices, we concede that musical humor cannot be reduced to simple musical features. There are aspects that may not be visible in the pitch and time structures of the music, such as the cultural context. Humor results from a complex combination of stimulants, and we do not yet have a robust way of automatically detecting musical humor.

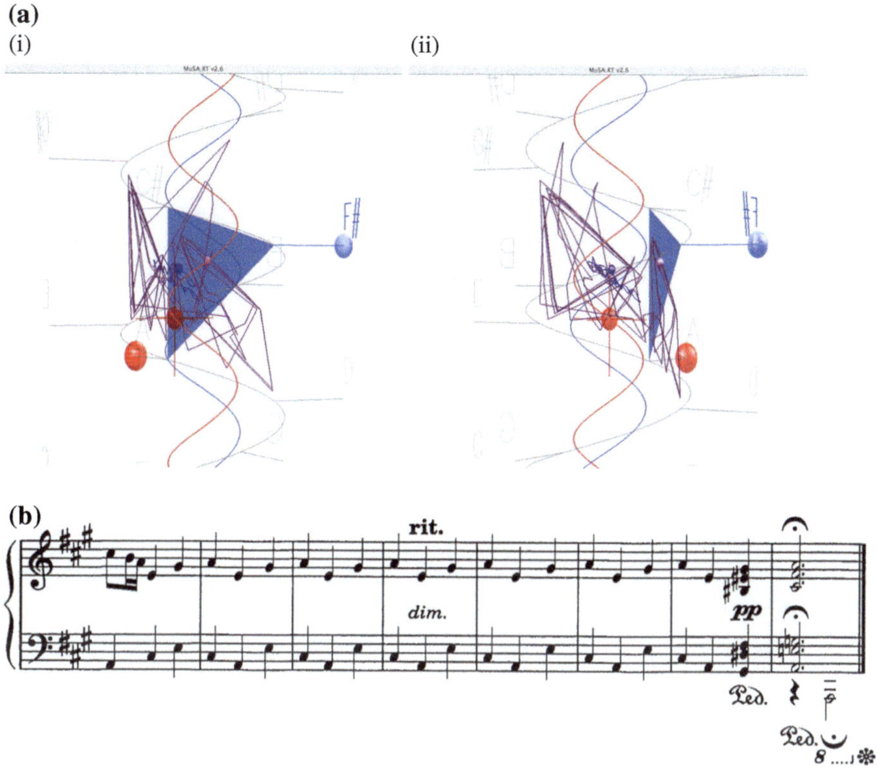

Fig. 10.9 MuSA.RT visualization of P. D. Q. Bach's Prelude No. 10 in A major. **a** CE trajectory of entire piece. (i) view from center; (ii) view from right. **b** Fugue No. 6 excerpt showing excessive repetition: bars 21–28

Acknowledgments This material is based upon work, and made use of Integrated Media Systems Center Shared Facilities, supported by the National Science Foundation under grant No. 0347988 and Cooperative Agreement No. EEC-9529152. Any opinions, findings, and conclusions or recommendations expressed in this material are those of the authors, and do not necessarily reflect the views of the National Science Foundation.

References

1. Chew, E.: Towards a mathematical model of tonality. Ph.D. thesis, Massachusetts Institute of Technology, Cambridge (2000)
2. Chew, E., François, A.R.J.: Interactive multi-scale visualizations of tonal evolution in MuSA.RT Opus 2. Comput. Entertain. **3**(4), 1–16 (2005)
3. François, A.R.J.: A hybrid architectural style for distributed parallel processing of generic data streams. In: Proceedings of the International Conference on Software Engineering, pp. 367–376 (2004)

4. François, A. R. J.: MuSA_RT https://itunes.apple.com/ca/app/musa-rt/id506866959?mt=12 (2012). Accessed 30 Aug 2013
5. Huron, D.: Music-engendered laughter: an analysis of humor devices in PDQ Bach. In: Proceedings of the International Conference on Music Perception and Cognition, pp. 93–95 (2004)
6. Schickele, P.: The Short-tempered Clavier (Preludes and Fugues in all the Major and Minor Keys Except for the Really Hard Ones), S. 3.14159, easy as edition, pp. 1807–1742?, Theodore Pressor Company, Bryn Mawr (1998)

Part VII
Extensions

Chapter 11
Sensitivity Analysis

Abstract The goal of this chapter is to present a systematic analysis of audio key-finding to determine the factors important to system design, and the selecting and evaluating of solutions. First, we present a basic audio key-finding system consisting of a Fuzzy Analysis technique and the Spiral Array Center of Effect Generator algorithm, with three key determination policies: the nearest-neighbor, the relative distance, and the average distance policy. An evaluation on 15-second excerpts of 410 classical pieces from a variety of stylistic periods shows that the average-distance key determination policy achieves a 79 % correct key identification rate, at least 8 % higher than the systems employing other key determination policies. Two examples are given to illustrate why audio key-finding could outperform symbolic key-finding, where precise pitch information is known. An analysis of the results by period shows that pieces from the romantic period may be the most challenging for key-finding. We next propose three extensions to the basic key-finding system—the modified Spiral Array approach, fundamental frequency identification, and post-weight balancing—to improve the performance at different stages of the system. The three extensions are evaluated using 15-second excerpts of recordings of Chopin's *Preludes* for piano. Quantitative analyses of the results show that fundamental frequency identification provides the greatest system improvement in the first 8s, while modifying the representation model using audio frequency features leads to the best performance after 8s. Two case studies show detailed analyses of an example where all extended systems give the correct answer, and another where all systems were incorrect.

This is a minor revision of the work "Audio key-finding: Considerations in System Design, and Case Studies on Chopin's 24 Preludes" by Ching-Hua Chuan and Elaine Chew in the *Journal on Advances in Signal Processing*2007:056561, doi:10.1155/2007/56561, published by EURASIP

E. Chew, *Mathematical and Computational Modeling of Tonality*, International Series in Operations Research & Management Science 204,
DOI: 10.1007/978-1-4614-9475-1_11,

11.1 Introduction

Our goal in this chapter is to present a systematic analysis of audio key-finding in order to determine the factors important to system design, and to explore the strategies for selecting and evaluating solutions. In this chapter we present a basic audio key-finding system, the Fuzzy Analysis technique with the Spiral Array Center of Effect Generator (CEG) algorithm [1, 2], also known as FACEG, first proposed in [6]. We propose three different policies, the nearest-neighbor (NN), the relative distance (RD), and the average distance (AD) policy, for key determination. Based on the evaluation of the basic system (FACEG), we provide three extensions at different stages of the system, the modified Spiral Array model (mSA), fundamental frequency identification (F0), and post-weight balancing (PWB). Each extension is designed to improve the system from different aspects. Specifically, the modified Spiral Array model is built with the frequency features of audio, the fundamental frequency identification scheme emphasizes the bass line of the piece, and post-weight balancing uses knowledge of music theory to adjust the pitch-class distribution. In particular, we consider several alternatives for determining pitch classes, for representing pitches and keys, and for extracting key information. The alternative systems are evaluated not only statistically, using average results on large datasets, but also through case studies of score-based analyses.

The problem of key-finding, that of determining the most stable pitch in a sequence of pitches, has been studied for more than two decades [2, 13, 12, 17]. In contrast, audio key-finding, determining the key from audio information, has gained interest only in recent years. Audio key-finding is far from simply the application of key-finding techniques to audio information with some signal processing. When the problem of key-finding was first posed in the literature, key-finding was performed on fully disclosed pitch data. Audio key-finding presents several challenges that differ from the original problem: in audio key-finding, the system does not determine key based on deterministic pitch information, but some audio features such as the frequency distribution; furthermore, full transcription of audio data to score may not necessarily result in better key-finding performance.

We aim to present a more nuanced analysis of an audio key-finding system. Previous approaches to evaluation have simply reported one overall statistic for key-finding performance [6–8, 16], which fails to fully address the importance of the various components in the system, or the actual musical content, to system performance. We represent a solution to audio key-finding as a system consisting of several alternative parts in various stages. By careful analysis of system performance with respect to choice of components in each stage, we attempt to give a clearer picture of the importance of each component, as well as the choice of music data for testing, to key-finding. Our approach draws inspiration from multiple domains, from music theory to audio signal processing. The system components we introduce aim to solve the problem from different viewpoints. The modular design allows us to explore the strengths and weaknesses of each alternative option, so that the change in system performance due to each choice can be made clear.

The remainder of the chapter is organized as follows: Sect 11.1.1 provides a literature review of related work in audio key finding. Section 11.2 describes the overall system diagram, with new alternatives and extensions. The basic system, the FACEG system, and the three key determination policies–the nearest-neighbor (NN), relative distance (RD), and average distance (AD) policies–are introduced in Sect. 11.3. The evaluation of the FACEG system with the three key determination policies follows in Sect. 11.4. Two case studies based on the musical score are examined, to illustrate situations in which audio key-finding performs better than symbolic key-finding. Section 11.5 describes three extensions of the system: the modified Spiral Array (mSA) approach, fundamental frequency identification (F0), and post-weight balancing (PWB). Qualitative and quantitative analyses and evaluations of the three extensions are presented in Sect. 11.6. Section 11.7 concludes the chapter.

11.1.1 Related Work

Various state-of-the-art audio key-finding systems were presented in the audio key-finding contest for MIREX 2005 [14]. Six groups participated in the contest, comprising Chuan and Chew, Gómez, İzmirl, Pauws, Purwins and Blankertz, and Zhu (see abstracts posted at [15]). Analysis of the six systems reveals that they share a similar structure, consisting of: some signal processing method, audio characteristic analysis, key template construction, query formation, key-finding method, and key determination criteria. The major differences between the systems occur in: the audio characteristic analysis, key template construction, and key determination criteria. In Gómez's system, the key templates are precomputed, and are generated from the Krumhansl-Schmuckler pitch-class profiles [12], with alterations to incorporate harmonics characteristic of audio signals. Two systems employing different key determination strategies are submitted by Gómez: one using only the start of a piece, and the other taking the entire piece into account. In İzmirli's system, he constructs key templates from monophonic instrument samples, weighted by a combination of the Krumhansl-Schmuckler and Temperley's modified pitch-class profiles. İzmirli's system tracks the confidence value for each key answer, and the global key is then selected as the one having the highest sum of confidence values over the length of the piece. The key templates in Pauws' and Purwins and Blankertz's systems are completely data-driven. The parameters are learned from training data. In their systems, the key is determined based on some statistical measure, or maximum correlation. In contrast, Zhu builds a rule-based key-finding system; the rules are learned from the MIDI training data. Further details of our comparative analysis of the systems can be found in Chuan and Chew's extended abstract at [15].

11.2 System Description

Consider a typical audio key-finding system as shown schematically in the top part of Fig. 11.1. The audio key-finding system consists of four main stages: processing of the audio signal to determine the frequencies present, determination of the pitch class description, application of a key-finding algorithm, and key answer determination. Results from the key-finding algorithm can feed back to the pitch class generation stage to help constrain the pitch class description to a reasonable set. In this chapter, we shall consider several possible alternative methods at each stage.

For example, as the basis for comparison, we construct a basic system that first processes the audio signal using the Fast Fourier Transform (FFT) on the all-frequency signal, then generates pitch class information using a Fuzzy Analysis (FA) technique, calculates key results using the CEG algorithm with a Periodic Cleanup procedure, and applies a key determination policy to output the final answer. This basic system, shown in the gray area in Fig. 11.1, is described in detail in Sect. 11.3, followed by an evaluation of the system using 410 classical pieces in Sect. 11.4. In Sect. 11.5, we present the details of several alternative options for the different stages of the audio key-finding system. In the audio processing stage, the two alternatives we consider are: performing the FFT on the all-frequency signal, or separating the signal into low and high frequencies for individual processing. In the pitch class generation stage, the options are: to use the peak detection method with Fuzzy Analysis, to use peak detection with fundamental frequency identification, or to determine pitch classes using sound sample templates. In the key determination stage, we consider the direct application of the Spiral Array CEG Algorithm [1, 2], the CEG method with feedback to reduce noise in the pitch class information, and the CEG method with PWB. The lower part of Fig. 11.1 shows the various combinations possible, with the alternate modules proposed, in assembling a key-finding system. The in-depth evaluation and qualitative analysis of all approaches are given in Sect. 11.6.

11.3 Basic System

We first construct our basic audio key-finding system as the main reference for comparison. This system, shown in the shaded portions of Fig. 11.1, consists of first an FFT on the audio sample. Then, we use the peak detection method described in Sec. 11.3.1 and the fuzzy analysis technique proposed in Sec. 11.3.2 to generate a pitch-class description of the audio signal. Finally, we map the pitch classes to the Spiral Array model [1] and apply the CEG algorithm [2] to determine the key. Distinct from our earlier approach, we explore here three key determination policies: nearest-neighbor (NN), relative distance (RD), and average distance (AD). Each method is described in the subsections below. We provide an evaluation of the system in Sect. 11.4.

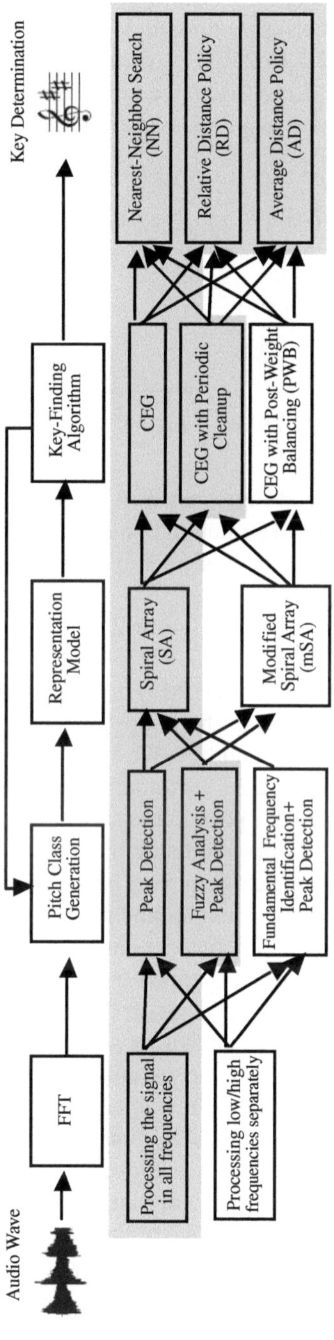

Fig. 11.1 Audio key-finding system (fundamental + extensions)

We synthesize audio wave files from MIDI at 44.1 kHz and with 16-bit precision. We process audio signals using FFT with nonoverlapped Hanning windows. The window size is set at 0.37 s, corresponding to $N = 2^{14}$ samples. Other sample sizes were tested in the range of 2^{10} to 2^{15} (i.e., window sizes of 0.0232 to 0.74 s), but these did not perform as well. Let $x(n)$ be the input signal, where $n = 0, \ldots, N-1$. The power spectrum is obtained using the equation:

$$\mathrm{X}(k) = \frac{1}{N}\sum_{n=0}^{N-1} x(n)\mathrm{W}_N^{kn}, \tag{11.1}$$

where $\mathrm{W}_N = e^{-j2\pi/N}$, and $k = 0, 1, \ldots, N-1$. We then calculate the magnitude from the power spectrum as follows:

$$\mathrm{M}(k) = ||\mathrm{X}(k)|| = \sqrt{\left(\mathrm{X}(k)_{real}^2 + \mathrm{X}(k)_{img}^2\right)}. \tag{11.2}$$

We set the reference fundamental frequency of A4 at 440 Hz. Let $h(\mathrm{p})$ be the number of half steps from the pitch A4 to a pitch p. For example, $h(\mathrm{p}) = -9$ when p = C4. The reference fundamental frequency of pitch p is then given by

$$\mathrm{F0}_{ref}(\mathrm{p}) = 440 \times 2^{h(\mathrm{p})/12}. \tag{11.3}$$

We employ a local maximum selection (LMS) method [7] to determine the presence of pitches and their relative weights. The midpoint between two adjacent reference fundamental frequencies forms a boundary. We examine $\mathrm{M}(k)$ in the frequency band between two such adjacent boundaries surrounding each pitch, p. The LMS method is based on two assumptions: (1) a peak value should be larger than the average to its left and to its right in the given frequency band; and (2) only one (the largest) peak value should be chosen in each frequency band. The value $\mathrm{M}(k)$ satisfying the above conditions for the frequency band around p, $\mathrm{M}^*(\mathrm{p})$, is chosen as the weight of that pitch. This method allows us to consider each pitch equally, so that the system is unaffected by the logarithmic scale of pitch frequencies.

We apply the FFT to the audio signals with two different setups. Under the first option, we process the signal as a whole, with a window size of 0.37 s, to generate the frequency magnitude for each pitch. In the second option, we partition the signals into two sub-bands, one for higher pitches (frequencies higher than 261 Hz, i.e., pitches higher than C4), and one for lower ones. We use the same window size to process the higher pitch signals, and use a larger and overlapped window size for the lower pitch signals. The window size is relatively large compared to the ones typically used in transcription systems. We give two main reasons for our choice of window size. First, a larger window more accurately captures lower pitches, which provide more valuable pitch information in key-finding. Second, a larger window smoothes the pitch information, allowing the method to be more robust to pitch variations less important to key identification such as grace notes, passing tones, non-chord tones, and chromatic embellishments.

11.3.1 Fuzzy Analysis Technique

We use the standard short-term FFT to extract frequency information for pitch identification. Music consists of streams of notes; each note has the properties pitch and duration. Pitch refers to the perceived fundamental frequency of the note. The peak values on the frequency spectrum correspond to the fundamental frequencies of the pitches present, and their harmonics. We use the frequency at the peak value to identify the pitch height, and map the peak spectral magnitude to the pitch weight. Pitches are defined on a logarithmic frequency scale. A range of frequencies, bounded by the midpoints between the reference frequencies, is deemed acceptable for the recognition of each pitch. We focus our attention on the pitches in the range between C_1 (32 Hz) and B_6 (1975 Hz), which covers most of the common pitches in our music corpus.

The peak detection method described above generates pitch-class distributions with limited accuracy. We design the Fuzzy Analysis technique to clarify the frequency magnitudes obtained from the FFT, in order to generate more accurate pitch class distributions for key-finding. The main idea behind the Fuzzy Analysis technique is that one can verify the existence of a pitch using its overtone series. Hence, we can emphasize the weight of a pitch that has been validated by its overtone series, and reduce the weight of a pitch that has been excluded due to the absence of its strongest overtones.

The problems stem from the fact that mapping of the frequency magnitude directly to pitch weight as input to a key-finding algorithm results in unbalanced pitch class distributions that are not immediately consistent with existing key templates. We have identified several sources of errors (see [6]) that include: uneven loudness of pitches in an audio sample, insufficient resolution of lower frequency pitches, tuning problems, and harmonic effects. In spite of the unbalanced pitch-class distributions, the key answer generally stays within the ballpark of the correct one, i.e., the answer given is typically a closely related key. Some examples of closely related keys are: the dominant major/minor, the relative minor/major, and the parallel major/minor keys.

The Fuzzy Analysis technique consists of three steps. The first step uses information about the overtone series to clarify the existence of the pitches in the lower frequencies. The second step, which we term adaptive level weighting, scales (multiplies) the frequency magnitudes by the relative signal density in a predefined range, so as to focus on frequency ranges containing the most information. After the frequency magnitudes have been folded into 12 pitch classes, we apply the third step to refine the pitch-class distribution. The third step sets all normalized pitch class values 0.2 and below to zero, and all values 0.8 and above to one. Details of each step are given below. After the three-part Fuzzy Analysis technique, we introduce the periodic cleanup procedure for preventing the accumulation of low-level noise over time.

Clarifying Lower Frequencies

In the first step, we use the overtone series to confirm the presence of pitches below 261 Hz (C4). Because of the logarithmic scale of pitch frequencies, lower pitches are more closely located on the linear frequency scale than higher ones. The mapping of lower frequencies to their corresponding pitch number is noisy and error prone, especially when using discrete frequency boundaries. There exists greater separation between the reference frequencies of higher pitches, and the mapping of higher frequencies to their corresponding pitches is a more accurate process. For lower pitches, we use the first overtone to confirm their presence and refine their weights.

We use the idea of the membership value in fuzzy logic to represent the likelihood that a pitch has been sounded. Assume that $P_{i,j}$ represents the pitch of class j at register i, for example, middle C (i.e., C4) is $P_{4,0}$. We consider the pitch range $i = 2, 3, 4, 5, 6$, and $j = 1, \ldots, 12$, which includes pitches ranging from C_2 (65 Hz) to B_6 (2000 Hz). The membership value of $P_{i,j}$ is defined as

$$mem(P_{i,j}) = \mathrm{M}^*(P_{i,j}) / \max_p \{\mathrm{M}^*(\mathrm{p})\}. \tag{11.4}$$

Next, we define the *membership negation value* for lower pitches, a quantity that represents the fuzzy likelihood that a pitch is not sounded. Let the membership negation value be:

$$\sim mem(P_{i,j}) = \max\{mem(P_{i,j+1}), mem(P_{i+1,j}), mem(P_{i+1,j+1})\}, \tag{11.5}$$

where $i = 2, 3$ and $j = 1, \ldots, 12$, because we consider only the lower frequency pitches, pitches below C_4. This value is the maximum of the membership values of the pitch one half step above ($P_{i,j+1}$), and the first overtones of the pitch itself ($P_{i+1,j}$), and that of the pitch one half step above the first overtone. The membership value of a lower frequency pitch is set to zero if its membership negation value is larger than its membership value:

$$mem(P_{i,j}) = \begin{cases} 0, & \text{if} \sim mem(P_{i,j}) > mem(P_{i,j}) \\ mem(P_{i,j}), & \text{if} \sim mem(P_{i,j}) \leq mem(P_{i,j}) \end{cases}, \tag{11.6}$$

where $i = 2, 3$ and $j = 1, \ldots, 12$. This step is based on the idea that, if the existence of the pitch a half step above, as indicated by $mem(P_{i,j+1})$ and $mem(P_{i+1,j+1})$, is stronger than that of the pitch itself, then the pitch itself is unlikely to have been sounded. And, if the signal for the existence of the pitch is stronger in the upper registers, then we can ignore the membership value of the present pitch.

Adaptive Level Weighting

The adaptive level weight for a given range, a scaling factor, is the relative density of signal in that range. We scale the weight of each pitch class by this adaptive level weight in order to focus on the regions with the greatest amount of pitch information. For example, the adaptive level weight for register i (which includes pitches C_i through B_i), Lw_i, is defined as

$$Lw_i = \frac{\sum_{j=1}^{12} M(P_{i,j})}{\sum_{k=2}^{6} \sum_{j=1}^{12} M(P_{k,j})}, \tag{11.7}$$

where $i = 2, \ldots, 6$. We generate the weight for each pitch class, $mem_C(C_j)$, by summing the membership values of that pitch over all registers, and multiplying the result by the corresponding adaptive level weight:

$$mem_C(C_j) = \sum_{i=2}^{6} Lw_i \times mem(P_{i,j}), \tag{11.8}$$

where $j = 1, \ldots, 12$.

Flatten High and Low Values

To reduce minor differences in the membership values of important pitch classes, and to eliminate low-level noise, we introduce the last step in this section. We set the pitch class membership values to 1 if they are greater than 0.8, and 0 if they are less than 0.2 (constants determined from held-out data). This flat output for high membership values prevents louder pitches from dominating the weight distribution. Finally, we normalize the membership values for all pitch classes to sum to one.

Periodic Cleanup Procedure

Based on our observations, errors tend to accumulate over time. To counter this effect, we implemented a periodic cleanup procedure that takes place every 2.5 s. In this cleanup step, we sort the pitch classes in ascending order and isolate the four pitches with the smallest membership values. We set the two smallest values to 0, a reasonable choice since most scales consist of only seven pitch classes. For the pitch classes with the third and fourth smallest membership values, we consult the current key assigned by the CEG algorithm; if the pitch class does not belong to the key, we set the membership value to 0 as well.

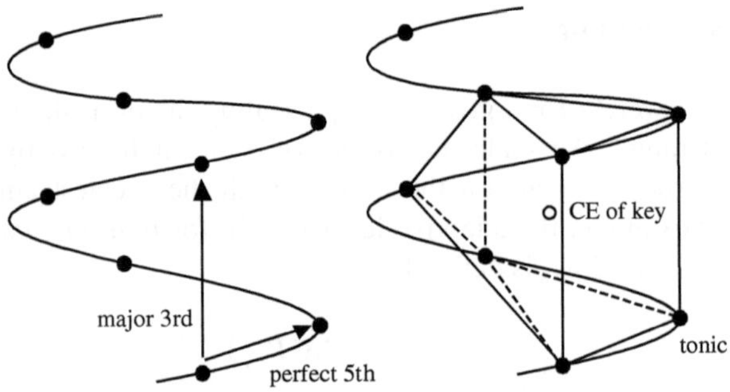

Fig. 11.2 Pitch spiral in the Spiral Array model (*left*), and the generating of a CE to represent the key (*right*)

11.3.2 Spiral Array Model and the Center of Effect Algorithm

The Spiral Array model, proposed by Chew in [1], is a three-dimensional model that represents pitches, and any pitch-based objects that can be described by a collection of pitches, such as intervals, chords, and keys, in the same three-dimensional space for easy comparison. On the Spiral Array, pitches are represented as points on a helix, and adjacent pitches are related by intervals of perfect fifths, while vertical neighbors are related by major thirds. The pitch spiral is shown on the left in Fig. 11.2. Central to the Spiral Array is the idea of the *center of effect* (CE), the representing of pitch-based objects as the weighted sum of their lower level components. The CE of a key is shown on the right in Fig. 11.2. Further details for the construction of the Spiral Array model are given in [1] and [2].

In the CEG algorithm, key selection is performed by a nearest-neighbor search in the Spiral Array space. We shall call this the nearest-neighbor (NN) policy for key determination. The pitch classes in a given segment of music are mapped to their corresponding positions in the Spiral Array, and their CE generated by a linear weighting of these pitch positions. The algorithm identifies the most likely key by searching for the key representation closest to the CE. The evolving CE creates a path that traces its dynamically changing relationships to the chord and key structures represented in the model [5].

Previous applications of the CEG algorithm have used the relative pitch durations as the CE weights, either directly [2] or through a linear filter [5]. Here, in audio key-finding, we use the normalized pitch-class distribution derived from the frequency weights to generate the CE.

One more step remains to map any numeric representation of pitch to its letter name for key analysis using the Spiral Array. The pitch-spelling algorithm, described in [3, 4], is applied to assign letter names to the pitches so that they can be mapped to their corresponding representations in the Spiral Array for key-finding. The pitch-spelling algorithm uses the current CE, generated by the past 5 s of music, as a proxy for the key context, and assigns pitch names through a nearest-neighbor search for the closest pitch-class representation. To initialize the process, all pitches in the first time chunk are spelt closest to the pitch class D in the Spiral Array, then the CE of these pitches is generated, and they are respelt using this CE.

11.3.3 Key Determination: Relative Distance Policy

In the audio key-finding systems under consideration, we generate an answer for the key using the cumulative pitch-class information (from time 0 until the present) at every analysis window, which eventually evolves into an answer for the global key for the whole duration of the music example. Directly reporting the key with the shortest distance to CE as the answer at each analysis window, i.e., the NN policy, does not fully reflect the extent of the tonal analysis information provided by the Spiral Array model. For example, at certain times, the CE can be practically equidistant from two different keys, showing strong ambiguity in key determination. Sometimes the first key answer (the one with the shortest distance to CE) may result from a local chord change, cadence, or tonicization, and the second answer is actually the correct global key. The next two key determination policies seek to address this problem.

We first introduce the relative distance key determination policy with distance threshold d, notated (RD, d). In the RD policy, we examine the first two keys with the shortest distances to the CE. If the distance difference between the first two keys is larger then the threshold, d, we report the first key as the answer. Otherwise, we compare the average distances of the two keys from the beginning to the current time chunk. The one with shorter average distance is reported as the answer.

Formally, let $d_{i,t}$ be the distance from the CE to key i at time t, where $i = 1, \ldots, 24$. At time t, assume that keys j and k are the closest keys to the CE with distances $d_{j,t}$ and $d_{k,t}$, respectively, and with $d_{j,t} \leq d_{k,t}$. Let $\bar{d}_{i,t}$ be the average distance between key i and each analysis window's CE throughout the piece up to time t. In pseudocode,

if $|d_{j,t} - d_{k,t}| \geq d$ **then**
  choose key j as the answer
else if $\bar{d}_{j,t} \leq \bar{d}_{k,t}$ **then**
  choose key j as the answer
else
  choose key k as the answer
end if
end if

The RD policy attempts to correct for tonal ambiguities introduced by local changes. The basic assumption underlying this method is that the NN policy is generally correct. In cases of ambiguity, which are identified as moments in time when the first and second closest keys are less than the threshold distance apart from each other, then we use the average distance policy to determine which among the two most likely candidates is the best choice. The next section describes the average distance policy in greater detail.

In this chapter, we test two values of d. The choice of d depends on the distance between keys in the Spiral Array. Assume d_1 denotes the shortest, and d_2 the second shortest, distance between any two keys in the Spiral Array model. Then, we constrain the value of d to the range:

$$\alpha d_1 \leq d \leq \beta d_2, \tag{11.9}$$

where $0 < \alpha,\ \beta \leq 0.5$. In this chapter, we set both α and β equal to 0.25. Intuitively, this means that the CE should lie in the center half of the line segment connecting two very close keys, if there is ambiguity between the two keys.

11.3.4 Key Determination: Average Distance Policy

The average-distance key determination policy (AD) is inspired by the method used by İzmirli in his winning submission to the MIREX 2005 audio key-finding competition [9, 15], where only the global key answer was evaluated. İzmirli's system tracks the confidence value for each key answer, a number based on the correlation coefficient between the query and key template. The global key was then selected as the one having the highest sum of confidence values over the length of the piece.

In the Spiral Array, the distance from each key to the current CE can serve as a confidence indicator for that key. In the AD policy, we use the average distance of the key to the CE at all time chunks to choose one key as the answer for the whole testing duration of the piece.

Formally, at time t, if $\bar{d}_{j,t} = \min_{i=1...24}(\bar{d}_{i,t})$, choose key j as the answer. We explore the advantages and the disadvantages of the (RD, d) and (AD) policies in the remainder of the chapter.

11.4 Evaluation of the Basic System

In this chapter we test the systems in two stages. In the first stage, we use 410 classical music pieces to test the basic systems described in Sect. 11.3, that is, the audio key-finding system using Fuzzy Analysis and the CEG algorithm, with the three key determination policies, (NN), (RD, d), and (AD). Both the local key answer (the result at each unit time) and the global key answer (one answer for each sample piece) are considered for the evaluation. The results are analyzed and classified by key relationships, as well as stylistic periods. At the second stage of the evaluation, we use audio recordings of 24 Chopin preludes to test the extensions of the audio key-finding system.

We choose excerpts from 410 classical music pieces by various composers across different time and stylistic periods, ranging from Baroque to Contemporary, to evaluate the methods. Table 11.1 shows the distribution of pieces across the various classical genres. Most of the chosen pieces are concertos, preludes, and symphonies,

Table 11.1 Results analysis of global key answers across periods obtained from fuzzy analysis technique and CEG algorithm

	Baroque	Classical	Early Romantic	Romantic	Late Romantic	Contemporary
CORR	80	95.7	72.4	76	72.9	82.8
DOM	16.8	0	25.3	8	5.9	0
SUBD	0	0.9	0	4	0	1
REL	0	0.9	0	6	5	3
PAR	2.1	1.7	0	2	1	1
Other	0	0.9	2.3	4	1	0
Num	**95**	**115**	**87**	**50**	**34**	**29**

Numbers indicate the percentage of answers that are correct (CORR), the dominant of the correct key (DOM), subdominant (SUBD), relative key (REL) or parallel key (PAR). Num indicates the number of pieces in each category

which consist of polyphonic sounds from a variety of instruments. We regard the key of each piece stated explicitly by the composer in the title as the ground truth for the evaluation. We use only the first 15 s of the first movement so that the test samples are highly likely to remain in the stated key for the entire duration of the sample.

In order to facilitate comparison of audio key-finding from symbolic and audio data, we collected MIDI samples from http://www.classicalarchives.com, and used the Winamp software with 44.1 kHz sampling rate to render MIDI files into audio (wave format). We concurrently tested four different systems on the same pieces. The first system applied the CEG algorithm with the nearest-neighbor policy, CEG(NN) to MIDI files, the second applied the CEG algorithm with the nearest-neighbor policy and Fuzzy Analysis technique, FACEG(NN), and the third and the fourth are similar to the second with the exception that they employ the relative distance policy in key determination, FACEG(RD, d), with different distance thresholds. The last system, FACEG(AD), applies the relative distance policy with average distances instead.

Two types of results are shown in the following sections. Section 11.4.1 presents the averages of the results of all periods over time for the four systems. Each system reported a key answer every 0.37 s, and the answers are classified into five categories: correct, dominant, relative, parallel, and others. Two score-based analyses are given to demonstrate the examples in which the audiokey-finding system outperforms the MIDI key-finding system that takes explicit note information as input. In Sect. 11.4.2, the global key results given by the audio key-finding system with Fuzzy Analysis technique and CEG algorithm are shown for each stylistic period.

11.4.1 Overall Results Over Time

Figure 11.3a shows the average correct rates of the five systems over time on 410 classical music pieces. We can observe that in the second half of the testing period, from 8 to 15 s, four of the systems, all except FACEG(AD), achieve almost the same results by the percentage correct measure.

The relative-distance key determination policy using average distance FACEG (AD) performed best. Its correct percentage is almost 10 % higher than the other systems from 8 to 15 s. Notice that the improved correct rate of FACEG(AD) is mainly due to the reduction of dominant and relative errors shown in Fig. 11.3b, c. The relative distance policy using threshold distance (RD, *d*) slightly outperforms the systems with only the nearest-neighbor (NN) policy in audio key finding. The results of the systems with the RD and AD policies maintain the same correct rates from 5 s to the end. The long-term stability of the results points to the advantage of the RD and AD policies for choosing the global key.

The CEG(NN) system outperforms all four audio systems in the first 5 s. The RD policy even lowers the correct rate of the FACEG(NN) audio key-finding system. The results show that audio key-finding system requires more time at the beginning to develop a clearer pitch-class distribution. The RD policy may change correct answers for the incorrect ones at the beginning if the pitch-class information at the first few seconds is ambiguous.

Figure 11.3b–e shows the results in dominant, relative, parallel, and others categories. Most differences between the CEG(NN) system and the FACEG audio key finding systems can be explained in the dominant and parallel errors, shown in Fig. 11.3b, d. We can use music theoretic counterpoint rules to explain the errors. In a composition, doubling of a root or the fifth of a chord is preferred over doubling the third. The third is the distinguishing pitch between major and minor chords. When this chord is the tonic, the reduced presence of thirds may cause a higher incidence of parallel major/minor key errors in the first 4 s.

For audio examples, the third becomes even weaker because the harmonics of the root and the fifth are more closely aligned, which explains why audio key-finding systems have more parallel errors than the MIDI key-finding system CEG(NN). The ambiguity between parallel major and minor keys subsides once the system gathers more pitch-class information.

In the relative and other error categories, shown in Fig. 11.3c and e, the audio key-finding systems perform slightly better than the MIDI key-finding system. We present two examples with score analysis in Figs. 11.4 and 11.5 to demonstrate how the audio key-finding systems—FACEG (NN), FACEG (RD, 0.1), FACEG (RD, 0.17), FACEG (AD)—outperform the MIDI key-finding system.

11.4.2 When Audio Outperforms Symbolic key-finding

Figure 11.4 shows the first four measures of Bach's *Double Concerto in D minor* for two violins, BWV1043. For the whole duration of the four measures, all audio systems give the correct key answer, D minor. In contrast, the MIDI key-finding system returns the answer F major in the first two measures, then changes the answer to G major at the end. We can explain the results by studying the pitch-class distributions for both the MIDI and audio systems at the end of the second and fourth measures.

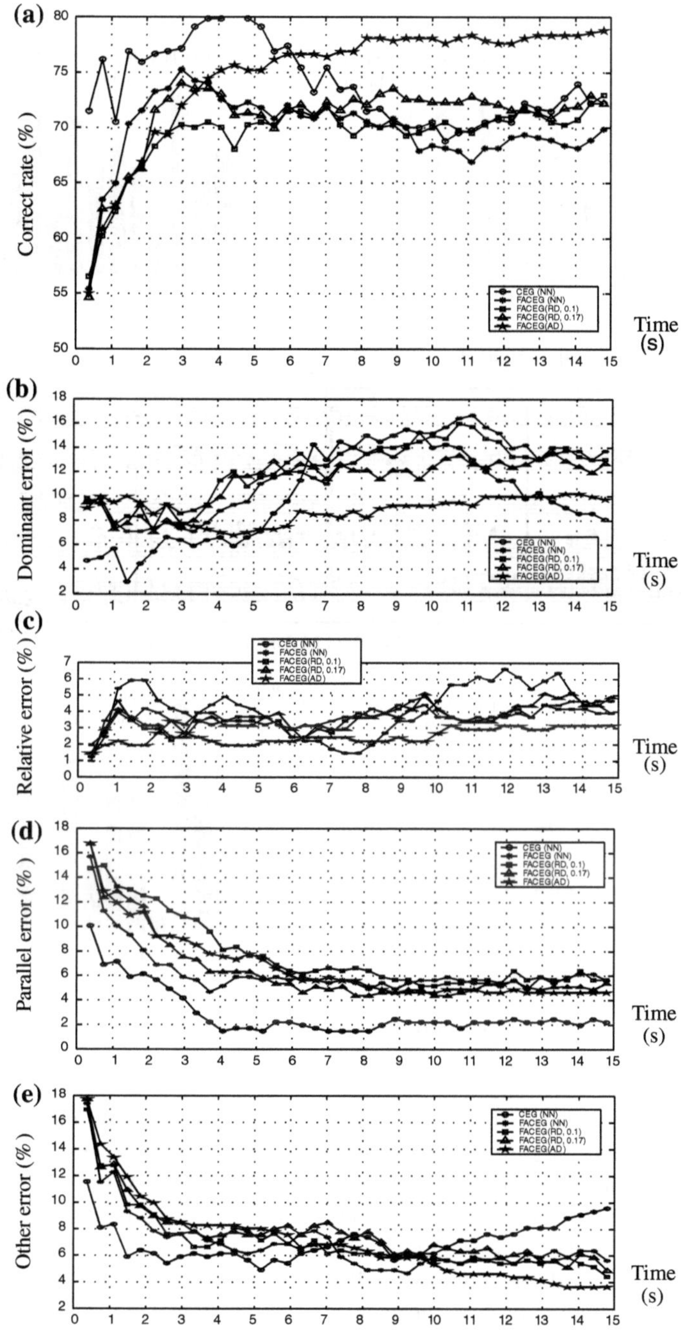

Fig. 11.3 Results for first 15 s of five categories of classical pieces, 410 in total

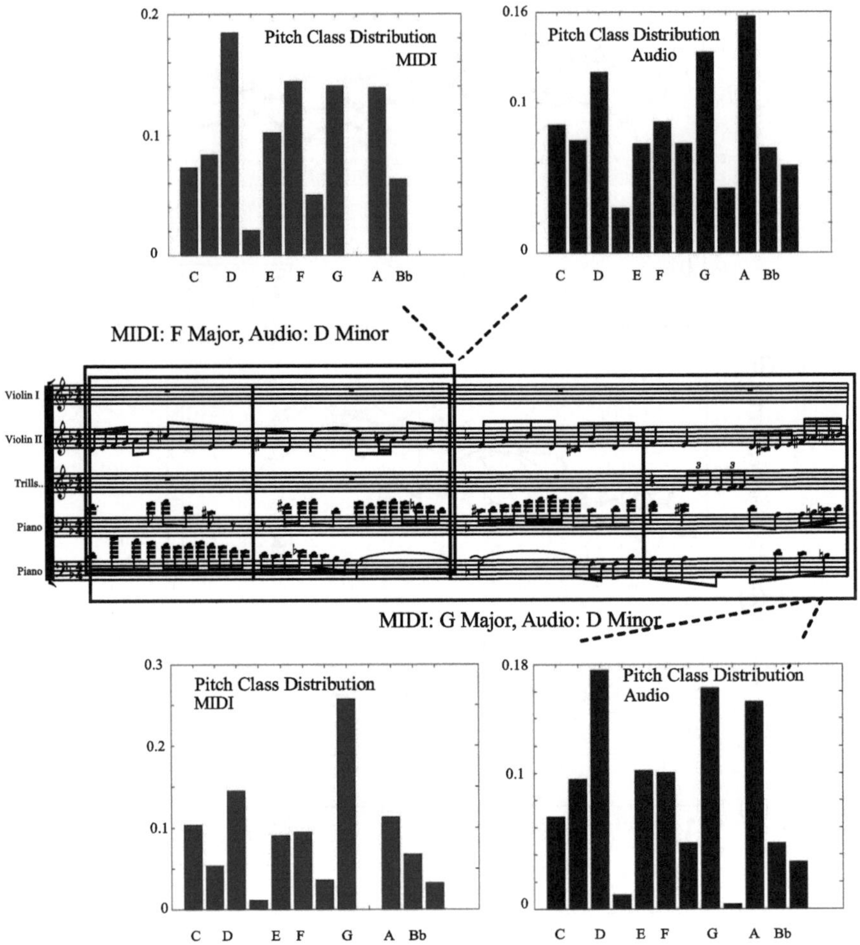

Fig. 11.4 Pitch-class distributions for Bach's *Double Concerto in D minor*

The pitch-class distribution of the MIDI system at the second measure does not provide sufficiently significant differences between the pitch sets belonging to F major and D minor; however, the high weight on pitch class A, the second harmonic of the pitch D, in the corresponding distribution derived from audio helps to break the tie to result in the answer, D minor. At the end of the second measure and the beginning of the third, there are two half note G's in the bass line of the piano part. These relatively long notes bias the answer toward G major in the MIDI key-finding system. The audio key-finding systems are not affected by these long notes because the effect of the overlapping harmonics results in a strong D, and a not-as-high weight on G in the pitch-class distribution.

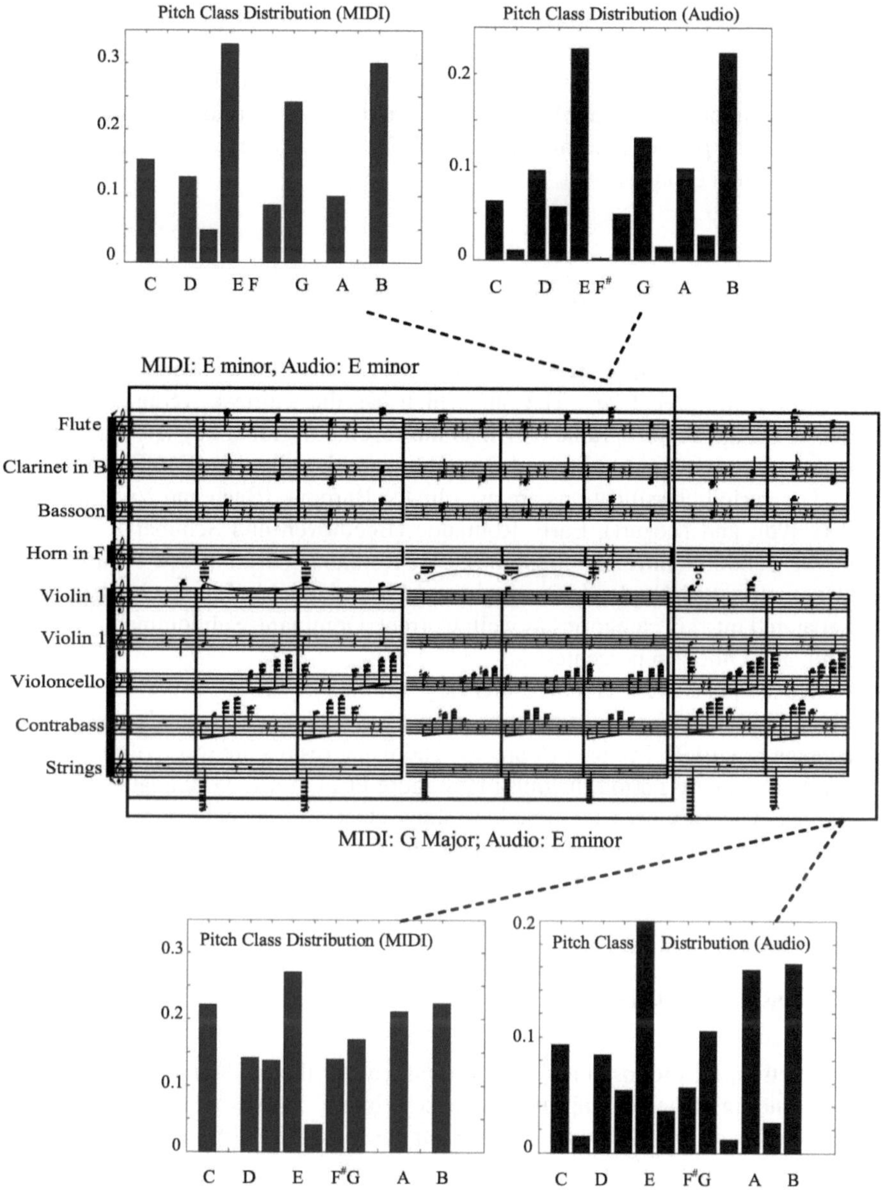

Fig. 11.5 Pitch-class distributions for Brahms' *Symphony No. 4 in E minor*

We give another example in Fig. 11.5, which shows the first eight measures of Brahms' *Symphony No. 4 in E minor*, Op. 98. Both the MIDI and audio key-finding systems report correct answers for the first six measures. At measures six through eight, the chords progress from vi (pitches C, E, G) to III (pitches G, B, D) to VII (pitches D, F♯, A) in E minor, which correspond to the IV, I, and V chords in G major. After these two measures the answer of the MIDI key-finding system becomes G major. This example shows that having explicit information about only the fundamental pitches present makes the MIDI key-finding system more sensitive to the local tonal changes.

11.4.3 Results of Global Key Across Periods

We use the average of the distances between the CE and the key over all time chunks to determine the global key. The one which has the shortest average distance is chosen to be the answer. Table 11.1 lists the results of global key answers, broken down by stylistic periods, obtained from the audio key-finding, FACEG(AD), systems. The period classifications are as follows: Baroque (Bach and Vivaldi), Classical (Haydn and Mozart), Early Romantic (Beethoven and Schubert), Romantic (Chopin, Mendelssohn, and Schumann), Late Romantic (Brahms and Tchaikovsky), and Contemporary (Copland, Gershwin and Shostakovich). The results themselves are separated into six categories as well: Correct, Dominant, Subdominant, Relative, Parallel, and Other (in percentages).

Notice that, in Table 11.1, the results vary significantly from one period to another. The best results are those of the Classical period, which attains the highest correct percentage rate of 95.7 % on 115 pieces. The worst results are those of pieces from the Early Romantic period, having many more errors on the dominant and others categories. The variances in Table 11.1 show clearly the dependency between the system performance and the music style. Lower correct rates could be interpreted as an index of the difficulty of the test data.

11.5 System Extensions

In this section, we propose three new alternatives for the pitch-class generation and the key-finding stages to improve audio key finding as was first presented in the system outline given in Fig. 11.1. These methods include modifying the Spiral Array model using sampled piano audio signals, fundamental frequency identification, and post-weight balancing. The three approaches affect different stages in the prototypical system, and use different domains of knowledge. In the first alternative, we modify the Spiral Array model so that the positions of the tonal entities reflect the frequency features of audio signals. The second alternative affects pitch class generation; we use the information from the harmonic series to identify the fundamental frequencies. The third method of post-weight balancing is applied after the key finding algorithm; it uses the key-finding answer to refine the pitch-class distribution. Each of the three approaches is described in the subsections to follow.

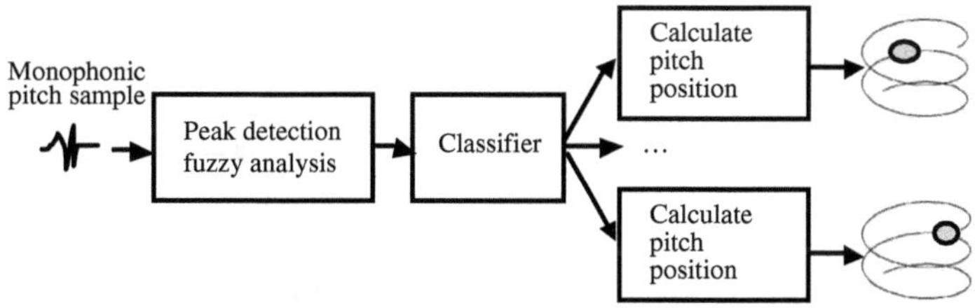

Fig. 11.6 System diagram of reconstructing pitches in spiral array model

11.5.1 Modified Spiral Array with Piano Signals

Since the pitch-class distribution for each audio sample is constructed using the frequency magnitudes derived from the FFT, in order to compare the CE of this distribution to an object of the same type, we propose to prepare the Spiral Array to also generate tonal representations based on audio signal frequency features. In this section, we describe how we modify the major and minor key spirals so that the positions are constructed according to the frequency features of the audio signals. The advantages of the proposed modification are that the modified Spiral Array can manage the diversity of the frequency features of audio signals, and tolerate the errors from the pitch detection method. A similar idea is proposed by İzmirli to modify the Krumhansl-Schmuckler key-finding method to address audio signals in [15], the Audio Key-Finding contest results.

Figure 11.6 shows the sequence of steps for remapping the Spiral Array representations for audio. The mapping uses the frequency distribution of monophonic pitch samples to first classify pitches into subclasses based on their harmonic profile, then calculates the new position of each pitch for each subclass. The monophonic pitch samples, piano sounds from B♭0 to C8, are obtained from the University of Iowa Musical Instrument Samples online [18]. The classification step is essential because tone samples from different registers exhibit different harmonic characteristics. Hence, the representations are regenerated for each subclass.

Formally, for each monophonic pitch sample, we apply the peak detection method and Fuzzy Analysis technique to generate a pitch-class distribution for that pitch, $mem(C_j)$, $i = 1, 2, \ldots, 12$. Each pitch then is classified into several subclasses according to the pitch-class distribution. The classification can be done by any existing classifiers, such as k-nearest neighbors. The classification must satisfy the constraint that each class consists of pitches that are close to one another. This constraint is based on the assumption that pitches in the same range are likely to have similar pitch-class distributions. For the purposes of the tests in this chapter, we classify the pitches into five classes manually.

The new position of the pitch representation in the Spiral Array, for each subclass, is recomputed using these weights. Assume $\mathbf{P}_i$ represents the original position of pitch class i in the Spiral Array model. The new position of pitch class i, $\mathbf{P}'_i$, is defined as

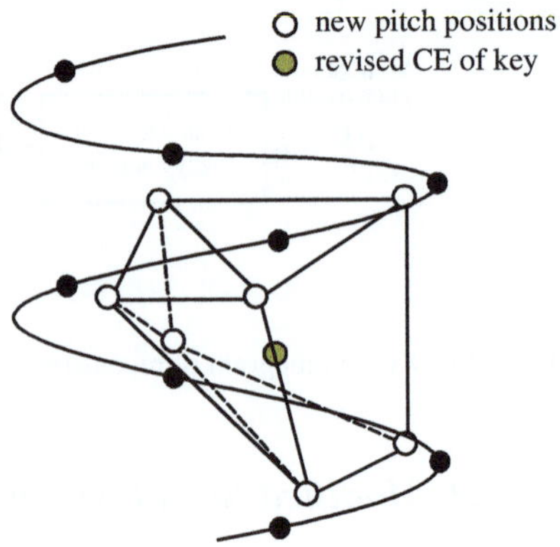

Fig. 11.7 Recalculating pitch position using pitch-class distribution

$$P_i{}' = \frac{1}{n}\sum_{j=1}^{12} mem(C_j) \times P_j, \tag{11.10}$$

where $j = 1, \ldots, 12$ and n is the size of the subclass. Figure 11.7 shows conceptually the generating of the new position for pitch class C.

Once we obtain the new position of pitches, we can calculate the new position of keys for each subclass by a weighted linear combination of the positions of the triads. The composite key spirals are generated in real time as the audio sample is being analyzed. We weight the key representation from each subclass in a way similar to that for the level weights method described in Sect. 11.3.2. That is to say, the level weight for a given subclass is given by the relative density of pitches from that subclass. The position of each key in a key spiral is the sum of the corresponding key representations for each subclass, multiplied by its respective level weight. Assume $\mathbf{T}_i$ is the original position of key i in the Spiral Array, the new position of key i, $\mathbf{T}'_i$, is calculated by:

$$\mathbf{T}'_i = Lw_i \times \mathbf{T}''_i, \tag{11.11}$$

where Lw_i is the level weight for subclass i and $\mathbf{T}_j{}''$ is the composite position for key j in subclass i, $j = 1, \ldots, 24$ for 24 possible keys.

As the final step, we perform the usual nearest-neighbor search between the CE generated by the pitch-class distribution of the audio sample and the key representations to determine the key.

11.5.2 Fundamental Frequency Identification

Audio signals from music differ from speech signals in three main aspects: the frequency range, the location of the fundamental frequency, and the characteristic of the harmonic series. Compared to human voices, instruments can sound in a much

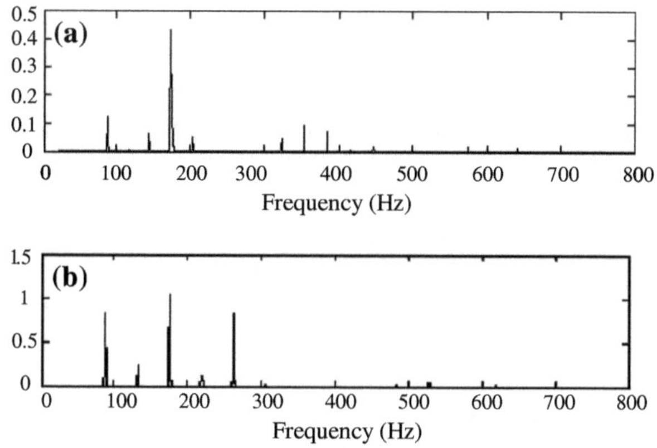

Fig. 11.8 Frequency responses of pitches **a** B♭0 and **b** B1 using FFT

wider range of frequencies. Furthermore, the lower pitches are typically organized in such a way as to highline the tonal structure of the music sample, while the higher pitches are less important structurally, and may contain many superfluous accidentals. However, the structurally more important lower pitches cannot always be detected using signal processing methods such as the FFT. Also, several lower pitches may generate similar distributions in the frequency spectrum. Missing information in the lower registers seriously compromises the results of key-finding. Figure 11.8 shows the FFT output for pitches B♭0 and F1. It is important to note that these two pitches have similar frequency distributions, yet neither of their fundamental frequencies appear in the FFT. In the case of pitch B♭0, none of the pitches in the pitch class B♭ is presented. This example reveals a key consideration as to why audio key-finding frequently suffers from dominant errors. The audio signals of each individual pitch are collected from the piano recordings on the Iowa University website [18].

Many systems for automatic transcription that use fundamental frequency to identify pitch have been proposed recently [10, 11]. The transcription problem requires the extraction of multiple fundamental frequencies of simultaneously sounding pitches. We, instead, are concerned with finding only the lowest pitch in the bass. We use the first seven harmonics to identify each fundamental frequency. The frequency ratio (multiple of the fundamental frequency) and the pitch relation of the harmonic structure are given in Table 11.2. We use this harmonic structure as a template for locating the fundamental frequencies as follows: Given an audio signal, first we

Table 11.2 Frequency and pitch relations of seven harmonics

Freq ratio	1	2	3	4	5	6	7
Pitch relation***	1	8va	8va+P5	16va	16va+M3	16va+P5	16va+m7
Semitone distance	0	12	19	24	28	31	34

***: P5: perfect fifth, M3: major third, m7: minor seventh

extract the frequencies with the largest and second largest frequency magnitudes. Then, we move the harmonic template so as to find all possible ways to cover the two frequencies, and calculate the total number of frequencies that are both in the harmonic template and the extracted frequency spectrum. The highest scoring option gives the location of the fundamental frequency. We employ a heuristic to break ties. Ties happen because not all the harmonics appear for a tone of a given fundamental pitch. When an octave pair is encountered, it is unclear if this represents the fundamental pitch class, or the fifth above the fundamental. The heuristic is based on our observations, and prefers the interpretation of the fifth when finding an octave pair in the lower registers.

Using a window size that is three times larger for low frequencies than higher ones (0.37 s), we tested the above method on monophonic piano samples of pitches ranging from B♭0 to B3 obtained from the Iowa University website [18]. For the 38 samples, we successfully identified the fundamental frequencies of 33 pitches, with 4 octave errors and 1 perfect fifth error. The octave error does not affect key-finding.

11.5.3 Post-weight Balancing

In audio key-finding, unbalanced pitch-class distributions are often obtained using the frequency spectrum derived from the FFT. One particularly problematic example occurs when the weight of a certain pitch class is much higher than the others. The pitch class dominates the weight distribution so much so that the CE is strongly biased by that pitch class, and cannot fairly represent the presence of the other pitch classes. The relative distance policy in key determination, in which the system compares the distance difference between the first two keys, cannot solve this unbalanced distribution problem. Similarly, one cannot readily eliminate the problem by simply examining the low-level features such as the frequency of the audio signal.

To solve the problem of unbalanced weight distributions, we design a PWB mechanism. We use high-level knowledge of the relations between keys to determine which pitch class has been weighted too heavily. The PWB mechanism is based on two principles: (1) if the three closest keys contain a *relative* major/minor pair, then the tonic of the other key is likely over weighted; (2) if the three closest keys contain a *parallel* major/minor pair, then the tonic of the pair is likely over weighted.

Once the over weighted pitch class is identified, we reduce its weight in the pitch-class distribution, and reapply the CEG algorithm to generate a new answer with the adjusted pitch-class distribution. The new answer is then verified again using the PWB mechanism to specifically disambiguate the relative or parallel major/minor answers. To differentiate between relative major/minor keys, we compare the weights of the respective tonic pitch classes. The one with larger weight is chosen as the answer. To differentiate between parallel major/minor keys, we examine the weights of the non-diatonic pitches in each candidate key. The PWB algorithm is summarized as follows:

Let $\mathbf{K}_3$ be the set of three closest keys and $\mathbf{K}_2$ is any subset with two keys of $\mathbf{K}_3$

if $\mathbf{K}_2$ contains a *relative* major/minor pair **then**
the tonic of $\mathbf{K}_3 \backslash \mathbf{K}_2$ is labeled as overweighted
else if $\mathbf{K}_2$ contains a *parallel* major/minor pair **then**
the tonic of $\mathbf{K}_2$ is labeled as overweighted
end if
Overweighted pitch class is assigned the average weight of pitches in $\mathbf{K}_3$

11.6 In-depth Analysis of Results

In order to explore the possible approaches for improving audio key-finding system outlined in Sect. 11.5, we test five systems on Evgeny Kissins CD recordings of Chopin's *Twenty-Four Preludes for Piano* (ASIN: B00002DE5F). We chose this test set for three reasons: (1) it represents one of the most challenging key finding datasets we have tested to date—in a previous study (see [6]), we discovered that the results for this test set was farthest from that for MIDI; (2) the audio recording created using an acoustic piano allows us to test the systems robustness in the presence of some timbral effects; and, (3) a minor but esthetic point–all 24 keys are represented in the collection.

The five systems selected for testing consist of combinations of the approaches described in Sects. 11.3 and 11.5, with a focus on the three new alternatives introduced in Sect. 11.5. We introduce a notation for representing the systems. The five systems are as follows:

(a) the basic system, sys(FA, SA/CEG);
(b) FACEG with post-weight balancing, sys(FA, SA/CEG, PWB);
(c) the modified Spiral Array with CEG, sys(mSA/CEG);
(d) FACEG with modified Spiral Array, sys(FA, mSA/CEG);
(e) fundamental frequency identification with CEG, sys(F0, SA/CEG).

We have ascertained in Sect. 11.4.1 that the best key determination policy was AD, the average distance policy, which is now employed in all five systems.

The first system, sys(FA, SA/CEG) serves as a reference, a basis for comparison. The second system, sys(FA, SA/CEG, PWB), tests the effectiveness of the PWB scheme applied to the basic system. The fourth system, sys(FA, mSA/CEG), tests the effectiveness of the modifications to the Spiral Array based on audio samples, in comparison to the basic system. To further test the power of the modified Spiral Array, we take away the Fuzzy Analysis layer from (d) to arrive at the third system, sys(mSA/CEG). The fifth system, sys(F0, SA/CEG), tests the effectiveness of the fundamental frequency identification scheme relative to the basic system.

The overall results for all five extended systems are shown in Fig. 11.9. The system employing fundamental frequency identification with the CEG algorithm, sys(F0, CEG), outperforms the others in the first 8 s. The systems using the modified Spiral Array model, sys(mSA/CEG), and sys(FA, mSA/CEG), achieve the best correct rates after 8 s, and significantly higher correct rates from 12 to 15 s. In comparison to the

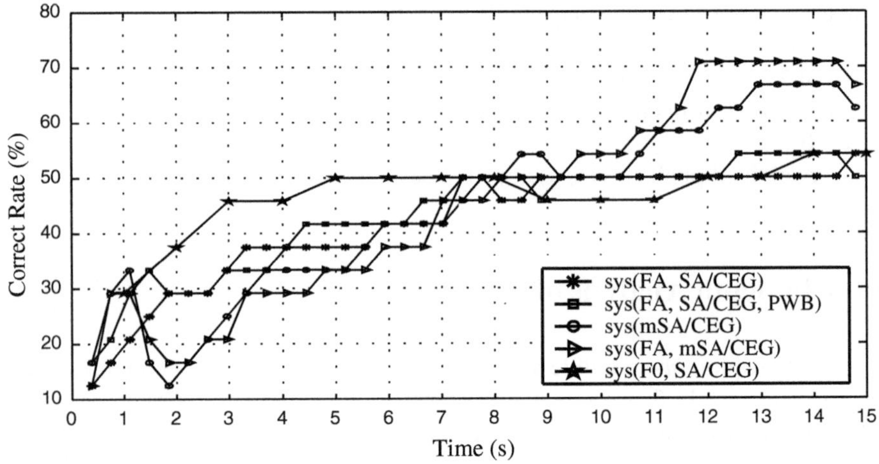

Fig. 11.9 Correct rate of five extended systems on Chopin's *24 Preludes*

(a)

SR	R	DR
S	T	D
SP	P	DP

(b)

0	2	2
0	12	6
0	1	1

(c)

0	1	4
1	13	2
0	2	0

(d)

1	2	1
0	16	3
0	0	0

(e)

1	2	1
0	17	3
0	0	0

(f)

1	3	1
0	13	4
0	0	0

Fig. 11.10 Results of five systems on Chopin's *24 Preludes* (T: Tonic, D: Dominant, S: Subdominant, R: Relative, P: Parallel). **a** gives the key map. The five systems are: **b** FA, SA/CEG; **c** FA, SA/CEG, PWB; **d** mSA/CEG; **e** FA, mSA/CEG; **f** F0, SA/CEG

original system, sys(FA, CEG), the PWB used in sys(FA, CEG, PWB) improves the results slightly.

Figure 11.10 shows the cross-section of results at 15 s for the five systems. Each result is represented in a key map, where the tonic (the ground truth) is in the center, and the dominant, subdominant, relative, parallel are to the right and left, and above and below, the tonic respectively. The number inside each grid is the number of the answers that fall in that category. Answers that are out of the range of the key map are not shown.

Table 11.3 Summary of the results of five systems categorizing the pieces by the number of the correct answers

No. of systems w/correct answer	5	4	3	2	1	0
Piece (Prelude No.)	6, 8, 17, 19, 20, 24	14, 15, 21, 23	2, 4, 7, 12	10, 22	1, 3, 11, 16	5, 9, 13, 18

Figure 11.10 (b) shows the results of the original system, the Fuzzy Analysis technique with Spiral Array CEG algorithm. The results of system extensions are given in (c) through (f). Notice that the FACEG system with modified Spiral Array model, (e), gets the most correct answers, which implies that including the frequency features within the model could be the most effective way to improve the system. This result is confirmed by Fig. 11.9.

In Table 11.3, we categorize the pieces by the number of systems that report correct answers. By analyzing pieces with the most or least number of correct answers, we can better understand which kinds of musical patterns pose the greatest difficulty for all systems, as well as the advantages of one system over the others for certain pieces. The results summarized in Table 11.3 suggest that the ease of arriving at a correct key solution is independent of the tempo of the piece.

We now focus on the analysis of some best- and worst-case answers for audio key-finding. Chopin's Prelude No. 20 in C minor is one of the pieces for which all of the systems gave a correct answer, even though it could be interpreted as being in four different keys in each of the first four bars. Figure 11.11 shows the first part of the score corresponding to the audio sample analyzed, the audio wave of the first 15 s, and the top two key answers over time, and the distances of these key representations to the CE of the audio sample. In the Spiral Array model, shorter distances correspond to more certain answers. The results over time presented in Fig. 11.11 show the two key representations closest to the current cumulative CE at each time sample. The closest answer is C minor for most of the time, except from 8.14 to 13.32 s, when the closest answer is E♭ major, while C minor is relegated to the second closest answer. The graph shows the chord sequences brief tonicization in E♭ major at the end of measure three, before reaching G major at the end of measure four, which acts as the dominant to the key of C minor.

Chopins Prelude No. 9 in E major is an example in which all five systems reported B major as the key, the dominant of E. Figure 11.12 shows the score for the first part of the piece, the audio wave of the first 15 seconds, and the top two key answers over time for the four audio systems. From Fig. 11.12a–d, we observe the following two common behaviors in all graphs: (1) for most of the time during the 15 s, the top ranked answer is B major, while the second answer traverses through related keys; and, (2) the first and second answers are almost equidistant from the CE in the middle part, between 7 and 8 s.

The test duration (first 15 s) corresponds to the first two and a half measures of the piece. From the score, the chord progression of the first measure is I-V-I-IV in

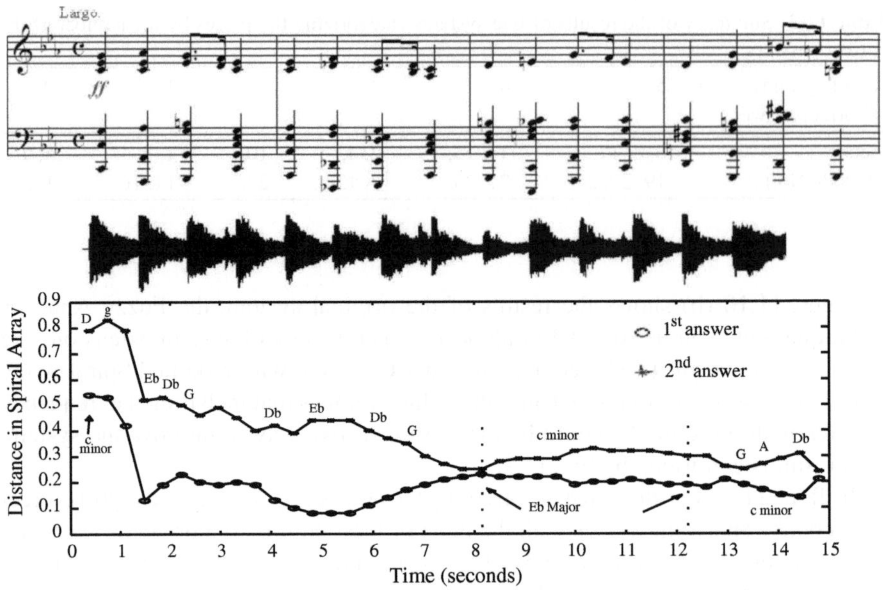

Fig. 11.11 Key answers over time for sys(FA, CEG) on Chopin's Prelude No. 20 in C minor

E major, which unequivocally sets up the key as E major. The dominant errors of all systems can be attributed to the audio attributes of Evgeny Kissins expressive performance. In the recording, Kissin emphasizes the pitch B3 at each beat as it is an important pitch that is repeated frequently in the melodic line. The emphasis on the pitch B3 and later the lowest note B1, results in the larger weight of B in the pitch-class distribution. Even though the notes in the score and the chords as early as the first bar clearly show E major as the key, the prominence of the pitch B in the performance overshadows the answer so that B major and B minor appear most frequently as top choices for this piece, even though the key of E major is ranked first or second at least once in the first two seconds for all systems.

The results in Fig. 11.12a and b show that PWB does not improve the answers in this case, where the pitches emphasized in the expressive performance dominate the results. The PWB method may be more effective if the parameters for the weight adjustments are optimized through learning. Compared with the other three systems, the F0 method employed in system (e) appears to have stabilized the results, and to react most sensitively to changing pitches in the lower registers. Sys(F0, SA/CEG) gets the correct answer within the first second, which is quicker than the others. In (c) and (d), sys (mSA/CEG) and sys(FA, mSA/CEG), the answers based on the modified Spiral Array model appear more random than the other systems, and are difficult to explain directly from the score. Using the frequency distribution of the pitches instead of the apparent duration does not appear to have been very helpful in this case.

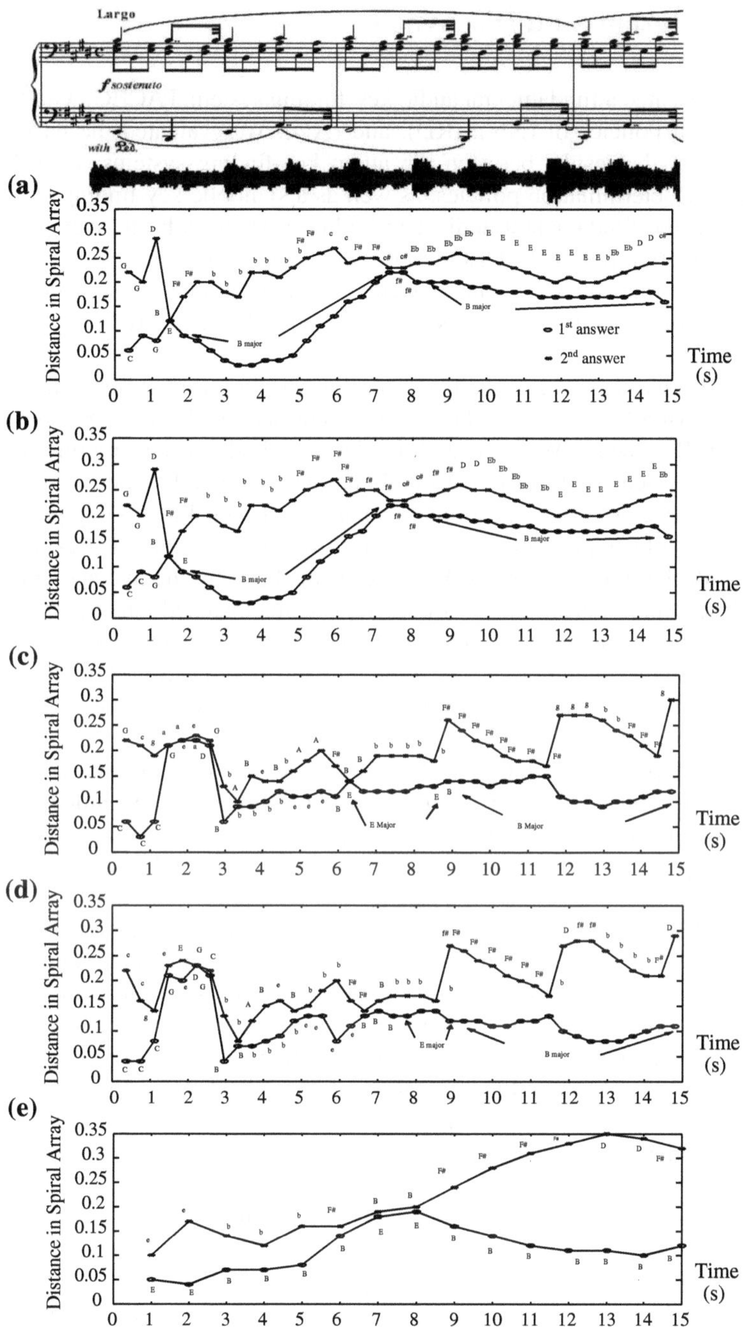

Fig. 11.12 Key answers over time for Chopin's Prelude No. 9 in E major. Results from the systems: **a** sys(FA, SA/CEG); **b** sys(FA, SA/CEG, PWB); **c** sys(mSA/CEG); **d** sys(FA, mSA/CEG); **e** sys(F0, SA/CEG)

11.7 Conclusions and Discussion

We have presented a fundamental audio key-finding system, FACEG, with three key determination policies of (NN), (RD), and (AD). We evaluated the basic system by comparing the results between the audio key-finding systems, with the three different key determination policies, as well as a symbolic key-finding system. We showed that the fundamental audio key-finding system with the average-distance key determination policy, FACEG (AD), is superior as it achieves generally 10 % higher correct rates than the other systems. We showed that the stylistic period of the classical pieces could be used as an indicator for the difficulty of key-finding for that test set.

We also presented three possible extensions to the basic system: the modified Spiral Array (mSA), fundamental frequency identification (F0), and post-weight balancing (PWB) scheme. We evaluated the three methods by constructing five audio key-finding systems with different combinations of possible extensions, and provided qualitative as well as quantitative analyses of their key-finding performance on recordings of Chopin's *Preludes* for piano. We observed that the Fuzzy Analysis system with the modified Spiral Array performs best, by the average correct rate metric. The overall performance of the five systems matched our expectations that identifying the fundamental frequencies is helpful in the first 8 s when fewer notes have been sounded; and the systems employing the modified Spiral Array, which incorporated audio frequency features in its weights, becomes the most effective after 8 s, when more notes have been played.

We further provided detailed analyses on the case when all five audio systems gave the correct answer and on another case when all five systems failed. The result of the case studies presented some evidence against the summary statistics as a metric for key-finding, and demonstrated the specific advantages and disadvantages of the system extensions.

Each of the improvements proposed to the specific system addressed in this chapter can also be employed in other audio key-finding systems. For example, modifying an existing model with instrument-specific signals (as in the modified Spiral Array method of Sect. 11.5.1) can be applied to any other key matching algorithm, such as the Krumhansl-Schmuckler method; fundamental frequency identification (Sect. 11.5.2) can be applied to the transcription of monophonic bass instruments, such as cello, with appropriate harmonic templates; PWB (Sect. 11.5.3) can be directly plugged into any audio key-finding system for refining pitch-class distributions based on the key answer.

In this chapter, our test data comprised audio synthesized from MIDI (mostly symphonies and concertos), and audio from acoustic piano. Methods such as the modified Spiral Array (Sect. 11.5.1), and post-weight balancing (Sect. 11.5.3), and analyses of when audio outperforms MIDI key-finding (Sect. 11.4.2) can be highly dependent on the instrumental timbre of the training and test data. Timbre is a variable that requires systematic study in the future.

The constant Q transform is another option for extracting frequency information from audio signal. The constant Q transform is closely related to the FFT, except the frequency-resolution ratio remains constant. This constant ratio confers two major advantages to the constant Q: first, with a proper choice of center frequency, the output of the constant Q transform corresponds to musical notes; second, the constant Q transform uses higher time resolutions for higher frequencies, which better models the human auditory system. One of the design choices for the present audio key-finding system was to concentrate on lower frequencies, because bass notes present strong and stable cues for the key; thus, our decision to use the proposed fuzzy analysis technique to confirm lower frequencies using the overtones. In the future, we plan to explore other techniques using the constant Q transform for audio key-finding.

References

1. Chew, E.: Towards a mathematical model of tonality. Ph.D. thesis, Massachusetts Institute of Technology, Cambridge, MA (2000)
2. Chew, E.: Modeling tonality: applications to music cognition. In: Moore, J.D., Stenning, K. (eds.) Proceedings of the 23rd Annual Mtg of the Cognitive Science Society, Lawrence Erlbaum Associates Publishers, Edinburgh, pp. 206–211 (2001)
3. Chew, E., Chen, Y.-C.: Mapping MIDI to the Spiral Array: disambiguating pitch spellings. In: Computational Modeling and Problem Solving in the Networked World—Proceedings of the 8th INFORMS Computer Society Conference, Kluwer, pp. 259–275 (2003)
4. Chew, E., Chen, Y.-C.: Real-time pitch spelling using the Spiral Array. Comput. Music J. **29**(2), 61–76 (2005). doi:10.1162/0148926054094378
5. Chew, E., François, A.R.J.: Interactive multi-scale visualization of tonal evaluation in MuSA.RT Opus 2. ACM Comput. Entertain. **3**(4), 1–16 (2005). doi:10.1145/1095534.1095539
6. Chuan, C.-H., Chew, E.: Fuzzy analysis in pitch class determination for polyphonic audio key-finding. In: Proceedings of the 6th International Conference on Music Information Retrieval, UK, London, pp. 296–303 (2005)
7. Chuan, C.-H., Chew, E.: Polyphonic audio key-finding using the Spiral Array CEG algorithm. In: Proceedings of the IEEE International Conference on Multimedia and Expo, The Netherlands, Amsterdam, pp. 21–24 (2005)
8. G6mez, E., Herrera, P.: Estimating the tonality of polyphonic audio files: cognitive versus machine learning modeling strategies. In: Proceedings of the 5th International Conference on Music Information Retrieval, Barcelona, Spain (2004)
9. İzmirli, Ö.: Template based key-finding from audio. In: Proceedings of the International Computer Music Conference, Barcelona, Spain (2005)
10. Klapuri, A.: Multiple fundamental frequency estimation by harmonicity and spectral smoothness. IEEE Trans. Speech Audio Process. **11**(6), pp. 804–816 (2003). doi:10.1109/TSA.2003.815516
11. Klapuri, A.: A perceptually motivated multiple-F0 estimation method. In: Proceedings of IEEE Workshop on Applications of Signal Processing to Audio and Acoustics, New Paltz, New York, pp. 291–294 (2005). doi:10.1109/ASPAA.2005.1540227
12. Krumhansl, C.L.: Cognitive Foundations of Musical Pitch. Chapter 2, pp. 16–49. Oxford University Press, Oxford (1990)
13. Longuet-Higgins, H.C., Steedman, M.J.: On interpreting Bach. Mach. Intell. **6**, 221–241 (1971)
14. First Annual Music Information Retrieval Evaluation eXchange, MIREX (2005). http://www.music-ir.org/mirex/wiki/2005. Cited 5 July 2013

15. Contest Results—Audio Key-Finding, MIREX (2005). http://www.music-ir.org/evaluation/mirex-results/audio-key. Cited 5 July 2013
16. Pauws, S.: Musical key extraction from audio. In: Proceedings of the 5th International Conference on Music Information Retrieval, Barcelona, Spain (2004)
17. Temperley, D.: What's key for key? The Krumhansl-Schmuckler key-finding algorithm reconsidered. Music Percept. **17**(1), 65–100 (1999)
18. Electronic Music Studios in the University of Iowa http://theremin.music.uiowa.edu/MIS.html. Cited 8 July 2013

Appendix A
Model Calibration

Abstract This appendix describes a calibration of the Spiral Array model parameters. The parameters include the aspect ratio of the pitch class helix, the pitch weights in the major/minor chord definitions, and the chord weights in the major/minor key definitions. The parameters are progressively selected in accordance to principles of cognition of musical interval (pitch-to-pitch), pitch-to-chord, chord-to-chord, key-to-pitch, and interval-to-key relations. The constraints corresponding to the desired comparative distances are derived and solved analytically, with the exception of the final examples, which used a computational heuristic. While solutions are presented for a particular set of conditions, parameters can be generated in a similar or alternate fashion to conform to a different set of constraints.

The goal of this chapter is to present an approach to the selection of parameters for the spatial representations in the Spiral Array so that relations between represented tonal entities have direct counterparts in the geometric structure.

Distance in the model is used as a indicator of "closeness" between the tonal entities as interpreted by the mind when apprehending tonal music. Note that closeness between pitches in the Spiral Array is different from closeness between pitches sorted by frequency, for example, like the pitches on a piano keyboard. Proximity in the Spiral Array can be the result of, amongst other things, shared pitches, shared intervals, or tonal elements within a perfect fifth, major third or minor third interval of each other.

Because pitches, chords, and keys are represented in the Spiral Array, the model allows one to relate musical entities from different hierarchical levels, for example, pitches to pitches, pitches to chords, and keys to chords. Figure A.1 shows the relationships that can be assigned to, and inferred from, the Spiral Array.

This chapter is a minor revision of Distances (Chap. 4) of "Towards a Mathematical Modeling of Tonality" by E. Chew, an MIT PhD dissertation, Cambridge, Massachusetts (2000) https://dspace.mit.edu/handle/1721.1/9139.

E. Chew, *Mathematical and Computational Modeling of Tonality*, International Series in Operations Research & Management Science 204, DOI: 10.1007/978-1-4614-9475-1,

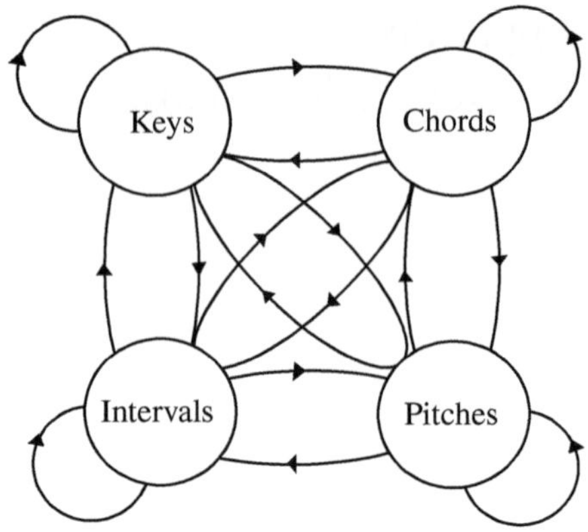

Fig. A.1 Relations that can be assigned to and inferred from the model

The Objects of Study
In this chapter, the objects of study are the parameters:

$$
\begin{aligned}
\mathrm{r} &= \text{radius of the spiral}\\
\mathrm{h} &= \text{height gain per quarter turn of the spiral}\\
w &= [w_1, w_2, w_3], \text{ weights on major chord pitches}\\
u &= [u_1, u_2, u_3], \text{ weights on minor chord pitches}\\
\omega &= [\omega_1, \omega_2, \omega_3], \text{ weights on major key chords}\\
\upsilon &= [\upsilon_1, \upsilon_2, \upsilon_3], \text{ weights on minor key chords}\\
\alpha &= \text{preference for V versus v chord in minor keys}\\
\beta &= \text{preference for iv versus IV chord in minor keys.}
\end{aligned}
$$

The formal definitions of chord and key representations in the Spiral Array space are presented in Chap. 3. The definitions constrain, in a useful manner, some of the relations between the tonal elements. In order for the model to reflect the cognition of tonal relations more precisely, a few more constraints have to be added.

In this chapter, I demonstrate the selection of model parameters based on one possible set of basic criteria. By assigning parameters that satisfy the constraints that correspond to these criteria, the Spiral Array aims to model tonal concepts in accordance with conceptual tonal pitch relations in western classical music.

A.1 Interval Relations

In this section, I examine the distance relations between pairs of pitch representations, and calibrate the model so that pitch proximity reflects one set of perceived interval relations in western tonal music.

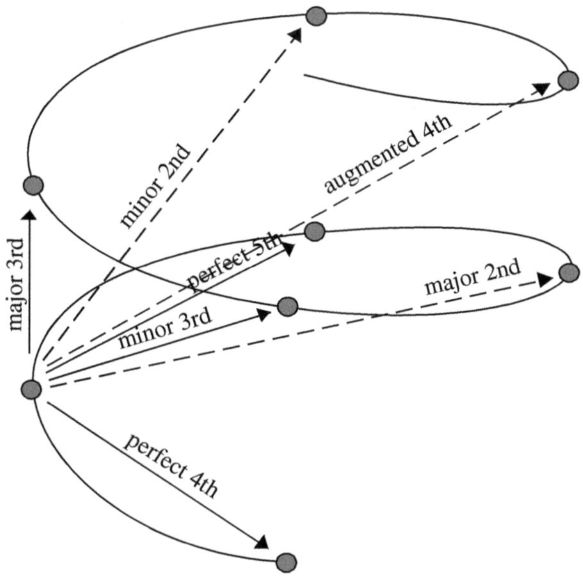

Fig. A.2 Intervals represented on the Spiral Array

The Aspect Ratio

The pitch positions are completely defined in the Spiral Array structure by the radius of the spiral and the vertical elevation at each step; these parameters are denoted by the variables r and h respectively. As a result, the aspect ratio h/r determines the distance relationships among the pitch positions. The design goal is that perceptually close intervals should be represented by shorter inter-pitch distances.

Scalar Versus Vector Representation of Intervals

In the Spiral Array, scalar distance serves as the analogue for conceptual distance between pitch pairs, which form intervals. As mentioned in Sect. 3.1, intervals could be represented by vectors, where vectors identical under rotation about the central axis and vertical translation would belong to the same equivalence class. In this section, I consider only the scalar representation of intervallic distance. Figure A.2 shows, geometrically, some interval representations. Next, I describe how the parameters can be chosen so that inter-pitch distances in the model preserve a particular understanding of interval preferences common to listeners of western tonal music.

A.1.1 Interval Preferences

In western tonal music, the interval of a fifth is considered the closest harmonic relationship between distinct pitches. Thus, the model requires that pitches related by a perfect fifth should be nearest each other. Other considerations are: pitches a

Table A.1 Distances associated with interval relations

Interval		Distance^2
Name	Class	
Perfect fifth, perfect fourth	(P4, P5)	$2r^2 + h^2$
Major third, minor sixth	(M3, m6)	$16h^2$
Minor third, major sixth	(m3, M6)	$2r^2 + 9h^2$
Major second, minor seventh	(M2, m7)	$4r^2 + 4h^2$
Minor second, major seventh	(m2, M7)	$2r^2 + 25h^2$
Diminished fifth, augmented fourth	(d5, A4)	$4r^2 + 36h^2$

third apart are more closely related than those a second apart; the major is regarded to be closer than the minor for both thirds and seconds; and, the tritone (diminished fifth or augmented fourth) is perceived to be more distant than these other interval classes.[1] These are the conditions against which I will calibrate the model.

Lemma 1 *Consider the interval classes: {(P4, P5), (M3, m6), (m3, M6), (M2, m7), (m2, M7), (d5, A4)}. Inherent in the model are the following interval distance relationships: [1] the (P4, P5) interval class is represented by distances closer than all but the (M3, m6) interval class; [2] (M3/m6) and (m3/M6) are both closer than the (m2/M7) interval class; and, [3] (d5/A4) is the farthest of all.*

[Proof]: The relationships follow naturally from the distances representing each interval class as given in Table A.1. □

Theorem 5 *In order that: [1] the closest distance between any two pitch positions denote a perfect fifth; [2] pitches a third apart are closer than those a second apart; [3] a major interval is closer than a minor; and [4] a tritone is more distant than any of the above, the aspect ratio h/r should be constrained so that:*

$$\frac{2}{15} < a < \frac{2}{7} \quad \text{where } a = \frac{h^2}{r^2}. \tag{A.1}$$

[Proof]: Condition [4] is always satisfied, according to Condition [3] in Lemma 1. The following inequality summarizes the remaining conditions:

$$\text{dist(P4, P5)} < \text{dist(m3, M6)} < \text{dist(M2, m7)} < \text{dist(m2, M7)}. \tag{A.2}$$

The first inequality in Eq. A.2 becomes:

[1] By *interval classes* I mean intervals which are equivalent under transformation. For example, a minor sixth interval is the same as a major third interval under inversion; and, a compound interval such as a major tenth, when transposed to the same octave, is equivalent to a major third. The naming convention used here is: P = perfect, M = major, m = minor, d = diminished, A = augmented.

$$\begin{aligned}
\text{dist}(\text{P4}, \text{P5}) &< \text{dist}(\text{M3}, \text{m6}) \\
\Rightarrow 2\text{r}^2 + \text{h}^2 &< (4\text{h})^2 \\
\Rightarrow 2\text{r}^2 &< 15\text{h}^2,
\end{aligned}$$

and, the second inequality implies that:

$$\begin{aligned}
\text{dist}(\text{M3}, \text{m6}) &< \text{dist}(\text{m3}, \text{M6}) \\
\Rightarrow (4\text{h})^2 &< 2\text{r}^2 + 9h^2 \\
\Rightarrow 7\text{h}^2 &< 2\text{r}^2.
\end{aligned}$$

The next two inequalities follow from the first two:

$$\begin{aligned}
\text{dist}(\text{m3}, \text{M6}) &= 2\text{r}^2 + 9\text{h}^2 = 2\text{r}^2 - 7\text{h}^2 + 16\text{h}^2 \\
&< 2\text{r}^2 - 7\text{h}^2 + 2\text{r}^2 + 9\text{h}^2 = 4\text{r}^2 + 2\text{h}^2, \text{ by the second inequality} \\
&< 4\text{r}^2 + 4\text{h}^2 = \text{dist}(\text{M2}, \text{m7}),
\end{aligned}$$

$$\begin{aligned}
\text{dist}(\text{M2}, \text{m7}) &= 4\text{r}^2 + 4\text{h}^2 \\
&< 2\text{r}^2 + 15\text{h}^2 + 4\text{h}^2, \text{ by the first inequality} \\
&< 2\text{r}^2 + 25\text{h}^2 = \text{dist}(\text{m2}, \text{M7}).
\end{aligned}$$

Hence, the inequalities combine to give the desired constraint on the aspect ratio:

$$0.3651 = \sqrt{\frac{2}{15}} < \frac{\text{h}}{\text{r}} < \sqrt{\frac{2}{7}} = 0.5345.$$

□

A.2 Major Chord–Chord Pitch Relations

In this section, I derive the conditions on the aspect ratio and the triad weights so that chord pitch to chord distances in the Spiral Array correspond to a set of simple relations.

Desired Distance Relationship
Consider the three pitches that constitute a major triad: its root, fifth and third. In conventional tonal harmony, theorists consider the root the most important pitch in the chord, followed by the fifth, then the third. The goal is to restrict the parameters so that each chord center is closest to its root, followed by the fifth, then the third.

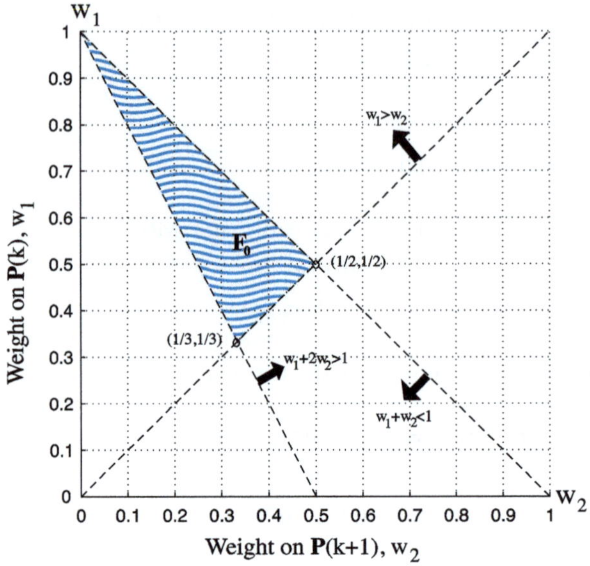

Fig. A.3 Feasible values for major chord weights, $w_1, w_2, w_3 (= 1 - w_1 - w_2)$

Major triad weights are given priority in the selection process since the behavior of major modes is better understood; the minor mode is considered by many to be derived from the major.

Definition of Major Chord Weights

Definition 3 in Chap. 3 specifies the mathematical equations for chords in the Spiral Array. By design, as reflected in the weights, the pitch with greatest influence on each chord representation is its root, followed by the fifth, then the third. Consider the constraints on the major chord weights as given by the definition in Eq. (3.4):

$$w_1 + w_2 + w_3 = 1$$

and $$w_1 > w_2 > w_3 = 1 - w_1 - w_2 > 0 \text{ (from the above line).}$$

Since w_3 can be written as a function of w_1 and w_2, the above expressions imply the following system of three inequalities on (w_1, w_2):

$$w_1 > w_2, \tag{A.3}$$

$$w_1 > -2w_2 + 1, \tag{A.4}$$

and $$w_1 < -w_2 + 1, \tag{A.5}$$

which are shown graphically on Fig. A.3. The shaded area, $\mathbf{F}_0$, shows the intersection of the three constraints. A similar set of inequalities applies to the weights on minor chords.

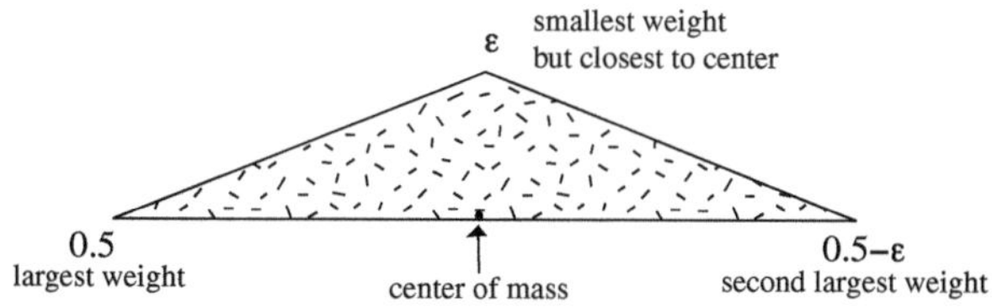

Fig. A.4 An example in which the point with least weight is closest to the center of effect (ε is a small number)

Reason for Further Analysis
Having the heaviest assigned weight does not guarantee that the root will be nearest to the chord representation. See Fig. A.4 for an example in which the point with least weight is actually closest to the center of effect. Thus, new constraints have to be derived so that proximity corresponds to cognition of tonal relations.

A.2.1 Constraints on Major Chord Weights

Allowing the major chord weights their full range of values within $\mathbf{F}_0$, shown in Fig. A.3, I derive the conditions on the aspect ratio so that the representations of the major triad's root, fifth and third are closer than other pitches to the triad's center.

First, I show that the condition from Theorem 5, $\sqrt{2/15} < \mathrm{h/r} < \sqrt{2/7}$, is a necessary and sufficient condition for the major chord's representation to be closest to its root, followed by the fifth, then the third (Theorem 6). Surprisingly, the constraint on the aspect ratio is identical for both Theorem 5 and Theorem 6. Thus, in the Spiral Array, restricting the perfect fifth to be the closest relation, followed by the major third, minor third, major second and minor second, is tantamount to constraining the major chord representation to be closest to the root, followed by the fifth, then the third.

In the next section, I derive the new constraints on the aspect ratio so that no other pitches are as close to the major chord center as its three constituent pitches.

Theorem 6 *A major chord's representation is closest to its root, followed by its fifth, then its third for all $(w_1, w_2) \in \mathbf{F}_0$ if and only if*

$$\frac{2}{15} < a < \frac{2}{7}, \quad where \quad a = \frac{h^2}{r^2}.$$

[Proof]: The goal is to find the aspect ratio values that satisfy the theorem conditions for all possible choice of (w_1, w_2) in the feasible region, $\mathbf{F}_0$. The theorem states that: for all $(w_1, w_2) \in \mathbf{F}_0$,

$$\frac{2}{15} < a < \frac{2}{7} \iff \text{dist(chord, root)} < \text{dist(chord, fifth)} < \text{dist(chord, third)},$$
$$\text{i.e. } \|\mathbf{C}_{\mathrm{M}}(k) - \mathbf{P}(k)\| < \|\mathbf{C}_{\mathrm{M}}(k) - \mathbf{P}(k+1)\| < \|\mathbf{C}_{\mathrm{M}}(k) - \mathbf{P}(k+4)\|.$$

Due to the model's symmetry, it is sufficient to simply consider the case of the C major chord, $\mathbf{C}_{\mathrm{M}}(0)$. The related fifth and third are $\mathbf{P}(1)$ and $\mathbf{P}(4)$ respectively. Their spatial positions are given as

$$\mathbf{P}(0) = \begin{bmatrix} 0 \\ \mathrm{r} \\ 0 \end{bmatrix}, \mathbf{P}(1) = \begin{bmatrix} \mathrm{r} \\ 0 \\ \mathrm{h} \end{bmatrix}, \text{ and } \mathbf{P}(4) = \begin{bmatrix} 0 \\ \mathrm{r} \\ 4\mathrm{h} \end{bmatrix}. \tag{A.6}$$

By definition (see Eq. 3.4), the C major chord position is located at:

$$\mathbf{C}_{\mathrm{M}}(0) = w_1 \cdot \mathbf{P}(0) + w_2 \cdot \mathbf{P}(1) + w_3 \cdot \mathbf{P}(4)$$
$$= w_1 \cdot \begin{bmatrix} 0 \\ \mathrm{r} \\ 0 \end{bmatrix} + w_2 \cdot \begin{bmatrix} \mathrm{r} \\ 0 \\ \mathrm{h} \end{bmatrix} + w_3 \cdot \begin{bmatrix} 0 \\ \mathrm{r} \\ 4\mathrm{h} \end{bmatrix} \tag{A.7}$$
$$= \begin{bmatrix} w_2 \cdot \mathrm{r} \\ (w_1 + w_3) \cdot \mathrm{r} \\ (w_2 + 4w_3) \cdot \mathrm{h} \end{bmatrix}. \tag{A.8}$$

Define a new variable for the standardized distance between $\mathbf{C}_{\mathrm{M}}(0)$ and a pitch:

$$y_{\mathrm{M}}(s) \stackrel{\mathrm{def}}{=} \frac{1}{\mathrm{r}^2} \|\mathbf{C}_{\mathrm{M}}(0) - \mathbf{P}(s)\|^2 \tag{A.9}$$

The standardized distances of each chord pitch from the chord are given by:

$$y_{\mathrm{M}}(0) = \frac{1}{\mathrm{r}^2} \|\mathbf{C}_{\mathrm{M}}(0) - \mathbf{P}(0)\|^2$$
$$= w_2^2 + (w_1 + w_3 - 1)^2 + (w_2 + 4w_3)^2 \cdot a$$
$$= 2 \cdot w_2^2 + (4 - 3w_2 - 4w_1)^2 \cdot a, \tag{A.10}$$

$$y_{\mathrm{M}}(1) = \frac{1}{\mathrm{r}^2} \|\mathbf{C}_{\mathrm{M}}(0) - \mathbf{P}(1)\|^2$$
$$= (w_2 - 1)^2 + (w_1 + w_3)^2 + (w_2 + 4w_3 - 1)^2 \cdot a$$
$$= 2 \cdot (1 - w_2)^2 + (3 - 3w_2 - 4w_1)^2 \cdot a, \tag{A.11}$$

and

$$y_{\mathrm{M}}(4) = \frac{1}{\mathrm{r}^2} \|\mathbf{C}_{\mathrm{M}}(0) - \mathbf{P}(4)\|^2$$
$$= w_2^2 + (w_1 + w_3 - 1)^2 + (w_2 + 4w_3 - 4)^2 \cdot a$$
$$= 2 \cdot w_2^2 + (3w_2 + 4w_1)^2 \cdot a. \tag{A.12}$$

The desired proximity conditions with respect to the chord center, $\mathbf{C}_{\mathrm{M}}(0)$, are that the root, $\mathbf{P}(0)$, should be closer than the fifth, $\mathbf{P}(1)$, which is in turn closer than the third, $\mathbf{P}(4)$. Mathematically, this means that:

$$\|\mathbf{C}_{\mathrm{M}}(0) - \mathbf{P}(0)\|^2 < \|\mathbf{C}_{\mathrm{M}}(0) - \mathbf{P}(1)\|^2 \iff y_{\mathrm{M}}(0) < y_{\mathrm{M}}(1) \quad \text{(A.13)}$$

$$\text{and} \quad \|\mathbf{C}_{\mathrm{M}}(0) - \mathbf{P}(1)\|^2 < \|\mathbf{C}_{\mathrm{M}}(0) - \mathbf{P}(4)\|^2 \iff y_{\mathrm{M}}(1) < y_{\mathrm{M}}(4). \quad \text{(A.14)}$$

These inequalities, simplified, yield the following constraints:

[1] Substituting Eqs. A.10 and A.11 into Eq. A.13, one gets

$$2 \cdot w_2^2 - 2 \cdot (1 - w_2)^2 + (4 - 3w_2 - 4w_1)^2 \cdot a - (3 - 3w_2 - 4w_1)^2 \cdot a < 0$$
$$-2 + 7a + (4 - 6a) \cdot w_2 - 8aw_1 < 0$$

$$\iff w_1 > \left(\frac{4-6a}{8a}\right) \cdot w_2 - \left(\frac{2-7a}{8a}\right)$$
$$= a_0 \cdot w_2 + \frac{1}{2}(1 - a_0), \quad \text{where } a_0 = \frac{1}{2a} - \frac{3}{4}. \quad \text{(A.15)}$$

[2] The result of substituting Eqs. A.11 and A.12 into Eq. A.14 is

$$2 \cdot (1 - w_2)^2 - 2 \cdot w_2^2 + (3 - 3w_2 - 4w_1)^2 \cdot a - (3w_2 + 4w_1)^2 \cdot a < 0$$
$$(2 + 9a) - (4 + 18a) \cdot w_2 - 24a \cdot w_1 < 0$$

$$\iff w_1 > -\left(\frac{4+18a}{24a}\right) \cdot w_2 + \left(\frac{2+9a}{24a}\right)$$
$$= -a_1 \cdot w_2 + \frac{1}{2}a_1, \quad \text{where } a_1 = \frac{3}{4} + \frac{1}{6a}. \quad \text{(A.16)}$$

Hence, the distance constraints given in Eqs. A.13 and A.14 are satisfied if and only if Eqs. A.15 and A.16 are true. The two latter inequalities (A.15 and A.16) relate the aspect ratio to the chord weights so that the chord representations are closest to the root, followed by the fifth, then the third.

Graphically, these conditions are satisfied if $\mathbf{F}_1$, the feasible region defined by Eqs. A.15 and A.16, contains $\mathbf{F}_0$. This relationship is illustrated in Fig. A.5. The line generated by Eq. A.15 has a fixed point at $w_2 = 1/2, w_1 = 1/2)$, a vertex of the triangular region $\mathbf{F}_0$, as shown in Fig. A.5. The line passes through this point for all values of a_0, the slope of the line. This line forms one of the boundaries for the feasible region, $\mathbf{F}_1$. The corresponding line from Eq. A.16 has a fixed point at $(w_2 = 1/2, w_1 = 0)$, passing through this point for all values of a_1. a_1 is the slope of the line that pivots on this fixed point. This line forms the other boundary for the feasible region, $\mathbf{F}_1$.

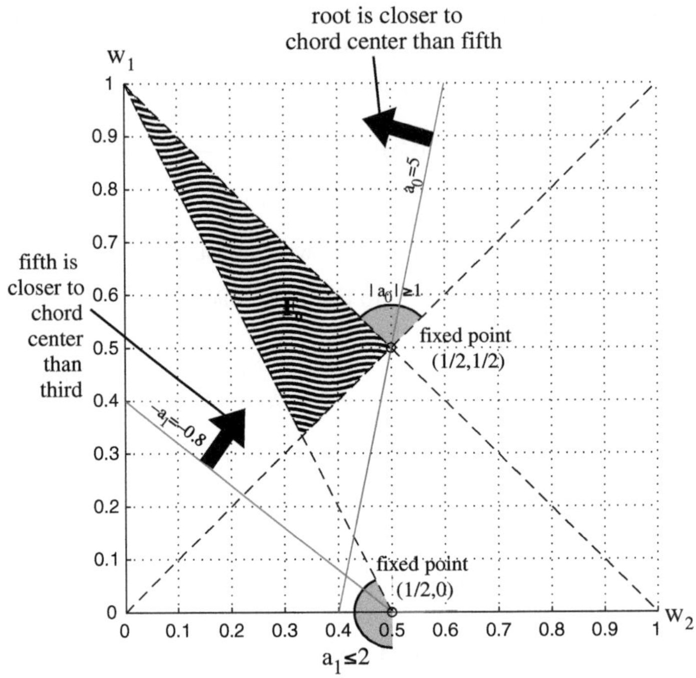

Fig. A.5 Feasible values for major chord weights, $w_1, w_2, w_3 (= 1 - w_1 - w_2)$, based on the desired proximity relations between a major chord and its component pitches

$\mathbf{F}_0$ will always be contained in $\mathbf{F}_1$ if and only if

$$|a_0| \geq 1 \quad \text{and} \quad a_1 \leq 2.$$

Figure A.5 gives a graphical interpretation. Solving for a:

$$\begin{aligned} a_0 = \frac{1}{2a} - \frac{3}{4} \geq 1 \text{ or } a_0 &= \frac{1}{2a} - \frac{3}{4} \leq -1 \\ \frac{1}{2a} \geq \frac{7}{4} \text{ or } \frac{1}{2a} &\leq -\frac{1}{4} \\ a \leq \frac{2}{7} \text{ or } a &\leq -2 \text{ (an impossibility)} \\ a &\leq \frac{2}{7}, \end{aligned} \tag{A.17}$$

and

$$\begin{aligned} a_1 &= \frac{3}{4} + \frac{1}{6a} \leq 2 \\ \frac{1}{6a} &\leq \frac{5}{4} \\ a &\geq \frac{2}{15}. \end{aligned} \tag{A.18}$$

Combining Eqs. A.17 and A.18:

$$\mathbf{F}_0 \text{ is a subset of } \mathbf{F}_1 \iff \frac{2}{15} \leq a \leq \frac{2}{7},$$

which is the very same constraint as that given in Theorem 5! □

A.3 Major Chord-All Pitch Relations

The theorem in this section derives the conditions on the aspect ratio so that no other pitches are as close to the major chord center as its three constituent pitches. The theorem is preceded by four lemmas that consider the proximity, to the C major triad $\mathbf{C}_M(0)$, of each of the four types of pitches, $\mathbf{P}(4n)$, $\mathbf{P}(4n+1)$, $\mathbf{P}(4n+2)$ and $\mathbf{P}(4n+3)$, where n is an integer.

Lemma 2 *Consider pitches of the type* $\mathbf{P}(4n)$*, which correspond to { ..., F♭, A♭, C, E, G♯, ...} and so on.* $\mathbf{P}(0)$ *is closest to* $\mathbf{C}_M(0)$*, i.e. pitch C is closest to chord C, followed by* $\mathbf{P}(4)$ *(pitch E) for all* $(w_1,w_2) \in \mathbf{F}_0$ *if and only if* $a < 2/7$.

[Proof]: By Eqs. 3.2 and A.8:

$$\begin{aligned} y_M(4n) &= \frac{1}{r^2}\|\mathbf{C}_M(0) - \mathbf{P}(4n)\|^2 \\ &= 2 \cdot w_2^2 + (4w_1 + 3w_2 + 4n - 4)^2 \cdot a \\ &= 2 \cdot w_2^2 + 16a \cdot \left(n - \frac{4 - 4w_1 - 3w_2}{4}\right)^2. \end{aligned} \tag{A.19}$$

Consider the function $(4w_1 + 3w_2)$. For all $(w_1, w_2) \in \mathbf{F}_0$, as shown in Fig. A.3, the function is minimized at $(1/3, 1/3)$ and maximized at $(1, 0)$ to give the range $7/3 < 4w_1 + 3w_2 < 4$.

Relaxing the integer constraint on n, the quadratic function where n is not constrained to be integer, $\tilde{y}(4n)$ has a minimum at $\tilde{n}^* = \frac{1}{4}(4 - 4w_1 - 3w_2)$. But $0 < \frac{1}{4}(4 - 4w_1 - 3w_2) < 5/12$.

Now, let $n_{(k)}$ denote the kth least value of $y_M(s)$. Observe, in Fig. A.6, that because $0 < \frac{1}{4}(4 - 4w_1 - 3w_2) < 5/12$, $\tilde{n}^*$ lies between 0 and 5/12. Hence, the minimum value for $y_M(4n)$ occurs when $n_{(1)} = 0$ at $y_M(0)$, followed by the second least value at $n_{(2)} = 1$ at $y_M(4)$.

In general, $y_M(4n) < y_M(-4n)$ and $y_M(4n) < y_M(4(n+1))$ for non-negative n as shown in Fig. A.6. Therefore,

$$\|\mathbf{C}_M(0) - \mathbf{P}(0)\|^2 < \|\mathbf{C}_M(0) - \mathbf{P}(4)\|^2 < \|\mathbf{C}_M(0) - \mathbf{P}(4n)\|^2, \quad \forall n \in \mathbb{Z} \text{ and } n \neq 0, 1.$$

□

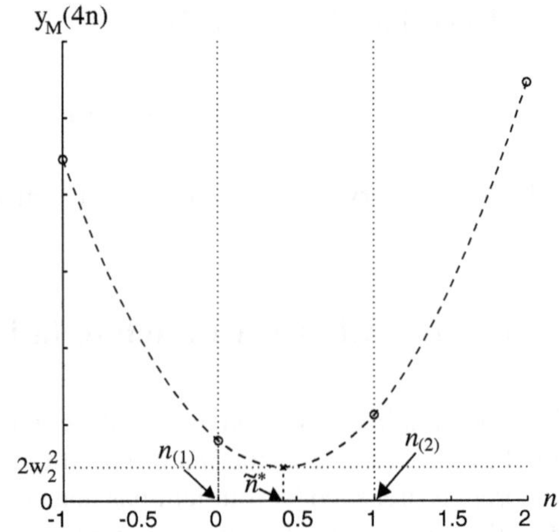

Fig. A.6 A graph explaining the choice of the integer $n = n_{(1)}$ that minimizes the function $y_M(4n) = \frac{1}{r^2}\|\mathbf{C}_M(0) - \mathbf{P}(4n)\|^2$; r = 1 in the plot

Lemma 3 *Consider pitches of the type* **P**$(4n+1)$*, which correspond to* { ..., C♭,E♭, G, B, D♯, ... }. **P**(1) *is closest to* $\mathbf{C}_M(0)$*, i.e. pitch G is closest to chord C. The second closest pitch is farther than* **P**(4) *(pitch E, the third in the triad) for all* $(w_1, w_2) \in \mathbf{F}_0$ *if and only if* $a < 2/7$*. When* $4w_1 + 3w_2 > 3$*, this second closest pitch is* $\mathbf{P}(-3)$ *(pitch E♭); when* $4w_1 + 3w_2 < 3$*, the second closest pitch is* **P**(5) *(pitch B), which is always farther than* **P**(4).

[Proof]: By Eqs. 3.2 and A.8:

$$\begin{aligned} y_M(4n+1) &= \frac{1}{r^2}\|\mathbf{C}_M(0) - \mathbf{P}(4n+1)\|^2 \\ &= 2\cdot(1-w_2)^2 + (4w_1 + 3w_2 - 3 + 4n)^2 \cdot a. \end{aligned} \tag{A.20}$$

Getting the expression for $y_M(4)$ from Eq. A.19,

$$\begin{aligned} &y_M(4n+1) - y_M(4) \\ &= 2\cdot(1-w_2)^2 + (4w_1+3w_2-3+4n)^2\cdot a - 2\cdot w_2^2 - (4w_1+3w_2)^2\cdot a \\ &= 2 - 4w_2 + a\cdot(-3+4n)(8w_1+6w_2-3+4n) \\ &= 2 - 4w_2 + 16a\cdot\left(n - \frac{3}{4}\right)\left(n - \frac{3-8w_1-6w_2}{4}\right). \end{aligned}$$

Because the weights are constrained to be in the feasible region $\mathbf{F}_0$ in Fig. A.3, we know that $-15/12 < \frac{1}{4}(3 - 8w_1 - 6w_2) < -5/12$, that is to say, $-15/12 < \tilde{n}^* < -5/12$. Consequently, $n_{(1)} = 0$. According to Theorem 6, $y_M(0) < y_M(1) < y_M(4)$.

Depending on the weights (w_1, w_2), the second closest pitch can be indexed either $n_{(2)} = -1$ or $n_{(2)} = 1$, that is to say, the pitch can be either $\mathbf{P}(-3)$ or **P**(5). The boundary for this change occurs at $\frac{1}{4}(3 - 8w_1 - 6w_2) = -\frac{3}{4}$. This boundary is shown in Fig. A.7.

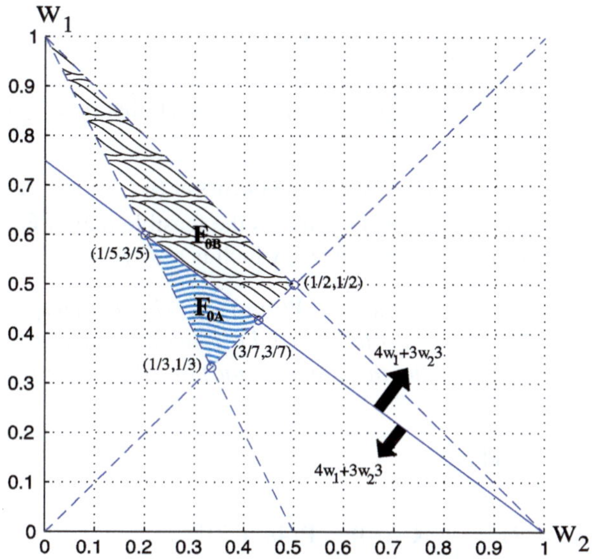

Fig. A.7 Feasible values for major chord weights, $w_1, w_2, w_3(= 1 - w_1 = w_2)$, and the boundary, $4w_1 + 3w_2 = 3$, between two subsets of weights in the analysis of $y_M(4n+1) - y_M(4)$

When $\frac{1}{4}(3 - 8w_1 - 6w_2) \leq -\frac{3}{4}$, $n_{(2)} = -1$:

$$y_M(-3) - y_M(4) = 2 - 4w_2 + 16a \cdot \left(-\frac{7}{4}\right)\left(-1 - \frac{3 - 8w_1 - 6w_2}{4}\right)$$
$$= 2 + 49a - (4 + 42a) \cdot w_2 - 56a \cdot w_1 > 0$$

$$\Longleftrightarrow \quad w_1 < -\left(\frac{4 + 42a}{56a}\right) \cdot w_2 + \left(\frac{2 + 49a}{56a}\right)$$
$$= -a_3 \cdot w_2 + \frac{1}{2}(a_3 + 1), \qquad \text{where } a_3 = \frac{1}{14a} + \frac{3}{4}.$$

Relating this inequality to the graph in Fig. A.7, the line $w_1 = -a_3 \cdot w_2 + \frac{1}{2}(a_3 + 1)$ has a fixed point at $(w_1, w_2) = (1/2, 1/2)$. The inequality $w_1 < -a_3 \cdot w_2 + \frac{1}{2}(a_3 + 1)$ contains the feasible region $\mathbf{F}_{0B}$ if and only if $a_3 > 1$, which means that $a < \frac{2}{7}$.

When $\frac{1}{4}(3 - 8w_1 - 6w_2) > -\frac{3}{4}$, $n_{(2)} = 1$.

$$y_M(5) - y_M(4) = 2 - 4w_2 + 16a \cdot \left(\frac{1}{4}\right)\left(1 - \frac{3 - 8w_1 - 6w_2}{4}\right)$$
$$= 2 + a + 8a \cdot w_1 - (4 - 6a) \cdot w_2$$

$$\iff w_1 > \left(\frac{4-6a}{8a}\right)\cdot w_2 - \left(\frac{2+a}{8a}\right)$$
$$= a_0 \cdot w_2 - \frac{1}{2}(a_0+1), \qquad \text{recall that } a_0 = \frac{1}{2a} - \frac{3}{4}.$$

The fixed point for $w_1 = a_0 \cdot w_2 - \frac{1}{2}(a_0+1)$ occurs at $(w_1, w_2) = (1/2, -1/2)$. The inequality $w_1 > a_0 \cdot w_2 - \frac{1}{2}(a_0+1)$ contains the feasible region $\mathbf{F}_{0A}$ if and only if $a_0 < -11/3$. This means that $a > -6/35$, which is always true. □

Lemma 4 *Consider pitches of the type* $\mathbf{P}(4n+2)$*, which correspond to { ..., G♭, B♭, D, F♯, A♯, ...}.* $\mathbf{P}(2)$ *is closest to* $\mathbf{C}_M(0)$*, i.e. pitch D is closest to chord C.* $\mathbf{P}(2)$ *is farther than* $\mathbf{P}(4)$ *(pitch E, the third in the triad) for all* $(w_1, w_2) \in \mathbf{F}_0$ *if and only if* $a < 3/15$.

[Proof]: By Eqs. 3.2 and A.8:

$$y_M(4n+2) = \frac{1}{r^2}\|\mathbf{C}_M(0) - \mathbf{P}(4n+2)\|^2$$
$$= 2w_2^2 - 4w_2 + 4 + (4w_1 + 3w_2 - 2 + 4n)^2 \cdot a, \qquad \text{(A.21)}$$

$$y_M(4n+2) - y_M(4)$$
$$= 2w_2^2 - 4w_2 + 4 + (4w_1 + 3w_2 - 2 + 4n)^2 \cdot a - 2w_2^2 - (4w_1 + 3w_2)^2 \cdot a$$
$$= 4(1-w_2) + 16a \cdot \left(n - \frac{1}{2}\right)\left(n - \frac{2 - 8w_1 - 6w_2}{4}\right).$$

Since $-3/2 < \frac{1}{4}(2 - 8w_1 - 6w_2) < -2/3$, the index of the distance-minimizing pitch, $n_{(1)} = 0$ for all $(w_1, w_2) \in \mathbf{F}_0$.

$$y_M(2) - y_M(4) = 4(1-w_2) + 16a \cdot \left(-\frac{1}{2}\right)\left(-\frac{2 - 8w_1 - 6w_2}{4}\right)$$
$$= 4(1+a) - (4+12a)\cdot w_2 - 16a \cdot w_1 > 0,$$

$$\iff w_1 < -\left(\frac{1+3a}{4a}\right)\cdot w_2 + \left(\frac{1+a}{4a}\right)$$
$$= -a_4 \cdot w_2 + \left(a_4 - \frac{1}{2}\right), \qquad \text{where } a_4 = \frac{1}{4a} + \frac{3}{4}.$$

The line $w_1 = -a_4 \cdot w_2 + (a_4 - \frac{1}{2})$ has a fixed point at $(w_1, w_2) = \left(-\frac{1}{2}, 1\right)$. In order that $w_1 < -a_4 \cdot w_2 + (a_4 - \frac{1}{2})$ contains $\mathbf{F}_0$, $a_4 > 2$. This is true if and only if $a < 3/15$. □

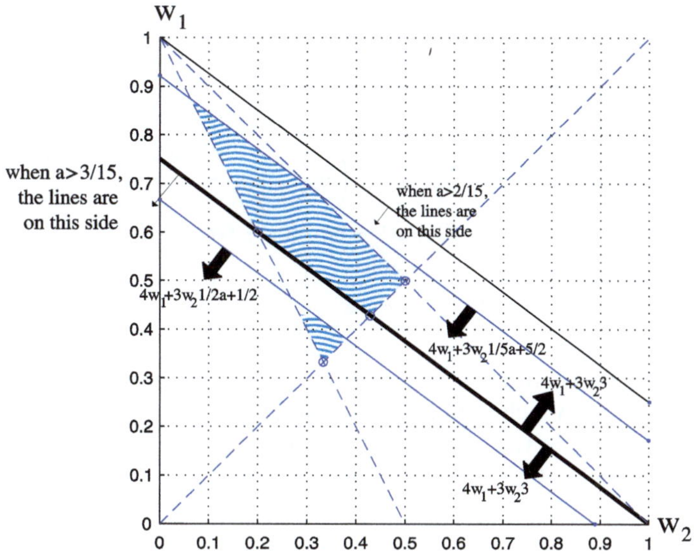

Fig. A.8 Feasible values for major chord weights, $w_1, w_2, w_3 (= 1 - w_1 - w_2)$, based on proximity relations between a major chord and all pitches

Lemma 5 *Consider pitches of the type* $\mathbf{P}(4n+3)$*, which correspond to* { ..., D♭, F, A, C♯, E♯, ... }. *For all* $(w_1, w_2) \in \mathbf{F}_0$*, when* $4w_1 + 3w_2 < 3$*,* $\mathbf{P}(3)$ *(pitch A) is closest to* $\mathbf{C}_\mathrm{M}(0)$*, the C major triad; and, when* $4w_1 + 3w_2 > 3$*,* $\mathbf{P}(-1)$ *(pitch F) is closest to* $\mathbf{C}_\mathrm{M}(0)$*. In either case, the closest pitch is still farther than* $\mathbf{P}(4)$ *(pitch E, the third in the triad) if and only if*

$$3 < 4w_1 + 3w_2 < \frac{1}{5a} + \frac{5}{2} \quad \textit{and} \quad 4w_1 + 3w_2 < \frac{1}{2a} + \frac{1}{2}.$$

[Proof]: By Eqs. 3.2 and A.8:

$$\begin{aligned} y_\mathrm{M}(4n+3) &= \frac{1}{\mathrm{r}^2} \| \mathbf{C}_\mathrm{M}(0) - \mathbf{P}(4n+3) \|^2 \\ &= 2 + 2w_2^2 + (4w_1 + 3w_2 - 1 + 4n)^2 \cdot a. \end{aligned} \tag{A.22}$$

$$\begin{aligned} & y_\mathrm{M}(4n+3) - y_\mathrm{M}(4) \\ &= 2 + 2w_2^2 + (4w_1 + 3w_2 - 1 + 4n)^2 \cdot a - 2w_2^2 - (4w_1 + 3w_2)^2 \cdot a \\ &= 2 + 16a \cdot \left(n - \frac{1}{4}\right)\left(n - \frac{1 - 8w_1 - 6w_2}{4}\right). \end{aligned}$$

In the possible range, $-7/4 < \frac{1}{4}(1-8w_1-6w_2) < 1/12$, when $\frac{1}{4}(1-8w_1-6w_2) > -5/4$, $n_{(1)} = 0$; and when $\frac{1}{4}(1-8w_1-6w_2) < -5/4$, $n_{(1)} = -1$.

When $\frac{1}{4}(1-8w_1-6w_2) > -5/4$,

$$\begin{aligned} y_M(3) - y_M(4) &= 2 + 16a \cdot \left(-\frac{1}{4}\right)\left(-\frac{1-8w_1-6w_2}{4}\right) \\ &= (2+a) - 8a \cdot w_1 - 6a \cdot w_2 > 0 \end{aligned}$$

$$\iff 4w_1 + 3w_2 < \frac{2+a}{2a} = \frac{1}{2a} + \frac{1}{2}.$$

Note that when $(2+a)/2a > 3$ ($\iff a < 3/15$), the gap between the two shaded areas in Fig. A.8 close to form one contiguous feasible region. In this case, $4w_1 + 3w_2 < 1/(2a) + 1/2$ is not a binding constraint.

When $\frac{1}{4}(1-8w_1-6w_2) < -5/4$,

$$\begin{aligned} y_M(-2) - y_M(4) &= 2 + 16a \cdot \left(-1-\frac{1}{4}\right)\left(-1-\frac{1-8w_1-6w_2}{4}\right) \\ &= 2 + 5a \cdot (5 - 8w_1 - 6w_2) > 0, \end{aligned}$$

$$\iff 4w_1 + 3w_2 < \frac{1}{5a} + \frac{5}{2}.$$

Note that when $a = 2/15$, this is not a binding constraint. Figure A.8 shows the two regions defined by these constraints. □

Theorem 7 *No other pitches are as close to a major chord's representation as its three constituent pitches if* $2/15 < a < 3/15$ *and* $4w_1 + 3w_2 < 1/(5a) + 5/2$.

[Proof]: Note that the chord pitch farthest from the chord representation is the third of the chord (by Theorem 6), so that for all $s \neq 0, 1, 4$:

$$\begin{aligned} &\|\mathbf{C}_M(\mathrm{k}) - \mathbf{P}(\mathrm{k}+\mathrm{s})\| \\ &> \max\{\|\mathbf{C}_M(\mathrm{k}) - \mathbf{P}(\mathrm{k})\|, \|\mathbf{C}_M(\mathrm{k}) - \mathbf{P}(\mathrm{k}+1)\|, \|\mathbf{C}_M(\mathrm{k}) - \mathbf{P}(\mathrm{k}+4)\|\} \\ &= \|\mathbf{C}_M(\mathrm{k}) - \mathbf{P}(\mathrm{k}+4)\|. \end{aligned}$$

Thus, one needs only prove that all non-chord pitches are farther from the chord than its third. By symmetry, one needs only show this for C major, i.e., $\mathbf{C}_M(0)$.

By Lemma 2, $a < 2/7$ is a necessary and sufficient condition for all $\mathbf{P}(4n)$ to be farther than $\mathbf{P}(4)$ from $\mathbf{C}_M(0)$, for $n \neq 0, 1$.

By Lemma 3, $a < 2/7$ is a necessary and sufficient condition for all $\mathbf{P}(4n+1)$ to be farther than $\mathbf{P}(4)$ from $\mathbf{C}_M(0)$, for $n \neq 0$.

By Lemma 4, $a < 3/15$ is a necessary and sufficient condition for all $\mathbf{P}(4n+2)$ to be farther than $\mathbf{P}(4)$ from $\mathbf{C}_M(0)$.

Given that $a < 3/15$, by Lemma 5, $4w_1 + 3w_2 < 1/(5a) + 5/2$ is a necessary and sufficient condition for all $\mathbf{P}(4n+3)$ to be farther than $\mathbf{P}(4)$ from $\mathbf{C}_{\mathrm{M}}(0)$. □

A.4 Minor Chord–Chord Pitch Relations

In this section, I derive the additional constraints on the minor chord weights so that, given the selected aspect ratio range, $\sqrt{2/15} < \mathrm{h/r} < \sqrt{3/15}$, the root is closest to the chord center, followed by the fifth, then the third; and, that all three are closer than other pitches to the chord center. In the following section, I derive the new constraints so that no other pitches are as close to the minor chord center as its three constituent pitches.

Definition of Minor Chord Weights
As in the case of the weights for the major triads, because of the definition of the minor triad representation (as given by Definition 3 in Chap. 3), the constraints on the minor chord weights can be summarized in the following three inequalities on (u_1, u_2):

$$u_1 > u_2, \tag{A.23}$$

$$u_1 > -2u_2 + 1, \tag{A.24}$$

$$\text{and} \qquad u_1 < -u_2 + 1. \tag{A.25}$$

These inequalities can be represented in a graph identical (except for the variables' names) to that in Fig. A.3. See Fig. A.9 for a graphical depiction of the feasible region for the minor triad weights, $\mathbf{F}_2$.

Theorem 8 *Given the aspect ratio constraint as given in Theorem 7, that is* $2/15 < a < 3/15$*, a minor chord's representation is closest to its root, followed by the fifth, then the third if and only if its weights,* $(u_1, u_2) \in \mathbf{F}_2$ *satisfy:*

$$a_0 \cdot u_1 > u_2 + \frac{1}{2}(a_0 - 1) \qquad \textit{where} \qquad a_0 = \frac{1}{2a} - \frac{3}{4}.$$

The sufficient condition, when $a = 2/15$ *is:*

$$u_1 > \frac{1}{3} \cdot u_2 + \frac{1}{3}.$$

[Proof]: The goal is to find the constraint on the weights $(u_1, u_2) \in \mathbf{F}_2$, for the aspect ratio range $\frac{2}{15} < a < \frac{3}{15}$, that satisfy the theorem conditions. The theorem states that: for $\frac{2}{15} < a < \frac{3}{15}$ and $(u_1, u_2) \in \mathbf{F}_2$,

$$\|\mathbf{C}_{\mathrm{m}}(k) - \mathbf{P}(k)\| < \|\mathbf{C}_{\mathrm{m}}(k) - \mathbf{P}(k+1)\| < \|\mathbf{C}_{\mathrm{m}}(k) - \mathbf{P}(k-3)\|$$

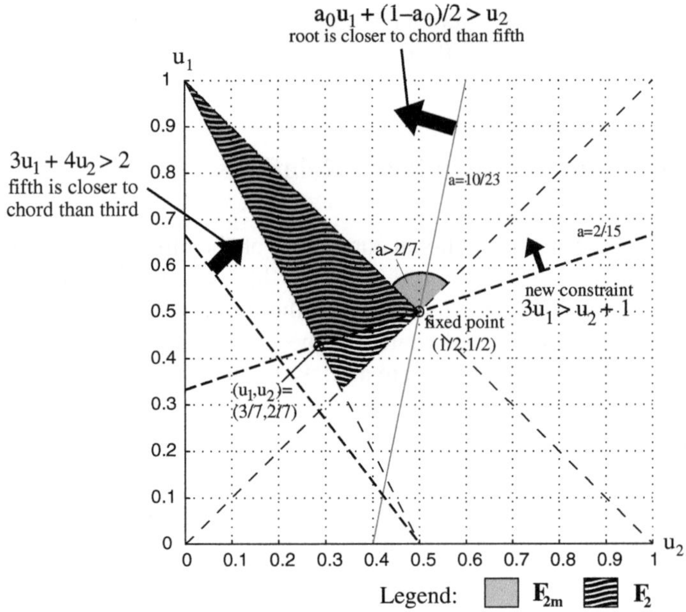

Fig. A.9 Feasible values for minor chord weights, $u_1, u_2, u_3 = (1 - u_1 - u_2)$, based on the desired proximity relations between a minor chord and its component pitches

$$\Longleftrightarrow \qquad a_0 u_1 > u_2 + \frac{1}{2}(a_0 - 1) \qquad \text{where} \qquad a_0 = \frac{1}{2a} - \frac{3}{4}.$$

Again, due to the model's symmetry, it is sufficient to simply consider the case of the C minor chord, $\mathbf{C}_\mathrm{m}(0)$. The related fifth and third are $\mathbf{P}(1)$ and $\mathbf{P}(-3)$ respectively. Their spatial positions are given by:

$$\mathbf{P}(0) = \begin{bmatrix} 0 \\ \mathrm{r} \\ 0 \end{bmatrix}, \mathbf{P}(1) = \begin{bmatrix} \mathrm{r} \\ 0 \\ \mathrm{h} \end{bmatrix}, \text{ and } \mathbf{P}(-3) = \begin{bmatrix} \mathrm{r} \\ 0 \\ -3\mathrm{h} \end{bmatrix}. \tag{A.26}$$

By definition (see Eq. 3.5), the C minor chord is spatially located at:

$$\begin{aligned} \mathbf{C}_\mathrm{m}(0) &= u_1 \cdot \mathbf{P}(0) + u_2 \cdot \mathbf{P}(1) + u_3 \cdot \mathbf{P}(-3) \\ &= u_1 \cdot \begin{bmatrix} 0 \\ \mathrm{r} \\ 0 \end{bmatrix} + u_2 \cdot \begin{bmatrix} \mathrm{r} \\ 0 \\ \mathrm{h} \end{bmatrix} + u_3 \cdot \begin{bmatrix} \mathrm{r} \\ 0 \\ -3\mathrm{h} \end{bmatrix} \\ &= \begin{bmatrix} (u_2 + u_3) \cdot \mathrm{r} \\ u_1 \cdot \mathrm{r} \\ (u_2 - 3u_3) \cdot \mathrm{h} \end{bmatrix}. \end{aligned} \tag{A.27}$$

Define a new variable for the standardized distance between $\mathbf{C}_{\mathrm{m}}(0)$ and a pitch:

$$y_{\mathrm{m}}(s) \stackrel{\text{def}}{=} \frac{1}{\mathrm{r}^2}\|\mathbf{C}_{\mathrm{m}}(0) - \mathbf{P}(s)\|^2 \tag{A.28}$$

The standardized distances of each chord pitch from the chords are given by:

$$\begin{aligned}
y_{\mathrm{m}}(0) &= \frac{1}{\mathrm{r}^2}\|\mathbf{C}_{\mathrm{m}}(0) - \mathbf{P}(0)\|^2 \\
&= (u_2+u_3)^2 + (u_1-1)^2 + (u_2-3u_3)^2 \cdot a \\
&= 2\cdot(1-u_1)^2 + (3u_1+4u_2-3)^2 \cdot a, \qquad \text{(A.29)}\\
y_{\mathrm{m}}(1) &= \frac{1}{\mathrm{r}^2}\|\mathbf{C}_{\mathrm{m}}(0) - \mathbf{P}(1)\|^2 \\
&= (u_2+u_3-1)^2 + u_1^2 + (u_2-3u_3-1)^2 \cdot a \\
&= 2u_1^2 + (3u_1+4u_2-4)^2 \cdot a, \qquad \text{(A.30)}\\
\text{and} \quad y_{\mathrm{m}}(-3) &= \frac{1}{\mathrm{r}^2}\|\mathbf{C}_{\mathrm{m}}(0) - \mathbf{P}(-3)\|^2 \\
&= (u_2+u_3-1)^2 + u_1^2 + (u_2-3u_3+3)^2 \cdot a \\
&= 2u_1^2 + (3u_1+4u_2)^2 \cdot a. \qquad \text{(A.31)}
\end{aligned}$$

The desired proximity conditions with respect to the chord center, $\mathbf{C}_{\mathrm{m}}(0)$, are:

$$\|\mathbf{C}_{\mathrm{m}}(0) - \mathbf{P}(0)\|^2 < \|\mathbf{C}_{\mathrm{m}}(0) - \mathbf{P}(1)\|^2 \iff y_{\mathrm{m}}(0) < y_{\mathrm{m}}(1), \tag{A.32}$$

$$\text{and } \|\mathbf{C}_{\mathrm{m}}(0) - \mathbf{P}(1)\|^2 < \|\mathbf{C}_{\mathrm{m}}(0) - \mathbf{P}(-3)\|^2 \iff y_{\mathrm{m}}(1) < y_{\mathrm{m}}(-3). \tag{A.33}$$

These inequalities, simplified, yield the following constraints:

[1] Substituting Eqs. A.29 and A.30 into Eq. A.32, one gets

$$\begin{aligned}
2\cdot(1-u_1)^2 - 2u_1^2 + (3u_1+4u_2-3)^2 \cdot a - (3u_1+4u_2-4)^2 \cdot a &< 0 \\
(2-7a) - (4-6a)\cdot u_1 + 8a \cdot u_2 &< 0 \\
\iff \frac{4-6a}{8a}\cdot u_1 - \frac{2-7a}{8a} &> u_2 \\
\iff a_0 u_1 + \frac{1}{2}\cdot(1-a_0) &> u_2, \qquad \text{(A.34)}
\end{aligned}$$

[2] The result of substituting Eqs. A.30 and A.31 into Eq. A.33 is

$$\begin{aligned}
(3u_1+4u_2-4)^2 \cdot a - (3u_1+4u_2)^2 \cdot a &< 0 \\
\iff 16 - 8\cdot(3u_1+4u_2) &< 0 \\
\iff 3u_1+4u_2 &> 2. \qquad \text{(A.35)}
\end{aligned}$$

Hence, the distance constraints given in Eqs. A.32 and A.33 are satisfied if and only if Eqs. A.34 and A.35 are true. The inequalities in Eqs. A.34 and A.35 are shown in Fig. A.9.

The inequality in Eq. A.35 always holds true since the region it defines always includes the values of $(u_1, u_2) \in \mathbf{F}_2$. The inequality in Eq. A.34 would not cut into $\mathbf{F}_2$ if and only if

$$\begin{array}{rcl} -1 < & \dfrac{4-6a}{8a} & < 1 \\ -8a < & 4-6a & < 8a \\ -2a < & 4 & < 14a \\ -a < & 2 & < 7a \\ & & \\ a > -2 & \text{and} & a > \dfrac{2}{7}. \end{array}$$

Since $2/15 < a < 3/15 < 2/7$, the line $a_0 \cdot u_1 + \frac{1}{2}(1 - a_0) = u_2$ will always cut into $\mathbf{F}_2$. Hence, the inequality in Eq. A.34 is only satisfied for a subset of $(u_1, u_2) \in \mathbf{F}_2$. The constraint in Eq. A.34 is now required to augment the inequality $u_1 > u_2$.

In Eq. A.34, by setting $a = 2/15$ (i.e. $a_0 = 3$), one gets:

$$u_2 < 3u_1 - \frac{1}{2}(1 - 3) = 3u_1 - 1.$$

Thus, this new constraint on the weights would always place the root closer to the chord's center than the fifth over the range of aspect ratio values $2/15 < a < 3/15$. □

Note that Theorem 8 effectively excludes the use of the weights $u = [\frac{1}{3}, \frac{1}{3}, \frac{1}{3}]$. This means that in the Spiral Array model, the minor triad requires a definite emphasis (as measured by the weight) on the root if it were to be more closely related (as reflected by distance in the model) to the chord than the fifth.

A.5 Minor Chord-All Pitch Relations

The theorem in this section defines the conditions on the minor chord weights so that all other pitches are farther from the minor chord center than its three constituent pitches. The theorem is preceded by four lemmas that consider the proximity, to the C minor triad, $\mathbf{C}_m(0)$, of each of the four types of pitches, $\mathbf{P}(4n)$, $\mathbf{P}(4n+1)$, $\mathbf{P}(4n+2)$ and $\mathbf{P}(4n+3)$, where n is an integer.

Lemma 6 *Consider pitches of the type* $\mathbf{P}(4n)$*, which correspond to { ..., F♭, A♭, C, E, G♯, ... }.* $\mathbf{P}(0)$ *is closest to* $\mathbf{C}_m(0)$*, i.e., pitch C is closest to minor chord C, followed by* $\mathbf{P}(-4)$ *(pitch A♭) for all* $(u_1, u_2) \in \mathbf{F}_{2m}$*, and* $\mathbf{P}(-4)$ *is always farther than* $\mathbf{P}(-3)$ *(pitch E♭, the third in the triad) if and only if* $a < 2/7$.

[Proof]: By Eqs. 3.2 and A.27:

$$\begin{aligned} y_m(4n) &= \frac{1}{r^2}\|\mathbf{C}_m(0) - \mathbf{P}(4n)\|^2 \\ &= 2\cdot(1-u_1)^2 + (4n - 3u_1 - 4u_2 + 3)^2 \cdot a \qquad \text{(A.36)} \\ &= 2\cdot(1-u_1)^2 + 16a\cdot\left(n - \frac{3u_1+4u_2-3}{4}\right)^2. \end{aligned}$$

Consider the function $\frac{1}{4}(3u_1+4u_2-3)$. For all $(u_1, u_2) \in \mathbf{F}_{2m}$, as shown in Fig. A.9, the function is minimized at $(u_1, u_2) = (3/7, 2/7)$ and maximized at $(\frac{1}{2}, \frac{1}{2})$. This means that $-1/7 < \frac{1}{4}(3u_1+4u_2-3) < 1/8$. Relaxing the integer constraint on n, the function $\tilde{y}_m(4n)$ is minimized at $\tilde{n}^* = \frac{1}{4}(3u_1+4u_2-3)$, which lies between $-1/7$ and $1/8$.

Using a parabola similar to that in Fig. A.6, it follows that $n_{(1)} = 0$ and $n_{(2)} = -1$. This means that the second farthest pitch of the type $\mathbf{P}(4n)$ is $\mathbf{P}(-4)$. Now, it remains to be proved that $y_m(-4) - y_m(-3) > 0$. Using Eqs. A.36 and A.31,

$$\begin{aligned} y_m(-4) - y_m(-3) &= 2\cdot(1-u_1)^2 - 2\cdot u_1^2 + a\cdot(3u_1+4u_2+1)^2 - a\cdot(3u_1+4u_2)^2 \\ &= (2+a) + (6a-4)\cdot u_1 + 8a\cdot u_2 > 0 \end{aligned}$$

$$\Longleftrightarrow \frac{1}{2}(a_0+1) + u_2 > a_0 u_1, \qquad \text{where } a_0 = \frac{1}{2a} - \frac{3}{4}.$$

The equation $a_0\cdot u_1 = \frac{1}{2}(a_0+1) + u_2$ has a fixed point at $(u_1, u_2) = (\frac{1}{2}, -\frac{1}{2})$. This line pivots around its fixed point. The inequality $a_0 u_1 < \frac{1}{2}(a_0+1) + u_2$, and does not interfere with the feasible region $\mathbf{F}_{2m}$ (or for that matter, $\mathbf{F}_2$), shown in Fig. A.9, if and only if $a_0 > 1$. $a_0 > 1$ simply means that $a < \frac{2}{7}$. □

Lemma 7 *Consider pitches of the type* $\mathbf{P}(4n+1)$*, which correspond to { ..., C♭, E♭, G, B, D♯, ... }.* $\mathbf{P}(1)$ *is closest to* $\mathbf{C}_m(0)$*, i.e. pitch G is closest to minor chord C. The* *second closest pitch* **P**(-3) *(pitch E♭, the third in the triad) for all* $(u_1, u_2) \in \mathbf{F}_{2m}$*; and, all other pitches* **P**$(4n+1)$ *are farther than* **P**(-3).

[Proof]: By Eqs. 3.2 and A.27,

$$\begin{aligned} y_{\mathrm{m}}(4n+1) &= \frac{1}{\mathrm{r}^2}\|\mathbf{C}_{\mathrm{m}}(0) - \mathbf{P}(4n+1)\|^2 \\ &= 2 \cdot u_1^2 + [4n+1-(3u_1+4u_2-3)]^2 \cdot a \\ &= 2 \cdot u_1^2 + 16a \cdot \left(n - \frac{3u_1+4u_2-4}{4}\right)^2. \end{aligned} \quad \text{(A.37)}$$

Consider the function $\frac{1}{4}(3u_1+4u_2-4)$. For all $(u_1, u_2) \in \mathbf{F}_{2m}$, as shown in Fig. A.9, since $-1/7 < \frac{1}{4}(3u_1+4u_2-3) < 1/8$, $-11/28 < \frac{1}{4}(3u_1+4u_2-4) < -1/8$, and $\tilde{n}^*$ lies between $-11/28$ and $-1/8$. It then follows that $n_{(1)} = 0$ and $n_{(2)} = -1$. In general, $y_{\mathrm{m}}(4n) < y_{\mathrm{m}}(-4n+1)$ for non-negative n. Therefore,

$$y_{\mathrm{m}}(0) < y_{\mathrm{m}}(-1) < y_{\mathrm{m}}(4n+1), \quad \forall n \in \mathbb{Z} \text{ and } n \neq 0, -1.$$

□

Lemma 8 *Consider pitches of the type* **P**$(4n+2)$*, which correspond to { ..., G♭, B♭, D, F♯, A♯, ...}. For all* $(u_1,u_2) \in \mathbf{F}_{2m}$*, when* $3u_1+4u_2 > 3$*,* **P**(2) *(pitch D) is closest to* $\mathbf{C}_{\mathrm{m}}(0)$*, the C minor chord; and when* $3u_1+4u_2 < 3$*,* **P**(-2) *(pitch B♭) is closest to* $\mathbf{C}_{\mathrm{m}}(0)$*. In either case, the closest pitch is still farther than* **P**(-3) *(E♭, the third in the triad) if and only if* $a < 3/15$.

[Proof]: By Eqs. 3.2 and A.8:

$$y_{\mathrm{m}}(4n+2) = \frac{1}{\mathrm{r}^2}\|\mathbf{C}_{\mathrm{m}}(0) - \mathbf{P}(4n+2)\|^2$$
$$= 2 \cdot (1-u_1)^2 + [4n+2-(3u_1+4u_2-3)]^2 \cdot a \quad \text{(A.38)}$$
$$= 2 \cdot (1-u_1)^2 + 16a \cdot \left(n - \frac{3u_1+4u_2-5}{4}\right)^2. \quad \text{(A.39)}$$

Consider the function $\frac{1}{4}(3u_1+4u_2-5)$. For all $(u_1, u_2) \in \mathbf{F}_{2m}$, as shown in Fig. A.9, since $-1/7 < \frac{1}{4}(3u_1+4u_2-3) < 1/8$, $-9/14 < \frac{1}{4}(3u_1+4u_2-5) < -3/8$, and $\tilde{n}^*$ lies between $-9/14$ and $-3/8$. When $\frac{1}{4}(3u_1+4u_2-5) > -\frac{1}{2}$, $n_{(1)} = 0$; and, when $\frac{1}{4}(3u_1+4u_2-5) < -\frac{1}{2}$, $n_{(1)} = -1$.

When $3u_1+4u_2 > 3$, $n_{(1)} = 0$:

$$\begin{aligned} y_{\mathrm{m}}(2) - y_{\mathrm{m}}(-3) &= 2 \cdot (1+u_1^2) - 2 \cdot u_1^2 + (3u_1+4u_2-5)^2 \cdot a - (3u_1+4u_2)^2 \cdot a \\ &= (2+25a) + 2(-5)(3u_1+4u_2) \cdot a > 0 \end{aligned}$$

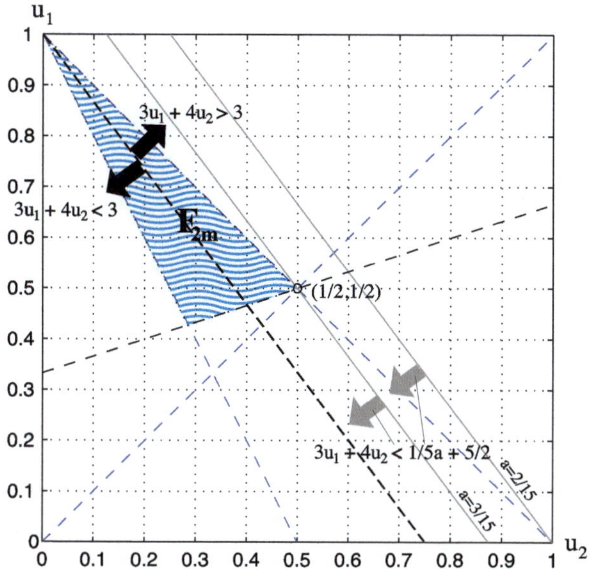

Fig. A.10 Feasible values for minor chord weights, $u_1, u_2, u_3 (= 1 - u_1 - u_2)$, and the boundary, $3u_1 + 4u_2 = 3$, between two subsets of weights in the analysis of $y_m(4n+2) - y_m(-3)$

$$\iff 3u_1 + 4u_2 < \frac{2+25a}{10a} = \frac{1}{5a} + \frac{5}{2}.$$

As shown in Fig. A.10, when $a < 3/15$, this is not a binding constraint. When $3u_1 + 4u_2 < 3$, $n_{(1)} = -1$:

$$\begin{aligned} y_m(-2) - y_m(-3) &= 2 \cdot (1 + u_1^2) - 2u_1^2 + (3u_1 + 4u_2 - 1)^2 \cdot a - (3u_1 + 4u_2)^2 \cdot a \\ &= (2 + a) + 2(-1)(3u_1 + 4u_2) \cdot a \\ &= (2 + a) + 6a \quad \text{since } 3u_1 + 4u_2 < 3, \\ &= 2 - 5a > 0 \quad \text{because } a < \frac{3}{15} < \frac{2}{5}. \end{aligned}$$

□

Lemma 9 *Consider pitches of the type* **P**$(4n+3)$*, which correspond to* { ..., D♭, F, A, C♯, E♯, ... }*. For all* $(u_1, u_2) \in \mathbf{F}_{2m}$*,* **P**$(-1)$ *(pitch F) is closest to* $\mathbf{C}_m(0)$*, the C minor triad. This closest pitch is still farther than* **P**(-3) *(E♭, the third in the triad) if and only if*

$$\left(a_5 + \frac{1}{2}\right) \cdot u_1 + u_2 < a_5, \quad \text{where} \quad a_5 = \frac{1}{8a} + \frac{1}{4}.$$

[Proof]: By Eqs. 3.2 and A.8:

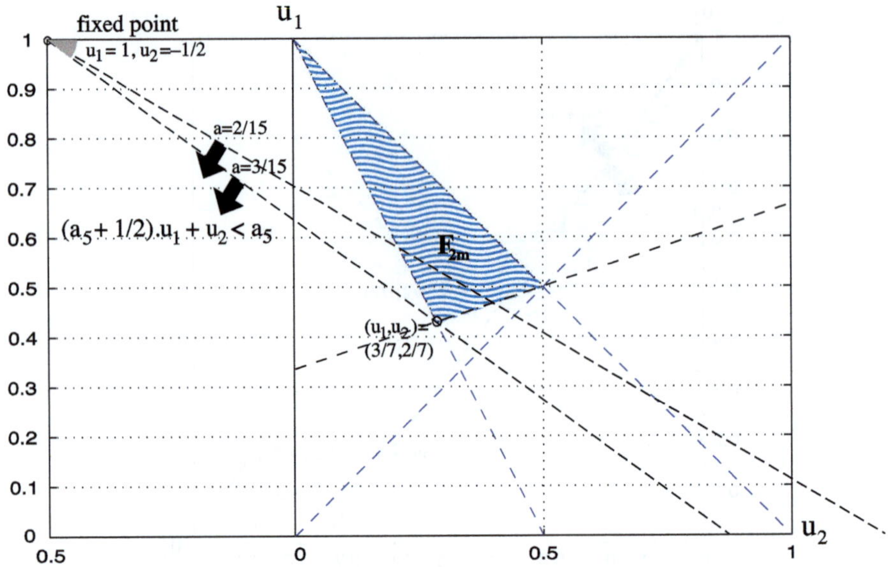

Fig. A.11 Feasible values for minor chord weights, u_1, u_2, $u_3(=1-u_1-u_2)$, based on the analysis of $y_{\mathrm{m}}(4n+3)-y_{\mathrm{m}}(-3)$

$$\begin{aligned} y_{\mathrm{m}}(4n+3) &= \frac{1}{\mathrm{r}^2}\|\mathbf{C}_{\mathrm{m}}(0)-\mathbf{P}(4n+3)\|^2 \\ &= 2\cdot(1-u_1+u_1^2)+[4n+3-(3u_1+4u_2-3)]^2\cdot a \qquad \text{(A.40)} \\ &= 2\cdot(1-u_1)^2+16a\cdot\left(n-\frac{3u_1+4u_2-6}{4}\right)^2. \qquad \text{(A.41)} \end{aligned}$$

Consider the function $\frac{1}{4}(3u_1+4u_2-6)$. For all $(u_1,u_2)\in\mathbf{F}_{2m}$, as shown in Fig. A.9, since $-\frac{1}{7}<\frac{1}{4}(3u_1+4u_2-3)<\frac{1}{8}$, $-25/28<\frac{1}{4}(3u_1+4u_2-6)<-5/8$, and $\tilde{n}^*$ lies between $-25/28$ and $-5/8$. It follows that $n_{(1)}=-1$.

$$\begin{aligned} y_{\mathrm{m}}(-1)-y_{\mathrm{m}}(-3) &= 2\cdot(1-u_1+u_1^2)-2\cdot u_1^2+(3u_1+4u_2-2)^2\cdot a \\ &\quad -(3u_1+4u_2)^2\cdot a \\ &= 2\cdot[(1-u_1)+2a+a(-2)(3u_1+4u_2)] \\ &= 2\cdot[(1+2a)-(1+6a)\cdot u_1+8a\cdot u_2]>0 \end{aligned}$$

$$\begin{aligned} &\Longleftrightarrow \frac{1+2a}{8a}-\left(\frac{1+6a}{8a}\right)\cdot u_1>u_2 \\ &\Longleftrightarrow \left(a_5+\frac{1}{2}\right)\cdot u_1+u_2<a_5, \quad \text{where } a_5=\frac{1+2a}{8a}=\frac{1}{8a}+\frac{1}{4}. \end{aligned}$$

The equation $(a_5 + \frac{1}{2}) \cdot u_1 + u_2 = a_5$ has a fixed point at $(u_1, u_2) = (1, -\frac{1}{2})$. Figure A.11 shows the range of the inequality $(a_5 + \frac{1}{2})u_1 + u_2 < a_5$ over $2/15 < a < 3/15$. □

Theorem 9 *No other pitches are as close to a minor chord center as its three constituent pitches if:*

$$\frac{2}{15} < a < \frac{3}{15}, \qquad 3u_1 > u_2 + 1, \quad \textit{and} \quad \left(a_5 + \frac{1}{2}\right) \cdot u_1 + u_2 < a_5,$$

where $a_5 = \frac{1}{8a} + \frac{1}{4}$.

[Proof]: Note that the chord pitch farthest from the chord representation is the third of the chord (by Theorem 8), so that for all $s \neq 0, 1, -3$:

$$\begin{aligned}&\|\mathbf{C}_\mathrm{m}(k) - \mathbf{P}(k+s)\| \\ &\quad > \max\{\|\mathbf{C}_\mathrm{m}(k) - \mathbf{P}(k)\|, \|\mathbf{C}_\mathrm{m}(k) - \mathbf{P}(k+1)\|, \|\mathbf{C}_\mathrm{m}(k) - \mathbf{P}(k-3)\|\}.\end{aligned}$$

Thus, one needs only prove that all non-chord pitches are farther from the chord than its third. By symmetry, one need only show this for C minor, i.e., $\mathbf{C}_\mathrm{m}(0)$.

By Lemma 6, $a < 2/7$ is a necessary and sufficient condition for all $\mathbf{P}(4n)$ to be farther than $\mathbf{P}(-3)$ from $\mathbf{C}_\mathrm{m}(0)$, for $n \neq 0$.

By Lemma 7, $\mathbf{P}(4n+1)$ are all farther than $\mathbf{P}(-3)$ from $\mathbf{C}_\mathrm{m}(0)$, for $n \neq 0, -1$.

By Lemma 8, $a < 3/15$ is a necessary and sufficient condition for all $\mathbf{P}(4n+2)$ to be farther than $\mathbf{P}(-3)$ from $\mathbf{C}_\mathrm{m}(0)$.

Given that $a > 2/15$, by Lemma 9, $(a_5 + \frac{1}{2})u_1 + u_2 < a_5$ where $a_5 = \frac{1}{8a} + \frac{1}{4}$, is necessary and sufficient to guarantee that all $\mathbf{P}(4n+3)$ are farther than $\mathbf{P}(-3)$ from $\mathbf{C}_\mathrm{m}(0)$. □

A.6 Summary of Aspect Ratio and Chord Weight Constraints

It is useful, at this point, to summarize the constraints on the aspect ratio and the chord weights generated by the basic condition: a chord's root should be the pitch closest to the chord representation, followed by its fifth, then the third.

For ease of transcription, instead of referring to the distance between a chord and a pitch, I shall use the standardized distances between a pitch of index s and a major or minor chord,

$$\begin{aligned} y_\mathrm{M}(s) &= \frac{1}{\mathrm{r}^2}\|\mathbf{C}_\mathrm{M}(0) - \mathbf{P}(s)\|^2, \\ \text{and} \quad y_\mathrm{m}(s) &= \frac{1}{\mathrm{r}^2}\|\mathbf{C}_\mathrm{m}(0) - \mathbf{P}(s)\|^2, \text{ respectively.}\end{aligned}$$

$y_M(s)$ is proportional to the distance from pitch **P**(s) to the C major chord representation, $\mathbf{C}_M(0)$; and, $y_m(s)$ is proportional to the distance from pitch **P**(s) to the C minor chord representation, $\mathbf{C}_m(0)$.

A.6.1 Summary of Chord to Pitch Relations

Recall that $a = (h/r)^2$. The results from the previous sections on major chords can be summarized as: for all $(w_1, w_2) \in \mathbf{F}_0$,

Theorem 6

$$y_M(0) < y_M(1) < y_M(4) \iff \frac{2}{15} < a < \frac{2}{7}.$$

Lemma 2

$$y_M(4) < y_M(4n), n \neq 0, 1 \iff a < \frac{2}{7}.$$

Lemma 3

$$y_M(4) < y_M(4n+1), n \neq 0 \iff a < \frac{2}{7}.$$

Lemma 4

$$y_M(4) < y_M(4n+2), \forall n \iff a < \frac{3}{15}.$$

Lemma 5

$$y_M(4) < y_M(4n+3), \forall n \iff 3 < 4w_1 + 3w_2 < \frac{1}{5a} + \frac{5}{2},$$
$$\text{and } 4w_1 + 3w_2 < \frac{1}{2a} + \frac{1}{2}.$$

Theorem 7

$$y_M(0) < y_M(1) < y_M(4) < y_M(s), \quad \forall s \neq 0, 1, 4$$

$$\text{if} \quad \frac{2}{15} < a < \frac{3}{15},$$

$$\text{and} \quad 4w_1 + 3w_2 < \frac{1}{5a} + \frac{5}{2}.$$

The results from the previous sections on minor chords can be summarized as: given that $\frac{2}{15} < a < \frac{3}{15}$ and $(u_1, u_2) \in \mathbf{F}_2$,

Theorem 8

$$y_m(0) < y_m(1) < y_m(-3) \iff u_1 > \frac{1}{3} \cdot u_2 + \frac{1}{3}.$$

Lemma 6

$$y_{\mathrm{m}}(-3) < y_{\mathrm{m}}(4n), n \neq 0 \iff a < \frac{2}{7}.$$

Lemma 7

$$y_{\mathrm{m}}(-3) < y_{\mathrm{m}}(4n+1), n \neq -1 \text{ is always true.}$$

Lemma 8

$$y_{\mathrm{m}}(-3) < y_{\mathrm{m}}(4n+2), \forall n \iff a < \frac{3}{15}.$$

Lemma 9

$$y_{\mathrm{m}}(-3) < y_{\mathrm{m}}(4n+3), \forall n \iff \left(a_5 + \frac{1}{2}\right) \cdot u_1 + u_2 < a_5,$$
$$\text{where } a_5 = \frac{1}{8a} + \frac{1}{4}$$

Theorem 9

$$y_{\mathrm{m}}(0) < y_{\mathrm{m}}(1) < y_{\mathrm{m}}(-3) < y_{\mathrm{m}}(s), \qquad \forall s \neq 0, 1, -3$$

$$\text{if} \quad 3u_1 > u_2 + 1,$$
$$\text{and} \quad \left(a_5 + \tfrac{1}{2}\right) \cdot u_1 + u_2 < a_5.$$

A.6.2 Results for Aspect Ratio, $h/r = \sqrt{2/15}$

When $a = 2/15$, the distance between pitches a perfect fifth apart and the distance between pitches a major third apart are the same. This is a boundary derived in Theorems 5 and 6. This value, 2/15, has been the aspect ratio of choice for the examples, illustrations, and applications in this book.

When $a = 2/15$, the results from the previous sections on major chords can be summarized as: for all $(w_1, w_2) \in \mathbf{F}_0$:

By Theorem 6: $y_{\mathrm{M}}(0) < y_{\mathrm{M}}(1) \leq y_{\mathrm{M}}(4)$.
By Lemma 2: $y_{\mathrm{M}}(4) < y_{\mathrm{M}}(4n), n \neq 0, 1$.
By Lemma 3: $y_{\mathrm{M}}(4) \leq y_{\mathrm{M}}(4n+1), n \neq 0$.
By Lemma 4: $y_{\mathrm{M}}(4) < y_{\mathrm{M}}(4n+2), \forall n$.
By Lemma 5: $y_{\mathrm{M}}(4) \leq y_{\mathrm{M}}(4n+3), \forall n$.
By Theorem 7: $y_{\mathrm{M}}(0) < y_{\mathrm{M}}(1) \leq y_{\mathrm{M}}(4) \leq y_{\mathrm{M}}(s), \forall s \neq 0, 1, 4$, if $4w_1 + 3w_2 < 4$.

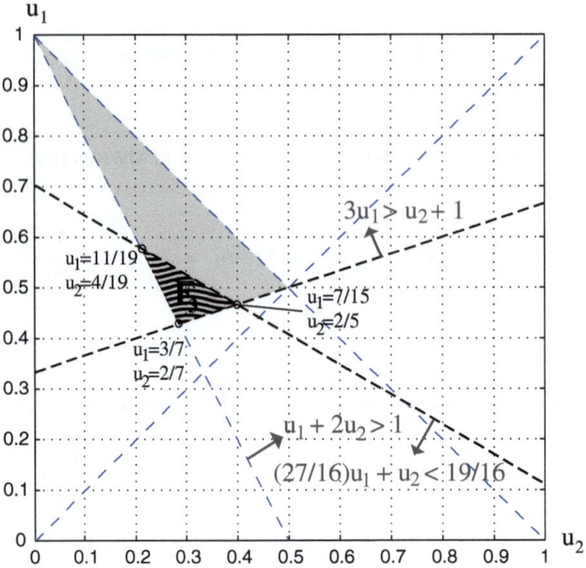

Fig. A.12 Feasible values for minor chord weights, $u_1, u_2, u_3 (= 1 - u_1 - u_2)$, when $a = 2/15$, based on the proximity relations between a minor chord and all pitches

The results from the previous sections on minor chords can be summarized as: given that $2/15 < a < 3/15$ and $(u_1, u_2) \in \mathbf{F}_2$:

By Theorem 8: $y_\mathrm{m}(0) < y_\mathrm{m}(1) < y_\mathrm{m}(-3) \iff 3u_1 > u_2 + 1.$
By Lemma 6: $y_\mathrm{m}(-3) < y_\mathrm{m}(4n), n \neq 0.$
By Lemma 7: $y_\mathrm{m}(-3) < y_\mathrm{m}(4n+1), n \neq -1.$
By Lemma 8: $y_\mathrm{m}(-3) < y_\mathrm{m}(4n+2), \forall n.$
By Lemma 9: $y_\mathrm{m}(-3) < y_\mathrm{m}(4n+3), \forall n \iff 27u_1 + 16u_2 < 19.$
By Theorem 9: $y_\mathrm{m}(0) < y_\mathrm{m}(1) < y_\mathrm{m}(-3) < y_\mathrm{m}(s), \forall s \neq 0, 1, -3$
if $3u_1 > u_2 + 1$ and $27u_1 + 16u_2 < 19.$

The intersection of $\mathbf{F}_2$ and the Theorem 8 constraint forms $\mathbf{F}_{2m}$ (shown in Fig. A.9). The intersection of $\mathbf{F}_{2m}$ and the Lemma 9 constraint forms $\mathbf{F}_3$ (shown in Fig. A.12).

A.7 Exploring Other Relationships Among Chords and Pitches

In this section, I shall explore the implications of the choices of weights on other relationships among major and minor chords and pitches. For example: what is the closest chord to any given pitch? what is the closest pitch to any given chord? what is the closest chord to any given chord?

In Chap. 3, as each chord and key representation were defined, I proved that, because of the definitions, the relationships among major chords, among minor chords, among major keys, and among minor keys, are identical to those among pitches. These relationships have yet to be explored across modes. This section will attempt to illustrate some of these major–minor mode relations for two sets of major triad and minor triad weights.

As shown in the previous section, for any choice of weights for a given chord—(w_1, w_2) for major chords and (u_1, u_2) for minor chords—if $a = 2/15$, and

$$3u_1 - u_2 > 1 \quad \text{and} \quad 27u_1 + 16u_2 < 19,$$

the chord is always closest to its root, followed by the fifth then the third. In the case of major triads, $\mathbf{C}_\mathrm{M}(\mathrm{k})$, the root is $\mathbf{P}(\mathrm{k})$, the fifth $\mathbf{P}(\mathrm{k}+1)$, and the third $\mathbf{P}(\mathrm{k}+4)$. For minor triads, $\mathbf{C}_\mathrm{m}(\mathrm{k})$, the root is $\mathbf{P}(\mathrm{k})$, the fifth $\mathbf{P}(\mathrm{k}+1)$, and the third $\mathbf{P}(\mathrm{k}-3)$.

This simple set of restrictions allows for the modeling of multiple relationships among chords and pitches, and also between pairs of chord representations.

A.7.1 Parameters Chosen for Illustrations

As mentioned before, the aspect ratio in all further examples are set at $a = 2/15$. There are two sets of weights which are used in the illustrations in this section. The major chord weights (w_1, w_2) are restricted to be within $\mathbf{F}_0$ (shown in Fig. A.3); and, the minor chord weights (u_1, u_2) are restricted to be within $\mathbf{F}_3$ (shown in Fig. A.12).

The first set of weights are:

$$w = \left[\frac{1}{3} + \varepsilon, \frac{1}{3}, \frac{1}{3} - \varepsilon\right], \quad \text{and} \quad u = \left[\frac{3}{7} + \varepsilon, \frac{2}{7}, \frac{2}{7} - \varepsilon\right].$$

These weights are chosen for their simplicity and because they are close to the vertices of the feasible spaces shown in Figs. A.3 and A.12 for illustrative purposes. This naïve choice of major triad weights reflects close to no preference among the relationships between the different chord pitches and major triad except for the model's own inbuilt preferences. The weights assigned the minor chord pitches shows little preference among the relationships between the minor chord and its fifth or third.

The second set of weights are:

$$w = \left[\frac{1}{2} + \varepsilon, \frac{1}{4}, \frac{1}{4} - \varepsilon\right], \quad \text{and} \quad u = \left[\frac{1}{2} + \varepsilon, \frac{1}{4}, \frac{1}{4} - \varepsilon\right].$$

In this case, the two sets of weights are chosen to be equal. The interpretation for these values is that emphasis is placed on the root of the chord, but close to equal preference is assigned to the fifth and the third. In addition, the relationships between

Table A.2 Example of pitch–chord relations when the major chord weights $w = [\frac{1}{3}+\varepsilon, \frac{1}{3}-\varepsilon, \frac{1}{3}-\varepsilon]$, and the minor chord weights $u = [\frac{3}{7}+\varepsilon, \frac{2}{7}-\varepsilon, \frac{2}{7}-\varepsilon]$, where $\varepsilon = 10^{-5}$

h/r	Pitch	Chord positions ranked by closeness					
$\sqrt{2/15}$	**P**(k)	$\mathbf{C}_M$(k)	$\mathbf{C}_m$(k)	$\mathbf{C}_m$(k - 1)	$\mathbf{C}_M$(k − 1)	$\mathbf{C}_M$(k - 4)	$\mathbf{C}_m$(k + 3)
	C	C (0.5926)[a]	c (0.6966)	f (0.6966)	F (0.9481)	A♭ (0.9482)	a (1.1538)
$\sqrt{3/15}$	**P**(*k*)	$\mathbf{C}_m$(k)	$\mathbf{C}_M$(k)	$\mathbf{C}_m$(k - 1)	$\mathbf{C}_M$(k-1)	$\mathbf{C}_M$(k - 4)	$\mathbf{C}_m$(k + 3)
	C	c (0.7183)	C (0.7778)	f (0.8612)	F (0.9778)	A♭ (1.3111)	a (1.5470)

[a]The number in the bracket denotes spatial distance between each chord representation and the reference pitch when radius r = 1 and h = the aspect ratio

Table A.3 Example of pitch–chord relations when the major and minor chord weights are both $[\frac{1}{2}+\varepsilon, \frac{1}{4}, \frac{1}{4}-\varepsilon]$, where $\varepsilon = 10^{-5}$

h/r	Pitch	Chord positions ranked by closeness					
	P(k)	$\mathbf{C}_M$(k)	$\mathbf{C}_m$(k)	$\mathbf{C}_m$(k − 1)	$\mathbf{C}_M$(k − 1)	$\mathbf{C}_M$(k − 4)	$\mathbf{C}_m$(k + 3)
$\sqrt{2/15}$	C	C (0.3333)	c (0.5333)	f (0.8000)	F (1.1334)	A♭ (1.1334)	a (1.1334)
$\sqrt{3/15}$	C	C (0.4374)	c (0.5499)	f (0.9500)	F (1.1375)	A♭ (1.6376)	a (1.7501)

minor chord pitches and the minor chord are assumed to be the same as those between major chord pitches and the major chord.

A.7.2 Implications for Chords with respect to Given Pitch

Because proximity in the Spiral Array indicates shared pitches and interval relations of fifths or thirds, the chords closest to a given pitch always has this pitch as its root, fifth or third. Changes in the weights only reorders the chords according to the distance gradations.

Tables A.2 and A.3 show examples of pitch–chord relations with respect to two different sets of the major and minor chord weights that satisfy the conditions given in Sect. A.2. In Table A.2, $w = \left[\frac{1}{3}+\varepsilon, \frac{1}{3}, \frac{1}{3}-\varepsilon\right]$, and $u = \left[\frac{3}{7}+\varepsilon, \frac{2}{7}, \frac{2}{7}-\varepsilon\right]$. Table A.3 shows the results when the weights are restricted to be both equal to $\left[\frac{1}{2}+\varepsilon, \frac{1}{4}, \frac{1}{4}-\varepsilon\right]$.

A.7.3 Implications for Pitches with respect to Given Chord

By design, the three closest pitch positions to a given major chord representation are its root, fifth and third. And, because pitches in a key occupy a compact space in the Spiral Array, the closest pitches to a chord, $\mathbf{C}_M$(k), will always include the pitches of the key whose tonic is **P**(k). For example, the pitches closest to the C major chord

Table A.4 Example of chord–pitch relations when the major chord weights $w = \left[\frac{1}{3} + \varepsilon, \frac{1}{3}, \frac{1}{3} - \varepsilon\right]$ and the minor chord weights $u = \left[\frac{3}{7} + \varepsilon, \frac{2}{7}, \frac{2}{7} - \varepsilon\right]$, where $\varepsilon = 10^{-5}$

h/r	Chord	Pitch positions ranked by closeness					
$\sqrt{2/15}$	$C_M(k)$	**P**(k)	**P**(k + 1)	**P**(k + 4)	**P**(k + 5)	**P**(k + 3)	**P**(k + 2)
	C	C (0.5926)	G (0.9481)	E (0.9482)	B (2.3704)	A (2.4593)	D (2.9037)
$\sqrt{3/15}$	$C_M(k)$	**P**(k)	**P**(k + 1)	**P**(k + 4)	**P**(k + 3)	**P**(k + 2)	**P**(k + 5)
	C	C (0.7778)	G (0.9778)	E (1.3111)	A (2.5778)	D (2.9111)	B (3.1112)
$\sqrt{2/15}$	$C_m(k)$	**P**(k)	**P**(k + 1)	**P**(k − 3)	**P**(k − 4)	**P**(k − 2)	**P**(k − 1)
	C	C (0.6966)	G (0.6966)	E♭ (1.1538)	A♭ (2.2204)	B♭ (2.6395)	F (2.6775)
$\sqrt{3/15}$	$C_m(k)$	**P**(k)	**P**(k + 1)	**P**(k − 3)	**P**(k − 1)	**P**(k − 2)	**P**(k − 4)
	C	C (0.7183)	G (0.8612)	E♭ (1.5470)	F (2.6898)	B♭ (2.7755)	A♭ (3.0041)

Table A.5 Example of chord-pitchs relation when the major and minor chord weights are both $\left[\frac{1}{2} + \varepsilon, \frac{1}{4}, \frac{1}{4} - \varepsilon\right]$, where $\varepsilon = 10^{-5}$

h/r	Chord	Pitch positions ranked by closeness					
$\sqrt{2/15}$	$C_M(k)$	**P**(k)	**P**(k + 1)	**P**(k + 4)	**P**(k + 3)	**P**(k − 1)	**P**(k + 5)
	C	C (0.3333)	G (1.1333)	E (1.1334)	A (2.5334)	F (2.8000)	B (3.0000)
$\sqrt{3/15}$	$C_M(k)$	**P**(k)	**P**(k + 1)	**P**(k + 4)	**P**(k + 3)	**P**(k − 1)	**P**(k + 2)
	C	C (0.4375)	G (1.1375)	E (1.6376)	A (2.7375)	F (3.1375)	D (3.2375)
$\sqrt{2/15}$	$C_m(k)$	**P**(k)	**P**(k + 1)	**P**(k − 3)	**P**(k − 4)	**P**(k − 1)	**P**(k − 2)
	c	C (0.5333)	G (0.8000)	E♭ (1.3334)	A♭ (2.1333)	F (2.5333)	B♭ (2.5333)
$\sqrt{3/15}$	$C_m(k)$	**P**(k)	**P**(k + 1)	**P**(k − 3)	**P**(k − 1)	**P**(k − 4)	**P**(k − 2)
	c	C (0.5500)	G (0.9500)	E♭ (1.7501)	F (2.5500)	A♭ (2.9500)	B♭ (2.9500)

center will always include those from the key of C major. And, the pitches closest to the C minor chord center will always include those from the key of C minor.

Table A.4 shows some sample results when the weights are restricted to be $\left[\frac{1}{3} + \varepsilon, \frac{1}{3}, \frac{1}{3} - \varepsilon\right]$ for major triads and $\left[\frac{3}{7} + \varepsilon, \frac{2}{7}, \frac{2}{7} - \varepsilon\right]$ for minor triads. Table A.5 shows similar results when the weights are restricted to be both equal to $\left[\frac{1}{2} + \varepsilon, \frac{1}{4}, \frac{1}{4} - \varepsilon\right]$.

A.7.4 Implications for Chords with respect to Given Chord

In a similar fashion, the choice of weights on the major and minor chords allow for different distance affiliations between chord centers. As examples, Table A.6 shows chord-chord relations when the major chord weights are $\left[\frac{1}{3} + \varepsilon, \frac{1}{3}, \frac{1}{3} - \varepsilon\right]$ and minor chord weights are $\left[\frac{3}{7} + \varepsilon, \frac{2}{7}, \frac{2}{7} - \varepsilon\right]$; and, Table A.7 shows the same relations when the weights are restricted to be both equal to $\left[\frac{1}{2} + \varepsilon, \frac{1}{4}, \frac{1}{4} - \varepsilon\right]$.

Table A.6 Example of chord–chord relations when the major chord weights $w = [\frac{1}{3}+\varepsilon, \frac{1}{3}, \frac{1}{3}-\varepsilon]$, and the minor chord weights $u = [\frac{3}{7}+\varepsilon, \frac{2}{7}, \frac{2}{7}-\varepsilon]$, where $\varepsilon = 10^{-5}$

h/r	Chord	Chord positions ranked by closeness					
$\sqrt{2/15}$	$C_M(k)$	$C_M(k)$	$C_m(k+4)$	$C_m(k+3)$	$C_m(k)$	$C_M(k+1)$	$C_M(k-1)$
	C	C (0.0000)	e (0.5273)	a (0.6670)	c (0.7812)	G (1.2444)	F (1.2444)
$\sqrt{3/15}$	$C_M(k)$	$C_M(k)$	$C_m(k+3)$	$C_m(k+4)$	$C_m(k)$	$C_M(k+1)$	$C_M(k-1)$
	C	C (0.0000)	a (0.7057)	e (0.7343)	c (1.1151)	G (1.3111)	F (1.3111)
$\sqrt{2/15}$	$C_m(k)$	$C_m(k)$	$C_M(k-4)$	$C_M(k-3)$	$C_M(k)$	$C_m(k-1)$	$C_m(k+1)$
	c	c (0.0000)	A♭ (0.5273)	E♭ (0.6670)	C (0.7812)	f (1.1537)	g (1.1537)
$\sqrt{3/15}$	$C_m(k)$	$C_m(k)$	$C_M(k-3)$	$C_M(k-4)$	$C_M(k)$	$C_m(k-1)$	$C_m(k+1)$
	c	c (0.0000)	E♭ (0.7057)	A♭ (0.7343)	C (1.1151)	f (1.2204)	g (1.2204)

Table A.7 Example of chord–chord relations when the major and minor chord weights are $[\frac{1}{2}+\varepsilon, \frac{1}{4}, \frac{1}{4}-\varepsilon]$, where $\varepsilon = 10^{-5}$

h/r	Chord	Chord positions ranked by closeness					
$\sqrt{2/15}$	$C_M(k)$	$C_M(k)$	$C_m(k-1)$	$C_m(k+4)$	$C_m(k+3)$	$C_M(k+1)$	$C_M(k-1)$
	C	C (0.0000)	c (0.5333)	e (0.8000)	a (0.8334)	G (1.3833)	F (1.3833)
$\sqrt{3/15}$	$C_M(k)$	$C_M(k)$	$C_m(k-1)$	$C_m(k+3)$	$C_m(k+4)$	$C_M(k+1)$	$C_m(k-1)$
	C	C (0.0000)	c (0.7374)	a (0.9376)	e (1.1376)	G (1.4500)	F (1.4500)
$\sqrt{2/15}$	$C_m(k)$	$C_m(k)$	$C_M(k)$	$C_M(k-4)$	$C_M(k-3)$	$C_m(k-1)$	$C_m(k+1)$
	c	c (0.0000)	C (0.5333)	A♭ (0.8000)	E♭ (0.8334)	f (1.1333)	g (1.1333)
$\sqrt{3/15}$	$C_m(k)$	$C_m(k)$	$C_M(k)$	$C_M(k-3)$	$C_M(k-4)$	$C_m(k-1)$	$C_m(k+1)$
	c	c (0.0000)	C (0.7374)	E♭ (0.9376)	A♭ (1.1376)	f (1.2000)	g (1.2000)

A.8 Desired Key–Interval–Pitch Relations

In this section, I examine three simple and desirable proximity relations between keys, pitches, and intervals, namely: the nearest key to a given pitch, the nearest key to a half-step interval, and the nearest key to a perfect fourth interval.

A.8.1 The Tonic-Key Relationship

The key representation closest to an isolated pitch should be the major key of the same name, which is to say, the major key with the pitch as its first degree. The interpretation is that: on hearing a single pitch, without further information, the single pitch is likely interpreted to be the first degree of the major key. For example, the key nearest the pitch C should be C major. The pitch may equally likely be the first degree of the minor key. For modeling purposes, the major key will be chosen to be the closer key. Some other keys close would then be: C minor, F major, F minor,

Fig. A.13 Example: "Londonderry Air" begins with a half step interval that forms a $(\hat{7}-\hat{1})$ transition

A minor, and A♭ major. A subtle difference in the weights can lead to any of these other keys being closer to the pitch C than C major.

Mathematically, this condition translates to:

$$\arg\min_{\mathbf{T}} d(\mathbf{T}, \mathbf{P}(\mathrm{k})) = \mathbf{T}_{\mathrm{M}}(\mathrm{k}), \text{ where } \mathbf{T} \in \{T_{\mathrm{M}}(\ell), T_{\mathrm{m}}(\ell) \;\; \forall \ell \in \mathbb{Z}\}. \quad \text{(A.42)}$$

A.8.2 Implications of the Half Step Interval

The half-step (also called a semitone) is an important element in defining a key. In an ascending diatonic major scale, there are only two half steps: between the mediant and subdominant $(\hat{3}-\hat{4})$, and the leading tone and tonic $(\hat{7}-\hat{1})$. All other intervals between consecutive scale steps are whole steps. As a result, an informed listener uses half steps as clues to the identity of a key.

In general, when taken out of context, the sounding of two pitches that forming the interval of a half step most strongly suggests a $(\hat{7}-\hat{1})$ transition, as in the beginning of "Londonderry Air" shown in Fig. A.13. Alternatively, this interval could suggest a $(\hat{3}-\hat{4})$ transition. Given no other information, I choose the former as the likelier interpretation. For example, BC would suggest C major, followed by G major.

In the model, the desired condition is:

$$\arg\min_{\mathbf{T}} d(\mathbf{T}, \mathrm{ave}(\mathbf{P}(\mathrm{k}+5), \mathbf{P}(\mathrm{k}))) = \mathbf{T}_{\mathrm{M}}(\mathrm{k}), \quad \text{(A.43)}$$
$$\text{where } \mathbf{T} \in \{T_{\mathrm{M}}(\ell), T_{\mathrm{m}}(\ell) \;\; \forall \ell \in \mathbb{Z}, \ell \neq \mathrm{k}+5\}.$$

A.8.3 Implications of the Perfect Fourth Interval

The other important element in assessing the key of a passage is the rising perfect fourth interval. This is usually interpreted to be the transition from a dominant (up) to the tonic $(\hat{5}-\hat{1})$ for both major and minor scales. A melody beginning with a rising fourth is used as the example in Chap. 4: "Simple Gifts", shown in Fig. 4.1. Two other examples (one a minor key, and one in a major key) are given in Fig. A.14.

Fig. A.14 Examples: two melodies that begin with a rising perfect fourth interval that form $(\hat{5}-\hat{1})$ transitions: Brahms' Piano Quintet is in F minor, and "The Ash Grove" is in F major

Less frequently, the rising fourth is a transition from the tonic (up) to the subdominant $(\hat{1}-\hat{4})$. The perfect fourth also exists as a relationship between many other scale degrees. For example, the intervals between the supertonic and dominant $(\hat{2}-\hat{5})$, and the mediant and submediant $(\hat{3}-\hat{6})$ are also perfect fourths. In the melodic minor scale, the relationship between the subdominant and leading tone $(\hat{4}-\hat{7})$ is also a perfect fourth.

Hence, the preference between interpretations is not as strong as that for the half-step. Nevertheless, it is still a strong indicator for key. Giving preference to the major mode, when given two pitches a perfect fourth apart, I select the major key of the upper pitch as the preferred key. The corresponding mathematical relationship is:

$$\arg\min_{\mathbf{T}} d(\mathbf{T}, \text{ave}(\mathbf{P}(\text{k}-1), \mathbf{P}(\text{k}))) = \mathbf{T}_{\text{M}}(\text{k}), \quad \text{(A.44)}$$
$$\text{where } \mathbf{T} \in \{T_{\text{M}}(\ell), T_{\text{m}}(\ell) \quad \forall \ell \in \mathbb{Z}, \ell \neq \text{k}-1\}.$$

A.9 Finding Solutions for Key–Interval–Pitch Relations

In this section, I discuss a selection of precise weights for all key and chord representations using computational methods, bearing in mind the constraints derived earlier in this chapter as summarized in Sect. A.6.

The problem of satisfying the conditions stated in Sect. A.8, translates to a mathematical one in the Spiral Array. The new problem is one of finding major chord weights w, minor chord weights, u, major key weights, ω, and minor chord weights υ that satisfy the conditions stated in Eqs. A.42, A.43, and A.44.

Even though the definitions for chords and keys—stated in Eqs. 3.4, 3.5, 3.10 and 3.12, and summarized in Sect. 3.5—contain simple linear relations, their interaction is nonlinear and not convex. Initial attempts at solving for the weights using the MATLAB nonlinear solver turned up no solutions. This does not mean that no solutions exist, just that the particular algorithm failed to find one.

Enumerating all possible values for each set of weights to do an exhaustive search would not be computationally viable. In response, I designed a Flip-Flop Heuristic to generate a set of weights that come close to satisfying all of the three conditions.

The Flip-Flop Heuristic

The Flip-Flop Heuristic is essentially an ad hoc perturbation technique that flip-flops between infeasible solutions that satisfy a subset of the conditions. Since it is difficult to find a solution to a nonconvex, nonlinear problem, this heuristic suggests beginning with an infeasible solution, and attempting to improve upon it by trying to satisfy alternating subsets of conditions. At worst, one could still use the original solution and not be any worse off; at best, one might find a feasible solution. More realistically, one would make incremental improvements on the infeasible solution.

The heuristic first begins with a solution that satisfies two out out of the three conditions. Rename the conditions given in Eqs. A.42, A.43, and A.44 A, B, and C, respectively. To summarize,

Condition A: The key nearest an isolated pitch, $\mathbf{P}(\mathrm{k})$, should be a major key of the same name, $\mathbf{T}_{\mathrm{M}}(\mathrm{k})$.

Condition B: A half step interval should be interpreted as the leading tone and tonic, i.e. scale degrees $\hat{7}$ and $\hat{1}$, of the major key.

Condition C: The lower and upper pitches of a perfect fourth interval should be interpreted as the dominant, $\hat{5}$, and tonic, $\hat{1}$, of the major key.

For example, two out of the three conditions satisfied might be A and B.

Such a weight is found by restricting the search space so that all weights, (w, u, ω, and υ) are the same, and allowed to take on every possible pair of values in $\mathbf{F}_0$ (shown in Fig. A.3). The equality of all weights is not an unreasonable assumption since, the cognition of the relative importance of the elements in each hierarchical level could to be similar.

By varying each set of weights in turn, and alternating between constraint subsets, the heuristic seeks a solution that comes closest to satisfying all three conditions given in Eqs. A.42 to A.44.

To formalize this ad-hoc algorithm, I list the steps of the Flip-Flop Heuristic:

Step 1: Restrict $w = u = \omega = \upsilon$.

Step 2: Find weights (w, u, ω, υ) that satisfy a subset of the conditions.

Step 3: Vary one set of weights over its feasible set of values (occasionally, allow infeasible weights).

Step 4: Find a boundary solution that satisfies another subset of the conditions.

Step 5: Choose another set of weights for Step 3.

Before giving an example of how one set of weights were found using this method, I first address the problem of choosing α (the preference for the dominant major, V, versus the dominant minor, v, chords in the minor key definition) and β (the preference for subdominant minor, iv, versus the subdominant major, IV, chords in the minor key definition) in the definition of a minor key representation, given in Definition 5 in Chap. 3.

Table A.8 Weights generated by the Flip-Flop Heuristic: minor keys using the harmonic definition

Iteration	Conditions	Major chord			Minor chord			Major key	Minor key
Initial	A, B	0.5353	0.2743	0.1904		like $\mathbf{C}_M$		like $\mathbf{C}_M$	like $\mathbf{C}_M$
1	B, C		as before		0.212	0.423	0.365	as before	as before
2	A, C	0.593	0.213	0.194		as before		like $\mathbf{C}_M$	like $\mathbf{C}_M$
3	B, C		as before		0.788	0.084	0.128	as before	as before
4	B, C	0.593	0.213	0.194		as before		like $\mathbf{C}_M$	like $\mathbf{C}_M$

Table A.9 Weights generated by the Flip-Flop Heuristic: minor keys using the "democratic" definition

Iteration	Conditions	Major chord			Minor chord			Major key	Minor key
Initial	A,B	0.536	0.274	0.19		like $\mathbf{C}_M$		like $\mathbf{C}_M$	like $\mathbf{C}_M$
1	A, C		as before		0.419	0.249	0.332	as before	as before
2	B, C	0.517	0.299	0.184		as before		like $\mathbf{C}_M$	like $\mathbf{C}_M$
3	A		as before		0.5247	0.1627	0.3126	as before	as before
4	B, C	0.5161	0.3155	0.1684		as before		like $\mathbf{C}_M$	like $\mathbf{C}_M$
5	A		as before		0.6011	0.2121	0.1868	as before	as before
6	B, C	0.6025	0.2930	0.1145		as before		as before	as before

A "Democratic" Definition of the Minor Key Representation

I propose the use of a "democratic" definition of the minor key in generating an alternative set of weights. Since the harmonic minor scale is such that the pitches of the dominant chord is major, and the melodic scale is such that the pitches of the dominant chord is major on its way up and minor on its way down, I set $\alpha = 0.75$ in Eq. 3.12, i.e., the V chord is used 75 % of the time, and the v chord 25 % of the time. For similar reasons, I set $\beta = 0.75$ in Eq. 3.12.

What the Flip-Flop Heuristic Found

The Flip-Flop Heuristic was implemented and run with two different definitions of minor keys: the harmonic minor definition which sets $\alpha = \beta = 1$; and the "democratic" definition of minor keys which sets $\alpha = \beta = 0.75$.

A numerical search using the harmonic minor definition for minor keys indicates that for r $=$ 1 and h $= \sqrt{2/15}$, $w = u = \omega = \upsilon = [0.5353, 0.2743, 0.1904]$ satisfies Conditions A and B. The Flip-Flop Heuristic was, unfortunately, not able to improve upon this set of initial weights, cycling between the last two infeasible values as shown in Table A.8.

Note that although this solution only satisfies A and B, it agrees with the constraints on w described in Sects. A.2 and A.3, so that the major chord pitches, more than any other pitches, are closest to the major chord. And the ordering, by closeness, is root, fifth, then third. The solution found, also agrees with the constraints on u described in Sects. A.4 and A.5, so a similar set of properties apply between the minor chord pitches and the minor chord.

In the "democratic" definition of minor keys scenario, the heuristic begins with the weights $w = u = \omega = \upsilon = [0.536, 0.274, 0.19]$. These weights clearly satisfy

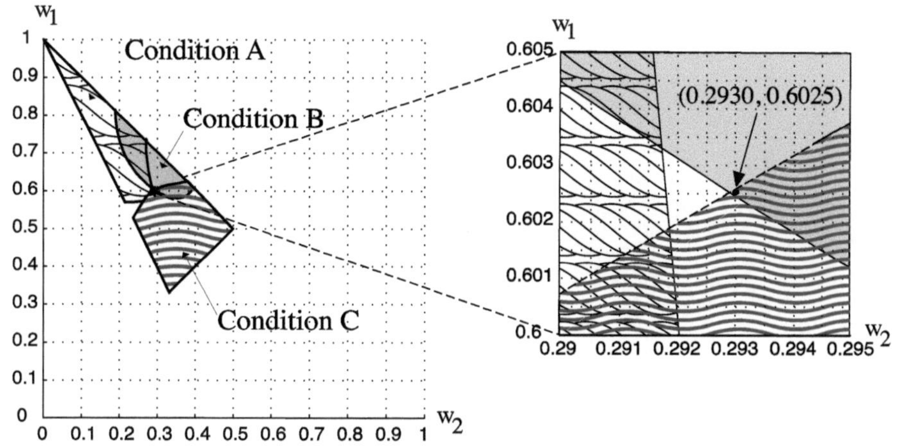

Fig. A.15 Venn diagram of the major chord weights, w, that satisfy different proximity conditions, given that $u = [0.6011, 0.2121, 0.1868]$, other weights ω and υ are restricted to be equal to w

Conditions A and B. The ensuing sets of weights generated using the Flip-Flop Heuristic are shown in Table A.9. These calculations also used $h = \sqrt{2/15}$ $(r = 1)$. This solution satisfies only two of the three conditions, and in addition, u is not in $\mathbf{F}_3$. However, it comes much closer to satisfying all three. By some measures, this solution is less infeasible than the one for the harmonic minor example.

Observe the weights in the final row of Table A.9. The minor chord weights have been assigned the values, $u = [0.6011, 0.2121, 0.1868]$. The other weights, (w, ω, υ), are constrained to be the same while being varied over the range $w \in \mathbf{F}_0$. The venn diagram in Fig. A.15 shows the parts of $\mathbf{F}_0$ that satisfy each of the three conditions, A, B, and C.

As can be seen in the magnified plot on the right half of Fig. A.15, while there is no intersection between the three sets of weights satisfying each of the conditions, the boundaries do come minutely close to one another. The final value $(w_1, w_2) = (0.6025, 0.2930)$ is chosen because it satisfies Conditions B and C, and is closest to satisfying Condition A. Recall that Condition A related the the closest key to an isolated pitch, which while the major key was chosen to be closest given no further information, could equally likely be the minor key, or one of the other closely related keys. Varying any of the weights do not lead to further improvements.

Chapters 4 through 11 uses these sets of weights in the various applications of the Spiral Array model.

Appendix B
CEG Key-Finding

Abstract This appendix contains the MATLAB source code and output of the Spiral Array Center of Effect Generator algorithm as implemented for the testing of the fugue subjects of Bach's *Well-Tempered Clavier* Book I. Each fugue subject is accompanied by a table documenting the pitch and duration of each note, and the three closest keys and their corresponding distances to the center of effect of the melody up to that point.

B.1 CEG MATLAB Source Code

This section contains the MATLAB code for the Center of Effect Generator (CEG) algorithm. Each subsection corresponds to a MATLAB source file of the name indicated.

Code for generating the Spiral Array model is embedded in these files: pitches2.m creates the pitch class representations on the pitch class helix; tonal_c2.m contains the formula for generating the major/minor chord helices as well as the major/minor keys. The code has been organized such that subsequent functions and procedures depend on prior ones, and the final one, BwtcI.m, produces the results documented and discussed in Sect. 5.

This particular implementation of the CEG algorithm is limited by the representation of pitch names only from F♭ through B♯, which impacts the definition of chords and keys spanning all possible letter names. Consequently, some of the fugue subjects have been transposed to keys with fewer sharps or flats (as noted) for testing.

Section B.2 is drawn from the final section of Finding Keys (Chap. 5) in "Towards a Mathematical Modeling of Tonality" by Elaine Chew, an MIT PhD dissertation, Cambridge, Massachusetts (2000) https://dspace.mit.edu/handle/1721.1/9139.

E. Chew, *Mathematical and Computational Modeling of Tonality*, International Series in Operations Research & Management Science 204,
DOI: 10.1007/978-1-4614-9475-1,

B.1.1 Code for Generating Pitches: pitches2.m

```
function [P] = pitches2(h,a,b);

% pitches.m : gives position of pitches from P(a) to P(b) given
%       h = aspect ratio
%       a = steps (in perfect 5ths) below C
%       b = steps (in perfect 5ths) above C
% returns
%       P = a 3 x (b-a+1) matrix of pitch positions
%
% created  by EC, 9 Nov 1998; updated 21 Jan 1999, 28 Aug 2013
%
% usage: [P] = pitches2(h,a,b)
% example: P = pitches2(sqrt(2/15),-8,12)  <- gives pitches (Fb-B#)

        k = a:b;

        t = pi*k./2;
        x = sin(t);
        y = cos(t);
        z = h*k;

        P = [x;y;z];
```

B.1.2 Code for Mapping Pitches to pc Spiral: position2.m

```
function [P] = position2(h,mypitches);

% position.m : gives pitch positions given
%       h = aspect ratio
%       mypitches = a list of pitch names from Fb-B#
% returns
%       P = a list of corresponding pc spiral positions
%
% created by EC, 20 Nov 1998; updated 28 Aug 2013
%
% usage:   [P] = position2(mypitches)
% example: position2(['Bb';'C-']);

M = mypitches';
[m,n] = size(M);

P = zeros(3,n);
pit = pitches2(h,-8,12);
N = ['Fb';'Cb';'Gb';'Db';'Ab';'Eb';'Bb';'F-';'C-';'G-';'D-';'A-';
```

```
        'E-';'B-';'F#';'C#';'G#';'D#';'A#';'E#';'B#']';

        for i = 1:n,
           I = (sum((N==M(1,i))+(N==M(2,i)))==2);
           P(:,i) = pit(:,I);
        end;
```

B.1.3 Code for Generating Keys : tonal_c2.m

```
function [T1,T2] = tonal_c2(h,w1,w2,t1,t2,alpha,beta);

% tonal_c2.m : gives position of keys (Db-F#,db-f#) given
%       h = aspect ratio
%       w1 = weights for major chords
%       w2 = weights for minor chords
%       t1 = weights for major keys
%       t2 = weights for minor keys
%       alpha = weight on V vs. v in minor keys
%       beta = preference for iv vs. IV in minor keys
% returns
%       T1 = a 3 x 12 matrix of major key positions (Db-F#)
%       T2 = a 3 x 12 matrix of minor key positions (db-f#)
%
% created by EC, 20 Nov 1998; updated 28 Aug 2013
%
% usage: [T1,T2] = tonal_c2

% generate pitch representations
P = pitches2(h,-9,11); % Bbb-E#

% generate chord representations
C1 = w1(1)*P(:,4:17) + w1(2)*P(:,5:18) + w1(3)*P(:,8:21); % Gb-C#
C2 = w2(3)*P(:,1:14) + w2(1)*P(:,4:17) + w2(2)*P(:,5:18); % gb-c#

% generate key representations
T1 = t1(3)*C1(:,1:12) + t1(1)*C1(:,2:13)
     + t1(2)*C1(:,3:14); % Db-F#
T2 = t2(3)*((1-beta)*C1(:,1:12) + beta*C2(:,1:12))
     + t2(1)*C2(:,2:13)
     + t2(2)*(alpha*C1(:,3:14) + (1-alpha)*C2(:,3:14)); % db-f#
```

B.1.4 Code for Computing the CE: wtd_pos.m

```
function [P] = wtd_pos(pitchpos, weights);

% wtd_pos.m : computes the c.e. given
%       pitchpos = a list of pitch positions
%       weights = weights on each pitch position
% returns
%       P = sum of weighted pitch positions
%
% created by EC, 29 Mar 1999
%
% usage: [P] = wtd_pos(position2(sqrt(2/15),['C-';'D-']),[1;2]);

[m,n] = size(pitchpos);

if (n==1), P = pitchpos;
   else weights = weights/(sum(weights));
        P = pitchpos*weights;
   end;
```

B.1.5 Code for Finding Nearest Key: rank_near2.m

```
function [nearP, dist] = rank_near2(T,N,P);

% nearestp.m : gives nearest centers T to positions P by rank given
%       T = an 3 x n1 matrix of target centers (e.g. keys)
%       N = corresponding names of the elements of T
%       P = an 3 x n2 matrix of spatial point from which to measure distance
% returns
%       nearP = a sorted list of target centers near each P
%       dist = the corresponding distances
%
% created by EC, 19 Nov 1998
%
% usage: R = rank_near2(T,N,[1,1,1]');

[m1,n1] = size(P);
[m2,n2] = size(T);
d = zeros(n1,n2);

for i = 1:n2,
   D = T(:,i*ones(1,n1)) - P;
   d(:,i) = (sum(D.*D))';
end;

[dist,I] = sort(d');

nearP = N(:,I);
```

B.1.6 Code for Analyzing a Melody: analyze_melody.m

```
function [R,D] = analyze_melody(Pit,Dur,T,N,h);

% analyze_melody.m: analyzes melody incrementally given
%       Pitches = string of pitches
%       Durations = string of durations
%       T = list of keys (positions)
%       N = corresponding list of key names
%       h = aspect ratio
% returns
%       R = list of closest keys at each step
%       D = list of distances at each step
%
% created by EC, 20 Oct 1999; updated 28 Aug 2013
%
% usage:
%       [T1,T2]=tonal_c2;
%       T = [T1,T2];
%       analyze_melody(['C-';'D-'],[1;1/2],T)

for i=1:length(Dur),

   pos(i,:) = (wtd_pos(position2(h,Pit(1:i,:)),Dur(1:i)))';
   [r,d] = rank_near2(T,pos(i,:)');
   R(i,:,:) = r;
   D(:,i) = d;

end;
```

B.1.7 Code for Analyzing Bach's WTC I Fugue Subjects: BwtcI.m

```
% BwtcI.m : analyzes Bach's fugue subjects from the WTC Bk.1
%       For comparison with the PTPM and SMA, the key is
%       analyzed incrementally with each additional note.
%
% created by EC, 19 Oct 1999; updated 28 Aug 2013.

clear;

% set spiral array model weights
% these are weights used for Bach WTC I
% for other weights see Tables A.8 and A.9 in Appendix A

h = sqrt(2/15);             % aspect ratio
```

```
w1 = [0.536, 0.274, 0.19];  % major chord weights
w2 = w1;                    % minor chord weights
t1 = w1;                    % major key weights
t2 = w1;                    % minor key weights

% select 'democratic' definition of key (see Appendix A)
% alternatively, alpha = beta = 1 (harmonic minor)

alpha = 0.75;               % alpha, use of V vs. v in minor key
beta = 0.75;                % beta, use of iv vs. IV in minor key

% generate key representations
[T1,T2] = tonal_c2(h, w1,w2,t1,t2,alpha,beta);  % Db-F#, db-f#
T = [T1,T2];         % list of major/minor key representations

N1 = ['Db';'Ab';'Eb';'Bb';'F-';'C-';'G-';'D-';'A-';'E-';'B-';'F#']';
N2 = ['db';'ab';'eb';'bb';'f-';'c-';'g-';'d-';'a-';'e-';'b-';'f#']';
N = [N1,N2];         % corresponding list of major/minor key names

% set note duration values
b = 1;               % semi-breve
dm = 3/4;            % dotted minim
m = 1/2;             % minim
dc = 3/8;            % dotted crotchet
c_s = 5/16;          % crotchet+semi-quaver
c = 1/4;             % crotchet
q = 1/8;             % quaver
dq = 3/16;           % dotted quaver
s = 1/16;            % semi-quaver
d = 1/32;            % demi-semi-quaver
% he = 1/64;         % hemi-demi-semi-quaver


% Bk.1 No.1 in C
Pit = ['C-';'D-';'E-';'F-';'G-';'F-';'E-';'A-';'D-';'G-';'A-';
'G-';'F-';'E-']; % pitches
Dur = [q;q;q;dq;d;d;q;q;q;dq;s;s;s;s]; % durations
[R,D] = analyze_melody(Pit,Dur,T,N,h); No1 = R(:,:,1)'
Top3R = [R(:,:,1)';R(:,:,2)';R(:,:,3)']
Top3D = D(1:3,:)
input('Type <Return> to continue ...');
clear Pit Dur R D;

% Bk.1 No.2 in c
Pit = ['C-';'B-';'C-';'G-';'Ab';'C-';'B-';'C-';'D-';'G-';'C-';
'B-';'C-';'D-';'F-';'G-';'Ab';'G-';'F-';'Eb'];
Dur = [s;s;q;q;q;s;s;q;q;q;s;s;q;q;s;s;c;s;s;s];
[R,D] = analyze_melody(Pit,Dur,T,N,h); No2 = R(:,:,1)'
Top3R = [R(:,:,1)';R(:,:,2)';R(:,:,3)']
Top3D = D(1:3,:)
input('Type <Return> to continue ...');
clear Pit Dur R D;

% Bk.1 No.3 in C# (transposed down a half-step to C)
```

```
% Pit = ['G#';'A#';'G#';'F#';'G#';'E#';'C#';'G#';'F#';'E#';'F#';
% 'D#';'E#';'C#';'D#';'B#';'C#']; <- original pitches
Pit = ['G-';'A-';'G-';'F-';'G-';'E-';'C-';'G-';'F-';'E-';'F-';
'D-';'E-';'C-';'D-';'B-';'C-'];
Dur = [q;s;s;s;s;q;q;q;s;s;q;q;q;q;q;q;q];
[R,D] = analyze_melody(Pit,Dur,T,N,h); No3trC = R(:,:,1)'
Top3R = [R(:,:,1)';R(:,:,2)';R(:,:,3)']
Top3D = D(1:3,:)
input('Type <Return> to continue ...');
clear Pit Dur R D;

% Bk.1 No.4 in c# (transposed to c)
% Pit = ['C#';'B#';'E-';'D#']; <- original pitches
Pit = ['C-';'B-';'Eb';'D-'];
Dur = [b;m;m;b];
[R,D] = analyze_melody(Pit,Dur,T,N,h); No4trc = R(:,:,1)'
Top3R = [R(:,:,1)';R(:,:,2)';R(:,:,3)']
Top3D = D(1:3,:)
input('Type <Return> to continue ...');
clear Pit Dur R D;

% Bk.1 No.5 in D
Pit = ['D-';'E-';'F#';'G-';'F#';'E-';'F#';'D-';'B-';'B-';'A-';
'G-';'F#'];
Dur = [s;s;s;s;s;s;s;s;dq;s;dq;s;dq];
[R,D] = analyze_melody(Pit,Dur,T,N,h); No5=R(:,:,1)'
Top3R = [R(:,:,1)';R(:,:,2)';R(:,:,3)']
Top3D = D(1:3,:)
input('Type <Return> to continue ...');
clear Pit Dur R D;

% Bk.1 No.6 in d
Pit = ['D-';'E-';'F-';'G-';'E-';'F-';'D-';'C#';'D-';'Bb';'G-';
'A-';'G-';'F-';'E-';'G-';'F-';'E-';'D-';'E-'];
Dur = [q;q;q;q;q;s;s;s;s;c;c;c_s;s;s;s;s;s;s;s;q];
[R,D] = analyze_melody(Pit,Dur,T,N,h); No6=R(:,:,1)'
Top3R = [R(:,:,1)';R(:,:,2)';R(:,:,3)']
Top3D = D(1:3,:)
input('Type <Return> to continue ...');
clear Pit Dur R D;

% Bk.1 No.7 in Eb
Pit = ['Bb';'G-';'F-';'G-';'Eb';'Ab';'G-';'Ab';'C-';'Bb';'A-';
'F-';'Eb';'D-';'C-';'Bb'];
Dur = [s;s;s;s;s;s;s;s;q;q;s;s;q;q;c;s];
[R,D] = analyze_melody(Pit,Dur,T,N,h); No7=R(:,:,1)'
Top3R = [R(:,:,1)';R(:,:,2)';R(:,:,3)']
Top3D = D(1:3,:)
input('Type <Return> to continue ...');
clear Pit Dur R D;

% Bk.1 No.8 in d# (transposed to d)
% Pit = ['D#';'A#';'B-';'A#';'G#';'F#';'G#';'A#';'D#';'G#';'F#';
% 'E#';'D#']; <- original pitches
```

```
Pit = ['D-';'A-';'Bb';'A-';'G-';'F-';'G-';'A-';'D-';'G-';'F-';
'E-';'D-'];
Dur = [c;dc;q;q;q;q;q;c;c;dc;q;c;dc];
[R,D] = analyze_melody(Pit,Dur,T,N,h); No8trd=R(:,:,1)'
Top3R = [R(:,:,1)';R(:,:,2)';R(:,:,3)']
Top3D = D(1:3,:)
input('Type <Return> to continue ...');
clear Pit Dur R D;

% Bk.1 No.9 in E
Pit = ['E-';'F#';'B-';'C#';'D#';'E-';'D#';'E-';'F#';'G#';'A-';
'B-';'G#';'E-';'D#';'E-';'F#';'E-';'D#';'E-';'F#';'D#';'C#';
'D#';'E-';'F#';'G#';'A-';'F#';'G#'];
Dur = [q;c;s;s;s;s;s;s;s;s;s;s;s;s;s;s;s;s;s;s;s;s;s;s;s;s;s;s;
q];
[R,D] = analyze_melody(Pit,Dur,T,N,h); No9=R(:,:,1)'
Top3R = [R(:,:,1)';R(:,:,2)';R(:,:,3)']
Top3D = D(1:3,:)
input('Type <Return> to continue ...');
clear Pit Dur R D;

% Bk.1 No.10 in e
Pit = ['E-';'G-';'B-';'E-';'D#';'E-';'D-';'E-';'C#';'E-';'C-';
'E-';'B-';'E-';'D#';'E-';'A#';'C#';'G-';'F#';'G-';'A#';'F#';
'E-';'D-';'B-'];
Dur = [s;s;s;s;s;s;s;s;s;s;s;s;s;s;s;s;s;s;s;s;s;s;s;s;q;q];
[R,D] = analyze_melody(Pit,Dur,T,N,h); No10=R(:,:,1)'
Top3R = [R(:,:,1)';R(:,:,2)';R(:,:,3)']
Top3D = D(1:3,:)
input('Type <Return> to continue ...');
clear Pit Dur R D;

% Bk.1 No.11 in F
Pit = ['C-';'D-';'C-';'Bb';'C-';'E-';'F-';'G-';'A-';'Bb';'C-';
'Bb';'A-';'G-';'A-';'G-';'F-';'G-';'A-';'Bb';'C-'];
Dur = [q;q;q;q;q;s;s;s;s;q;s;s;s;s;s;s;s;s;s;s;s];
[R,D] = analyze_melody(Pit,Dur,T,N,h); No11=R(:,:,1)'
Top3R = [R(:,:,1)';R(:,:,2)';R(:,:,3)']
Top3D = D(1:3,:)
input('Type <Return> to continue ...');
clear Pit Dur R D;

% Bk.1 No.12 in f
Pit = ['C-';'Db';'C-';'B-';'E-';'F-';'Bb';'A-';'Ab';'G-';'F-'];
Dur = [c;c;c;c;c;c;c;c;c;m;q];
[R,D] = analyze_melody(Pit,Dur,T,N,h); No12=R(:,:,1)'
Top3R = [R(:,:,1)';R(:,:,2)';R(:,:,3)']
Top3D = D(1:3,:)
input('Type <Return> to continue ...');
clear Pit Dur R D;

% Bk.1 No.13 in F# (transposed to F)
% Pit = ['C#';'F#';'E#';'F#';'E#';'D#';'C#';'B-';
% 'C#';'D#';'C#';'B-';'A#';'G#';'C#';'A#']; <- original pitches
```

```
Pit = ['C-';'F-';'E-';'F-';'E-';'D-';'C-';'Bb';'C-';'D-';'C-';
'Bb';'A-';'G-';'C-';'A-'];
Dur = [q;q;q;q;s;s;dq;d;d;c;q;q;q;q;q;q];
[R,D] = analyze_melody(Pit,Dur,T,N,h); No13trF=R(:,:,1)'
Top3R = [R(:,:,1)';R(:,:,2)';R(:,:,3)']
Top3D = D(1:3,:)
input('Type <Return> to continue ...');
clear Pit Dur R D;

% Bk.1 No.14 in f#
Pit = ['F#';'G#';'A-';'G#';'A#';'B-';'A#';'G#';'A#';'B#';'C#';
'B-';'A-';'C#';'B-';'A-';'G#';'F#'];
Dur = [c;c;b;q;q;m;q;q;q;q;dc;q;q;q;q;q;m;dm];
[R,D] = analyze_melody(Pit,Dur,T,N,h); No14=R(:,:,1)'
Top3R = [R(:,:,1)';R(:,:,2)';R(:,:,3)']
Top3D = D(1:3,:)
input('Type <Return> to continue ...');
clear Pit Dur R D;

% Bk.1 No.15 in G
Pit = ['G-';'A-';'G-';'F#';'G-';'A-';'B-';'A-';'G-';'A-';'B-';
'A-';'G-';'D-';'C-';'B-';'A-';'G-';'F#';'E-';'D-';'E-';'D-';'C-';
'B-';'A-';'C-';'B-';'A-';'G-';'F#'];
Dur = [q;s;s;s;s;q;s;s;s;s;q;q;q;q;c;q;q;q;q;c;q;s;s;s;s;q;s;s;s;
s;q];
[R,D] = analyze_melody(Pit,Dur,T,N,h); No15=R(:,:,1)'
Top3R = [R(:,:,1)';R(:,:,2)';R(:,:,3)']
Top3D = D(1:3,:)
input('Type <Return> to continue ...');
clear Pit Dur R D;

% Bk.1 No.16 in g
Pit = ['D-';'Eb';'G-';'F#';'G-';'A-';'Bb';'C-';'Bb';'A-';'Bb';
'G-'];
Dur = [q;q;q;c;c;s;s;q;s;s;c;c];
[R,D] = analyze_melody(Pit,Dur,T,N,h); No16=R(:,:,1)'
Top3R = [R(:,:,1)';R(:,:,2)';R(:,:,3)']
Top3D = D(1:3,:)
input('Type <Return> to continue ...');
clear Pit Dur R D;

% Bk.1 No.17 in Ab
Pit = ['Ab';'Eb';'C-';'Ab';'F-';'Db';'Eb'];
Dur = [q;q;q;q;q;q;c_s];
[R,D] = analyze_melody(Pit,Dur,T,N,h); No17=R(:,:,1)'
Top3R = [R(:,:,1)';R(:,:,2)';R(:,:,3)']
Top3D = D(1:3,:)
input('Type <Return> to continue ...');
clear Pit Dur R D;

% Bk.1 No.18 in g# (transposed to a)
% Pit = ['G#';'Fx';'G#';'A#';'B-';'A#';'G#';'Cx';'D#';'F#';'G#';
% 'G#';'A#';'A#';'D#']; <- original pitches
Pit = ['A-';'G#';'A-';'B-';'C-';'B-';'A-';'D#';'E-';'G-';'A-';
```

```
'A-';'B-';'B-';'E-'];
Dur = [c;q;s;s;q;q;q;q;q;q;q;q;q;q;q];
[R,D] = analyze_melody(Pit,Dur,T,N,h); No18tra=R(:,:,1)'
Top3R = [R(:,:,1)';R(:,:,2)';R(:,:,3)']
Top3D = D(1:3,:)
input('Type <Return> to continue ...');
clear Pit Dur R D;

% Bk.1 No.19 in A
Pit = ['A-';'G#';'C#';'A-';'D-';'B-';'E-';'C#';'F#';'E-';'A-';
'D-';'C#';'D#';'G#';'E-';'A-';'G#'];
Dur = [q;q;q;q;q;q;q;q;q;q;q;c;q;q;q;q;q;c];
[R,D] = analyze_melody(Pit,Dur,T,N,h); No19=R(:,:,1)'
Top3R = [R(:,:,1)';R(:,:,2)';R(:,:,3)']
Top3D = D(1:3,:)
input('Type <Return> to continue ...');
clear Pit Dur R D;

% Bk.1 No.20 in a
Pit = ['A-';'G#';'A-';'B-';'C-';'C-';'B-';'C-';'D-';'E-';'D-';
'C-';'D-';'E-';'F-';'G#';'E-';'A-';'B-';'C-';'A-';'B-';'C-';'D-';
'B-';'C-';'D-';'C-';'B-';'E-';'A-'];
Dur = [s;s;q;q;q;s;s;q;q;s;s;s;s;q;q;q;q;q;s;s;s;s;s;s;s;s;q;q;q;
q;c];
[R,D] = analyze_melody(Pit,Dur,T,N,h); No20=R(:,:,1)'
Top3R = [R(:,:,1)';R(:,:,2)';R(:,:,3)']
Top3D = D(1:3,:)
input('Type <Return> to continue ...');
clear Pit Dur R D;

% Bk.1 No.21 in Bb
Pit = ['F-';'G-';'F-';'Bb';'D-';'C-';'A-';'G-';'Bb';'A-';'G-';
'F-';'C-';'Eb';'D-';'Bb';'C-';'A-';'Bb';'C-';'D-';'Eb';'D-';'C-';
'Eb';'D-';'C-';'Bb';'C-';'A-';'Bb';'C-';'D-';'Eb';'D-';'C-';'Eb';
'D-'];
Dur = [q;q;q;q;q;q;s;s;s;s;s;s;q;q;q;s;s;s;s;s;s;s;s;s;s;s;s;s;s;
s;s;s;s;s;s;s;s;dq];
[R,D] = analyze_melody(Pit,Dur,T,N,h); No21=R(:,:,1)'
Top3R = [R(:,:,1)';R(:,:,2)';R(:,:,3)']
Top3D = D(1:3,:)
input('Type <Return> to continue ...');
clear Pit Dur R D;

% Bk.1 No.22 in bb (transposed to c)
% Pit = ['Bb';'F-';'Gb';'F-';'Eb';'Db';'C-';'Db';'Eb';'F-'];
% <- original pitches
Pit = ['C-';'G-';'Ab';'G-';'F-';'Eb';'D-';'Eb';'F-';'G-'];
Dur = [m;m;c;c;c;c;c;c;c;c];
[R,D] = analyze_melody(Pit,Dur,T,N,h); No22trc=R(:,:,1)'
Top3R=[R(:,:,1)';R(:,:,2)';R(:,:,3)']
Top3D=D(1:3,:)
input('Type <Return> to continue ...');
clear Pit Dur R D;
```

```
% Bk.1 No.23 in B (transposed to C)
% Pit = ['B-';'A#';'B-';'C#';'F#';'G#';'A#';'B-';'C#';'D#';'E-';
% 'D-';'C#';'B-']; <- original pitches
Pit = ['C-';'B-';'C-';'D-';'G-';'A-';'B-';'C-';'D-';'E-';'F-';
'E-';'D-';'C-'];
Dur = [q;q;q;c;q;s;s;q;s;s;q;q;m;dc];
[R,D] = analyze_melody(Pit,Dur,T,N,h); No23trC=R(:,:,1)'
Top3R = [R(:,:,1)';R(:,:,2)';R(:,:,3)']
Top3D = D(1:3,:)
input('Type <Return> to continue ...');
clear Pit Dur R D;

% Bk.1 No.24 in b
Pit = ['F#';'D-';'B-';'G-';'F#';'B-';'A#';'E-';'D#';'C-';'B-';
'F#';'E#';'D-';'C#';'B#';'C#';'A-';'F#';'G#';'F#'];
Dur = [q;q;q;q;q;q;q;q;q;q;q;q;q;q;q;q;q;q;q;m;dq];
[R,D] = analyze_melody(Pit,Dur,T,N,h); No24=R(:,:,1)'
Top3R = [R(:,:,1)';R(:,:,2)';R(:,:,3)']
Top3D = D(1:3,:)
input('Type <Return> to continue ...');
clear Pit Dur R D;
```

B.2 CEG WTC (Book I) Results Data

This section contains data generated by the CEG key-finding algorithm when analyzing the twenty-four fugue subjects of Bach's *Well-Tempered Clavier* Book I [1]. Each subsection comprises the score notation of a fugue subject, and the CEG algorithm's output.

Of the CEG's output, I have chosen to show only the top three (nearest three key) selections. Hence the primary result is to be read from the middle column labelled "First." The accompanying numbers in brackets record the distances between each key and the current aggregate position. The smaller the distance, the closer the fit.

The encoding for durations follows the convention as set out in Table 4.1. For convenience, the definitions are reproduced below:

Symbol	Description
b	semi-breve (whole note)
m	minim (half note)
dc	dotted crotchet (dotted quarter note)
c+s	crotchet+semi-quaver (quarter note + sixteenth)
c	crotchet (quarter note)
dq	dotted-quaver (dotted eighth-note)
q	quaver (eighth-note)
s	semi-quaver (sixteenth-note)

B.2.1 Bach WTC Book I: Fugue No. 1 in C major

Fig. B.1 Fugue No. 1 in C major

Table B.1 CEG key analysis of the fugue subject from the WTC Book I No. 1

Fugue subject		Key choice		
Pitches	Durations	First	Second	Third
C	q	C (0.6117)	c (0.6121)	f (0.6140)
D	q	C (0.1791)	g (0.2129)	F (0.2810)
E	q	C (0.1480)	a (0.2675)	G (0.4288)
F	dq	F (0.1050)	C (0.3154)	d (0.4503)
G	d	F (0.1143)	C (0.2479)	d (0.4449)
F	d	F (0.0860)	C (0.2918)	f (0.4285)
E	q	C (0.2226)	F (0.2541)	a (0.4044)
A	q	a (0.3107)	F (0.3162)	C (0.3624)
D	q	d (0.2109)	a (0.3350)	F (0.3382)
G	dq	C (0.1944)	d (0.2662)	G (0.3590)
A	s	d (0.2323)	C (0.2366)	F (0.3713)
G	s	C (0.1987)	d (0.2572)	G (0.3379)
F	s	C (0.2137)	d (0.2537)	F (0.3211)
E	s	C (0.2021)	d (0.2714)	F (0.3578)

B.2.2 Bach WTC Book I: Fugue No. 2 in C minor

Fig. B.2 Fugue No. 2 in C minor

Table B.2 CEG key analysis of the fugue subject from the WTC Book I No. 2

Fugue subject		Key choice		
Pitches	Durations	First	Second	Third
C	s	C (0.6117)	c (0.6121)	f (0.6140)
B	s	C (0.3579)	e (0.4277)	G (0.5384)
C	q	C (0.1515)	c (0.3720)	F (0.8444)
G	q	C (0.1036)	c (0.2008)	G (0.6396)
A♭	q	c (0.1165)	C (0.3010)	f (0.6037)
C	s	c (0.1436)	C (0.3075)	f (0.5769)
B	s	c (0.1009)	C (0.1604)	F (0.7513)
C	q	c (0.1387)	C (0.1882)	f (0.6905)
D	q	c (0.0435)	C (0.0498)	F (0.5010)
G	q	c (0.0362)	C (0.0604)	g (0.5622)
C	s	c (0.0378)	C (0.0606)	F (0.5881)
B	s	C (0.0441)	c (0.0715)	g (0.5612)
C	q	C (0.0453)	c (0.0700)	F (0.6247)
D	q	C (0.0231)	c (0.0685)	g (0.4523)
F	s	C (0.0216)	c (0.0602)	g (0.4425)
G	s	C (0.0268)	c (0.0575)	g (0.4238)
A♭	c	c (0.0180)	C (0.1387)	F (0.4138)
G	s	c (0.0111)	C (0.1329)	F (0.4416)
F	s	c (0.0174)	C (0.1388)	F (0.3846)
E♭	s	c (0.0138)	C (0.1693)	f (0.4077)

B.2.3 Bach WTC Book I: Fugue No. 3 in C♯ major

Fig. B.3 Fugue No. 3 in C♯ major

Table B.3 CEG key analysis of the fugue subject from the WTC Book I No. 3

Fugue subject		Key choice		
Pitches	Durations	First	Second	Third
G♯	q	G♯ (0.6117)	g♯ (0.6121)	c♯ (0.6140)
A♯	s	G♯ (0.0714)	g♯ (0.1559)	C♯ (0.1899)
G♯	s	G♯ (0.1120)	g♯ (0.1755)	C♯ (0.2389)
F♯	s	g♯ (0.1327)	C♯ (0.1360)	c♯ (0.2073)
G♯	s	g♯ (0.1238)	C♯ (0.1516)	c♯ (0.1862)
E♯	q	C♯ (0.0685)	G♯ (0.2412)	c♯ (0.3520)
C♯	q	C♯ (0.0118)	c♯ (0.2387)	G♯ (0.4658)
G♯	q	C♯ (0.0230)	c♯ (0.1872)	G♯ (0.3760)
F♯	s	C♯ (0.0041)	c♯ (0.1470)	G♯ (0.4080)
E♯	s	C♯ (0.0077)	c♯ (0.2140)	G♯ (0.4195)
F♯	q	C♯ (0.0244)	c♯ (0.1908)	F♯ (0.3136)
D♯	q	C♯ (0.0633)	c♯ (0.2393)	g♯ (0.3226)
E♯	q	C♯ (0.0531)	c♯ (0.3145)	G♯ (0.3691)
C♯	q	C♯ (0.0340)	c♯ (0.2717)	F♯ (0.3736)
D♯	q	C♯ (0.0613)	c♯ (0.3003)	G♯ (0.3521)
B♯	q	C♯ (0.0770)	G♯ (0.2517)	g♯ (0.3573)
C♯	q	C♯ (0.0469)	c♯ (0.3146)	G♯ (0.3219)

B.2.4 Bach WTC Book I: Fugue No. 4 in C♯ minor

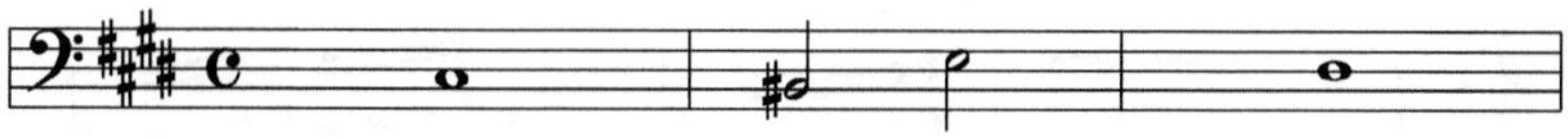

Fig. B.4 Fugue No. 4 in C♯ minor

Table B.4 CEG key analysis of the fugue subject from the WTC Book I No. 4

Fugue subject		Key choice		
Pitches	Durations	First	Second	Third
C♯	b	C♯ (0.6117)	c♯ (0.6121)	f♯ (0.6140)
B♯	m	C♯ (0.1462)	c♯ (0.4400)	G♯ (0.7930)
E	m	c♯ (0.0797)	C♯ (0.1542)	g♯ (0.8120)
D♯	b	g♯ (0.1238)	C♯ (0.1516)	c♯ (0.1862)

B.2.5 Bach WTC Book I: Fugue No. 5 in D major

Fig. B.5 Fugue No. 5 in D major

Table B.5 CEG key analysis of the fugue subject from the WTC Book I No. 5

Fugue subject		Key choice		
Pitches	Durations	First	Second	Third
D	s	D (0.6117)	d (0.6121)	g (0.6140)
E	s	D (0.1791)	a (0.2129)	G (0.2810)
F♯	s	D (0.1480)	b (0.2675)	A (0.4288)
G	s	G (0.1861)	D (0.2254)	e (0.4207)
F♯	s	D (0.2295)	b (0.2569)	G (0.4417)
E	s	D (0.2483)	e (0.2638)	b (0.2820)
F♯	s	b (0.1643)	D (0.2695)	e (0.3847)
D	s	D (0.1873)	b (0.2492)	G (0.4977)
B	dq	b (0.1349)	e (0.3289)	D (0.4862)
B	s	b (0.1313)	e (0.3092)	B (0.4696)
A	dq	b (0.2262)	D (0.2477)	e (0.2986)
G	s	D (0.2525)	b (0.2772)	e (0.2785)
F♯	dq	b (0.1531)	D (0.2805)	e (0.4150)

B.2.6 Bach WTC Book I: Fugue No. 6 in D minor

Fig. B.6 Fugue No. 6 in D minor

Table B.6 CEG key analysis of the fugue subject from the WTC Book I No. 6

Fugue subject		Key choice		
Pitches	Durations	First	Second	Third
D	q	D (0.6117)	d (0.6121)	g (0.6140)
E	q	D (0.1791)	a (0.2129)	G (0.2810)
F	q	d (0.1218)	F (0.3641)	a (0.4150)
G	q	C (0.1967)	g (0.2259)	G (0.2282)
E	q	C (0.1745)	a (0.2886)	G (0.3022)
F	s	C (0.1727)	d (0.3216)	a (0.3416)
D	s	C (0.2180)	d (0.2377)	G (0.3049)
C♯	s	d (0.1792)	a (0.2267)	D (0.3231)
D	s	d (0.1314)	D (0.2651)	a (0.2803)
B♭	c	d (0.1766)	g (0.1852)	G (0.3349)
G	c	g (0.0709)	G (0.1933)	C (0.3145)
A	c+s	d (0.1318)	g (0.2263)	G (0.2792)
G	s	d (0.1568)	g (0.1955)	G (0.2466)
F	s	d (0.1579)	g (0.2154)	G (0.2915)
E	s	d (0.1573)	g (0.2368)	G (0.2798)
G	s	d (0.1802)	g (0.2098)	G (0.2514)
F	s	d (0.1805)	g (0.2275)	G (0.2920)
E	s	d (0.1819)	g (0.2492)	G (0.2840)
D	s	d (0.1681)	g (0.2335)	G (0.2717)
E	q	d (0.1784)	G (0.2650)	C (0.2761)

B.2.7 Bach WTC Book I: Fugue No. 7 in E♭ major

Fig. B.7 Fugue No. 7 in E♭ major

Table B.7 CEG key analysis of the fugue subject from the WTC Book I No. 7

Fugue subject		Key choice		
Pitches	Durations	First	Second	Third
B♭	s	B♭ (0.6117)	b♭ (0.6121)	e♭ (0.6140)
G	s	E♭ (0.3579)	g (0.4277)	B♭ (0.5384)
F	s	B♭ (0.0505)	b♭ (0.3066)	E♭ (0.3467)
G	s	g (0.2740)	B♭ (0.2764)	c (0.3337)
E♭	s	E♭ (0.1573)	c (0.3743)	B♭ (0.3980)
A♭	s	E♭ (0.0844)	c (0.3847)	e♭ (0.4264)
G	s	E♭ (0.1508)	c (0.2742)	B♭ (0.5409)
A♭	s	E♭ (0.1451)	c (0.3427)	A♭ (0.3889)
C	q	c (0.2112)	E♭ (0.3438)	f (0.3710)
B♭	q	E♭ (0.1515)	c (0.3185)	f (0.3934)
A	s	c (0.2410)	E♭ (0.2488)	f (0.2794)
F	s	f (0.2144)	B♭ (0.2697)	c (0.2835)
E♭	q	E♭ (0.1693)	c (0.3210)	f (0.3337)
D	q	E♭ (0.2096)	B♭ (0.2370)	c (0.2883)
C	c	c (0.1636)	f (0.2383)	F (0.3429)
B♭	s	c (0.1979)	f (0.2542)	E♭ (0.3166)

B.2.8 Bach WTC Book I: Fugue No. 8 in D♯ minor

Fig. B.8 Fugue No. 8 in D♯ minor

Table B.8 CEG key analysis of the fugue subject from the WTC Book I No. 8

Fugue subject		Key choice		
Pitches	Durations	First	Second	Third
D♯	c	D♯ (0.6117)	d♯ (0.6121)	g♯ (0.6140)
A♯	dc	d♯ (0.1013)	D♯ (0.1907)	a♯ (0.6867)
B	q	d♯ (0.0772)	D♯ (0.3233)	g♯ (0.7262)
A♯	q	d♯ (0.0704)	D♯ (0.3027)	a♯ (0.8502)
G♯	q	d♯ (0.0167)	D♯ (0.2340)	g♯ (0.5266)
F♯	q	d♯ (0.0583)	D♯ (0.3826)	F♯ (0.5746)
G♯	q	d♯ (0.0589)	D♯ (0.3619)	g♯ (0.3787)
A♯	c	d♯ (0.0292)	D♯ (0.3066)	F♯ (0.5673)
D♯	c	d♯ (0.0213)	D♯ (0.2590)	g♯ (0.4859)
G♯	dc	d♯ (0.0861)	g♯ (0.2100)	G♯ (0.2710)
F♯	q	d♯ (0.0893)	g♯ (0.2389)	G♯ (0.3397)
E♯	c	d♯ (0.0825)	G♯ (0.2878)	g♯ (0.2888)
D♯	dc	d♯ (0.0662)	g♯ (0.2458)	D♯ (0.2548)

B.2.9 Bach WTC Book I: Fugue No. 9 in E major

Fig. B.9 Fugue No. 9 in E major

Table B.9 CEG key analysis of the fugue subject from the WTC Book I No. 9

Fugue subject		Key choice		
Pitches	Durations	First	Second	Third
E	q	E (0.6117)	e (0.6121)	a (0.6140)
F♯	c	b (0.0947)	f♯ (0.2101)	B (0.2319)
B	s	b (0.0308)	B (0.1483)	f♯ (0.3391)
C♯	s	b (0.1155)	f♯ (0.1789)	B (0.1867)
D♯	s	B (0.0722)	b (0.1234)	f♯ (0.2241)
E	s	B (0.1057)	b (0.1405)	f♯ (0.2662)
D♯	s	B (0.0495)	b (0.1748)	E (0.3133)
E	s	B (0.0938)	b (0.1993)	E (0.2040)
F♯	s	B (0.0721)	b (0.1579)	E (0.2699)
G♯	s	B (0.0949)	b (0.2402)	E (0.2420)
A	s	B (0.1464)	E (0.2160)	b (0.2301)
B	s	B (0.1194)	E (0.1942)	b (0.1979)
G♯	s	B (0.1403)	E (0.1805)	b (0.2681)
E	s	E (0.1324)	B (0.1822)	b (0.2967)
D♯	s	B (0.1453)	E (0.1623)	b (0.3080)
E	s	E (0.1215)	B (0.1843)	b (0.3333)
F♯	s	E (0.1504)	B (0.1508)	b (0.2855)
E	s	E (0.1151)	B (0.1872)	b (0.3107)
D♯	s	E (0.1410)	B (0.1570)	b (0.3199)
E	s	E (0.1097)	B (0.1908)	b (0.3423)
F♯	s	E (0.1324)	B (0.1603)	b (0.2997)
D♯	s	B (0.1380)	E (0.1580)	b (0.3116)
C♯	s	B (0.1511)	E (0.1723)	b (0.3277)
D♯	s	B (0.1344)	E (0.1984)	b (0.3415)
E	s	B (0.1580)	E (0.1653)	b (0.3541)
F♯	s	B (0.1355)	E (0.1842)	b (0.3201)
G♯	s	B (0.1555)	E (0.1861)	b (0.3637)
A	s	E (0.1687)	B (0.1708)	b (0.3482)
F♯	s	B (0.1511)	E (0.1870)	b (0.3186)
G♯	q	B (0.1864)	E (0.1909)	b (0.3968)

B.2.10 Bach WTC Book I: Fugue No. 10 in E minor

Fig. B.10 Fugue No. 10 in E minor

Table B.10 CEG key analysis of the fugue subject from the WTC Book I No. 10

Fugue subject		Key choice		
Pitches	Durations	First	Second	Third
E	s	E (0.6117)	e (0.6121)	a (0.6140)
G	s	C (0.3579)	e (0.4277)	G (0.5384)
B	s	e (0.1935)	G (0.5801)	E (0.6363)
E	s	e (0.1204)	E (0.4523)	a (0.6805)
D♯	s	e (0.1013)	E (0.1907)	b (0.6867)
E	s	e (0.0797)	E (0.1542)	b (0.8120)
D	s	e (0.0071)	E (0.1820)	b (0.5511)
E	s	e (0.0065)	E (0.1595)	A (0.5772)
C♯	s	e (0.0277)	E (0.0618)	A (0.3761)
E	s	e (0.0394)	E (0.0699)	A (0.3680)
C	s	e (0.0538)	E (0.1752)	a (0.3401)
E	s	e (0.0750)	E (0.1863)	a (0.3377)
B	s	e (0.0400)	E (0.1541)	a (0.4095)
E	s	e (0.0580)	E (0.1640)	a (0.4013)
D♯	s	e (0.0421)	E (0.0823)	A (0.4780)
E	s	e (0.0556)	E (0.0933)	A (0.4738)
A♯	s	E (0.0255)	e (0.0655)	A (0.4494)
C♯	s	E (0.0130)	e (0.1017)	A (0.3849)
G	s	E (0.0316)	e (0.0535)	A (0.3935)
F♯	s	E (0.0277)	e (0.0612)	A (0.3741)
G	s	e (0.0302)	E (0.0545)	A (0.3931)
A♯	s	E (0.0356)	e (0.0707)	b (0.3530)
F♯	s	E (0.0484)	e (0.0930)	b (0.2912)
E	s	E (0.0391)	e (0.0819)	b (0.3329)
D	q	e (0.0818)	E (0.1021)	b (0.2685)
B	q	e (0.0741)	E (0.1036)	b (0.2473)

B.2.11 Bach WTC Book I: Fugue No. 11 in F major

Fig. B.11 Fugue No. 11 in F major

Table B.11 CEG key analysis of the fugue subject from the WTC Book I No. 11

Fugue subject		Key choice		
Pitches	Durations	First	Second	Third
C	q	C (0.6117)	c (0.6121)	f (0.6140)
D	q	C (0.1791)	g (0.2129)	F (0.2810)
C	q	C (0.0714)	c (0.1559)	F (0.1899)
B♭	q	F (0.1791)	c (0.2129)	B♭ (0.2810)
C	q	c (0.1327)	F (0.1360)	f (0.2073)
E	s	c (0.1263)	C (0.1278)	F (0.1463)
F	s	F (0.0841)	c (0.1880)	C (0.1987)
G	s	c (0.1276)	F (0.1394)	C (0.1491)
A	s	F (0.1063)	C (0.1520)	c (0.1976)
B♭	q	F (0.1140)	c (0.2235)	f (0.2797)
C	s	F (0.1030)	c (0.1972)	f (0.2502)
B♭	s	F (0.1202)	c (0.2223)	f (0.2529)
A	s	F (0.0933)	c (0.2653)	f (0.2733)
G	s	F (0.1272)	c (0.2206)	C (0.2669)
A	s	F (0.1083)	c (0.2654)	C (0.2657)
G	s	F (0.1405)	c (0.2268)	C (0.2340)
F	s	F (0.1120)	c (0.2584)	C (0.2701)
G	s	F (0.1417)	c (0.2237)	C (0.2411)
A	s	F (0.1275)	C (0.2435)	c (0.2634)
B♭	s	F (0.1363)	c (0.2761)	C (0.2869)
C	s	F (0.1230)	c (0.2520)	C (0.2623)

B.2.12 Bach WTC Book I: Fugue No. 12 in F minor

Fig. B.12 Fugue No. 12 in F minor

Table B.12 CEG key analysis of the fugue subject from the WTC Book I No. 12

Fugue subject		Key choice		
Pitches	Durations	First	Second	Third
C	c	C (0.6117)	c (0.6121)	f (0.6140)
D♭	c	D♭ (0.3579)	f (0.4277)	A♭ (0.5384)
C	c	f (0.1935)	A♭ (0.5801)	F (0.6363)
B	c	c (0.1625)	f (0.2098)	F (0.2213)
E	c	C (0.1114)	c (0.2653)	F (0.3498)
F	c	F (0.1446)	C (0.2091)	f (0.2838)
B♭	c	F (0.0560)	f (0.1592)	c (0.2390)
A	c	F (0.0347)	f (0.2538)	C (0.2740)
A♭	c	F (0.0383)	f (0.1020)	c (0.3268)
G	m	c (0.1204)	F (0.1388)	C (0.1688)
F	q	F (0.1069)	c (0.1532)	f (0.2007)

B.2.13 Bach WTC Book I: Fugue No. 13 in F♯ major

Fig. B.13 Fugue No. 13 in F♯ major

Table B.13 CEG key analysis of the fugue subject from the WTC Book I No. 13

Fugue subject		Key choice		
Pitches	Durations	First	Second	Third
C♯	q	C♯ (0.6117)	c♯ (0.6121)	f♯ (0.6140)
F♯	q	f♯ (0.0797)	F♯ (0.1542)	c♯ (0.8120)
E♯	q	F♯ (0.3153)	C♯ (0.3851)	f♯ (0.5592)
F♯	q	F♯ (0.1227)	f♯ (0.3058)	C♯ (0.5699)
E♯	s	F♯ (0.2063)	f♯ (0.4668)	C♯ (0.4684)
D♯	s	F♯ (0.1418)	C♯ (0.3725)	f♯ (0.4680)
C♯	dq	F♯ (0.1693)	C♯ (0.3117)	f♯ (0.3857)
B	d	F♯ (0.1199)	C♯ (0.2946)	f♯ (0.3242)
C♯	d	F♯ (0.1285)	C♯ (0.2916)	f♯ (0.3202)
D♯	c	F♯ (0.1135)	C♯ (0.2288)	c♯ (0.4242)
C♯	q	F♯ (0.1008)	C♯ (0.1894)	c♯ (0.3653)
B	q	F♯ (0.0689)	C♯ (0.2417)	c♯ (0.3309)
A♯	q	F♯ (0.0809)	C♯ (0.2820)	f♯ (0.4096)
G♯	q	F♯ (0.1170)	C♯ (0.2065)	c♯ (0.3411)
C♯	q	F♯ (0.1040)	C♯ (0.1762)	c♯ (0.3012)
A♯	q	F♯ (0.1108)	C♯ (0.2125)	c♯ (0.3903)

B.2.14 Bach WTC Book I: Fugue No. 14 in F♯ minor

Fig. B.14 Fugue No. 14 in F♯ minor

Table B.14 CEG key analysis of the fugue subject from the WTC Book I No. 14

Fugue subject		Key choice		
Pitches	Durations	First	Second	Third
F♯	c	F♯ (0.6117)	f♯ (0.6121)	b (0.6140)
G♯	c	F♯ (0.1791)	B (0.2810)	f♯ (0.3056)
A	b	A (0.1397)	a (0.4397)	D (0.5415)
G♯	q	A (0.1123)	a (0.4569)	f♯ (0.5300)
A♯	q	A (0.1891)	f♯ (0.3182)	a (0.6482)
B	m	A (0.1676)	f♯ (0.2826)	E (0.3209)
A♯	q	f♯ (0.1851)	A (0.2669)	b (0.3486)
G♯	q	f♯ (0.1870)	A (0.2899)	E (0.3110)
A♯	q	f♯ (0.1259)	b (0.3589)	E (0.3633)
B♯	q	f♯ (0.1250)	F♯ (0.3296)	B (0.3493)
C♯	dc	f♯ (0.0962)	F♯ (0.2942)	B (0.4615)
B	q	f♯ (0.1066)	F♯ (0.3012)	B (0.4013)
A	q	f♯ (0.1070)	F♯ (0.3381)	E (0.4315)
C♯	q	f♯ (0.1005)	F♯ (0.3287)	E (0.4653)
B	q	f♯ (0.1099)	F♯ (0.3341)	E (0.4167)
A	q	f♯ (0.1130)	F♯ (0.3691)	E (0.4212)
G♯	m	f♯ (0.1881)	E (0.3761)	F♯ (0.3844)
F♯	dm	f♯ (0.0853)	F♯ (0.2521)	B (0.3460)

B.2.15 Bach WTC Book I: Fugue No. 15 in G major

Fig. B.15 Fugue No. 15 in G major

Table B.15 CEG key analysis of the fugue subject from the WTC Book I No. 15

Fugue subject		Key choice		
Pitches	Durations	First	Second	Third
G	q	G (0.6117)	g (0.6121)	c (0.6140)
A	s	G (0.0714)	g (0.1559)	C (0.1899)
G	s	G (0.1120)	g (0.1755)	C (0.2389)
F♯	s	G (0.0118)	g (0.2387)	D (0.4658)
G	s	G (0.0199)	g (0.2091)	C (0.5158)
A	q	G (0.0738)	D (0.2425)	d (0.2590)
B	s	G (0.0702)	D (0.2772)	g (0.3669)
A	s	G (0.1276)	D (0.1889)	d (0.2789)
G	s	G (0.0720)	D (0.2713)	g (0.3378)
A	s	G (0.1243)	D (0.2029)	d (0.2560)
B	q	G (0.1293)	D (0.2698)	e (0.3552)
A	q	D (0.1696)	G (0.2029)	a (0.3045)
G	q	G (0.1117)	D (0.2576)	d (0.3658)
D	q	G (0.0878)	D (0.2040)	d (0.3014)
C	c	G (0.1215)	C (0.2706)	g (0.3295)
B	q	G (0.1033)	C (0.3347)	g (0.3746)
A	q	G (0.1424)	D (0.3177)	a (0.3254)
G	q	G (0.1062)	C (0.3203)	g (0.3582)
F♯	q	G (0.1054)	D (0.2814)	d (0.3504)
E	c	G (0.1897)	a (0.2834)	D (0.3403)
D	q	G (0.1567)	D (0.2914)	a (0.3147)
E	s	G (0.1759)	a (0.2927)	D (0.3058)
D	s	G (0.1615)	D (0.2844)	a (0.3082)
C	s	G (0.1644)	a (0.2991)	D (0.3091)
B	s	G (0.1625)	a (0.3110)	D (0.3159)
A	q	G (0.1894)	a (0.2674)	D (0.2790)
C	s	G (0.1924)	a (0.2609)	D (0.3025)
B	s	G (0.1887)	a (0.2713)	D (0.3077)
A	s	G (0.2016)	a (0.2526)	D (0.2914)
G	s	G (0.1850)	a (0.2727)	D (0.3058)
F♯	q	G (0.1884)	D (0.2557)	a (0.2812)

B.2.16 Bach WTC Book I: Fugue No. 16 in G minor

Fig. B.16 Fugue No. 16 in G minor

Table B.16 CEG key analysis of the fugue subject from the WTC Book I No. 16

Fugue subject		Key choice		
Pitches	Durations	First	Second	Third
D	q	D (0.6117)	d (0.6121)	g (0.6140)
E♭	q	E♭ (0.3579)	g (0.4277)	B♭ (0.5384)
G	q	g (0.3114)	c (0.5077)	E♭ (0.5783)
F♯	c	G (0.1655)	g (0.2821)	D (0.4929)
G	c	G (0.0927)	g (0.1761)	D (0.7778)
A	s	G (0.0474)	g (0.1420)	D (0.6035)
B♭	s	G (0.0703)	g (0.0853)	d (0.6477)
C	q	G (0.0397)	g (0.0405)	C (0.4798)
B♭	s	g (0.0152)	G (0.0764)	C (0.5030)
A	s	g (0.0182)	G (0.0638)	C (0.4717)
B♭	c	g (0.0150)	G (0.2495)	d (0.5721)
G	c	g (0.0100)	G (0.2110)	c (0.5128)

B.2.17 Bach WTC Book I: Fugue No. 17 in A♭ major

Fig. B.17 Fugue No. 17 in A♭ major

Table B.17 CEG key analysis of the fugue subject from the WTC Book I No. 17

Fugue subject		Key choice		
Pitches	Durations	First	Second	Third
A♭	q	A♭ (0.6117)	a♭ (0.6121)	d♭ (0.6140)
E♭	q	a♭ (0.0797)	A♭ (0.1542)	e♭ (0.8120)
C	q	A♭ (0.1462)	a♭ (0.4400)	E♭ (0.7930)
A♭	q	A♭ (0.1515)	a♭ (0.3720)	D♭ (0.8444)
F	q	A♭ (0.1289)	f (0.4762)	a♭ (0.4887)
D♭	q	A♭ (0.1585)	D♭ (0.2718)	a♭ (0.4396)
E♭	c+s	A♭ (0.0012)	a♭ (0.1557)	D♭ (0.4487)

B.2.18 Bach WTC Book I: Fugue No. 18 in G♯ minor

Fig. B.18 Fugue No. 18 in G♯ minor

Table B.18 CEG key analysis of the fugue subject from the WTC Book I No. 18

Fugue subject		Key choice		
Pitches	Durations	First	Second	Third
G♯	c	G♯ (0.6117)	g♯ (0.6121)	c♯ (0.6140)
F♯♯	q	G♯ (0.1462)	g♯ (0.4400)	D♯ (0.7930)
G♯	s	G♯ (0.1401)	g♯ (0.3921)	e♯ (0.8755)
A♯	s	G♯ (0.0392)	g♯ (0.2912)	D♯ (0.5655)
B	q	g♯ (0.0396)	G♯ (0.0738)	d♯ (0.6068)
A♯	q	G♯ (0.0663)	g♯ (0.0799)	d♯ (0.2773)
G♯	q	G♯ (0.0343)	g♯ (0.0459)	d♯ (0.4246)
C♯♯	q	G♯ (0.0687)	D♯ (0.1979)	d♯ (0.2308)
D♯	q	G♯ (0.0870)	D♯ (0.1849)	d♯ (0.2142)
F♯	q	d♯ (0.1227)	G♯ (0.1444)	g♯ (0.1881)
G♯	q	G♯ (0.0858)	g♯ (0.1256)	d♯ (0.1994)
G♯	q	G♯ (0.0524)	g♯ (0.0889)	d♯ (0.2787)
A♯	q	G♯ (0.0892)	g♯ (0.1424)	d♯ (0.1905)
A♯	q	d♯ (0.1301)	G♯ (0.1359)	g♯ (0.2033)
D♯	q	d♯ (0.1238)	G♯ (0.1394)	g♯ (0.1923)

B.2.19 Bach WTC Book I: Fugue No. 19 in A major

Fig. B.19 Fugue No. 19 in A major

Table B.19 CEG key analysis of the fugue subject from the WTC Book I No. 19

Fugue subject		Key choice		
Pitches	Durations	First	Second	Third
A	q	A (0.6117)	a (0.6121)	d (0.6140)
G♯	q	A (0.3579)	E (0.5384)	f♯ (0.7211)
C♯	q	f (0.5077)	A (0.5783)	F♯ (0.8664)
A	q	A (0.3200)	f♯ (0.5749)	a (0.7980)
D	q	A (0.1289)	f♯ (0.4762)	a (0.4887)
B	q	A (0.0844)	f♯ (0.3847)	a (0.4264)
E	q	A (0.0381)	a (0.3099)	E (0.3333)
C♯	q	A (0.0833)	f♯ (0.3994)	E (0.4057)
F♯	q	A (0.1480)	f♯ (0.2675)	E (0.4288)
E	q	A (0.0975)	E (0.3351)	f♯ (0.3715)
A	q	A (0.0632)	a (0.3997)	E (0.4227)
D	c	A (0.0893)	D (0.2823)	a (0.3567)
C♯	q	A (0.0975)	D (0.3638)	f♯ (0.4166)
D♯	q	A (0.1473)	f♯ (0.3048)	b (0.4160)
G♯	q	A (0.1604)	f♯ (0.2765)	E (0.3455)
E	q	A (0.1297)	E (0.2950)	f♯ (0.3369)
A	q	A (0.0992)	E (0.3431)	f♯ (0.3633)
G♯	c	A (0.1588)	E (0.2565)	f♯ (0.3460)

B.2.20 Bach WTC Book I: Fugue No. 20 in A minor

Fig. B.20 Fugue No. 20 in A minor

Table B.20 CEG key analysis of the fugue subject from the WTC Book I No. 20

Fugue subject		Key choice		
Pitches	Durations	First	Second	Third
A	s	A (0.6117)	a (0.6121)	d (0.6140)
G♯	s	A (0.3579)	E (0.5384)	f (0.7211)
A	q	A (0.1515)	a (0.3720)	D (0.8444)
B	q	A (0.0609)	E (0.2310)	a (0.2921)
C	q	a (0.0382)	A (0.1597)	e (0.3078)
C	s	a (0.0487)	A (0.2877)	e (0.4066)
B	s	a (0.0935)	e (0.2504)	A (0.2833)
C	q	a (0.1455)	C (0.3809)	e (0.4537)
D	q	a (0.1388)	C (0.3313)	G (0.4051)
E	s	a (0.1218)	C (0.3641)	e (0.4150)
D	s	a (0.1365)	C (0.3595)	G (0.3779)
C	s	a (0.1549)	C (0.2923)	G (0.4026)
D	s	a (0.1723)	C (0.2985)	G (0.3332)
E	q	a (0.1330)	C (0.3404)	G (0.4229)
F	q	a (0.1524)	C (0.2758)	d (0.3953)
G♯	q	a (0.0689)	A (0.3452)	C (0.4410)
E	q	a (0.0639)	A (0.3306)	e (0.4133)
A	q	a (0.0366)	A (0.2841)	e (0.4657)
B	s	a (0.0453)	A (0.2756)	e (0.4027)
C	s	a (0.0542)	A (0.3161)	e (0.4351)
A	s	a (0.0426)	A (0.2960)	e (0.4585)
B	s	a (0.0508)	A (0.2884)	e (0.4021)
C	s	a (0.0591)	A (0.3253)	e (0.4321)
D	s	a (0.0626)	A (0.3243)	e (0.4349)
B	s	a (0.0718)	A (0.3188)	e (0.3862)
C	s	a (0.0788)	A (0.3517)	e (0.4141)
D	q	a (0.0912)	A (0.3557)	e (0.4273)
C	q	a (0.1055)	C (0.3637)	A (0.4157)
B	q	a (0.1194)	C (0.3894)	e (0.3955)
E	q	a (0.1102)	e (0.3737)	A (0.3875)
A	c	a (0.0674)	A (0.3215)	e (0.4311)

B.2.21 Bach WTC Book I: Fugue No. 21 in B♭ major

Fig. B.21 Fugue No. 21 in B♭ major

Table B.21 CEG key analysis of the fugue subject from the WTC Book I No. 21

Fugue subject		Key choice		
Pitches	Durations	First	Second	Third
F	q	F (0.6117)	f (0.6121)	b♭ (0.2810)
G	q	F (0.1791)	c (0.2129)	B♭ (0.1899)
F	q	F (0.0714)	f (0.1559)	B♭ (0.3236)
B♭	q	B♭ (0.0176)	b♭ (0.1723)	F (0.4106)
D	q	B♭ (0.0632)	b♭ (0.3931)	F (0.3423)
C	q	B♭ (0.1097)	F (0.1908)	f (0.3684)
A	s	F (0.1494)	B♭ (0.1682)	f (0.3735)
G	s	F (0.1729)	B♭ (0.1941)	g (0.3864)
B♭	s	B♭ (0.1369)	F (0.2224)	g (0.3872)
A	s	B♭ (0.1848)	F (0.1888)	g (0.3156)
G	s	B♭ (0.2105)	F (0.2113)	g (0.3798)
F	s	F (0.1804)	B♭ (0.1881)	g (0.3120)
C	q	F (0.1133)	B♭ (0.2453)	f (0.2838)
E♭	q	F (0.1756)	B♭ (0.2033)	c (0.3302)
D	q	B♭ (0.1995)	F (0.2168)	g (0.3400)
B♭	s	B♭ (0.1652)	F (0.2450)	g (0.3254)
C	s	B♭ (0.1864)	F (0.2151)	c (0.3464)
A	s	F (0.1873)	B♭ (0.2048)	c (0.3565)
B♭	s	B♭ (0.1722)	F (0.2117)	g (0.3382)
C	s	F (0.1871)	B♭ (0.1920)	c (0.3370)
D	s	B♭ (0.1941)	F (0.2064)	g (0.3319)
E♭	s	B♭ (0.1833)	F (0.2280)	c (0.3135)
D	s	B♭ (0.1864)	F (0.2457)	g (0.3212)
C	s	B♭ (0.2025)	F (0.2221)	c (0.3003)
E♭	s	B♭ (0.1953)	F (0.2442)	c (0.3085)
D	s	B♭ (0.1968)	F (0.2589)	g (0.2920)
C	s	B♭ (0.2120)	F (0.2376)	c (0.3149)
B♭	s	B♭ (0.1880)	F (0.2556)	c (0.2897)
C	s	B♭ (0.2030)	F (0.2360)	c (0.3025)
A	s	B♭ (0.2125)	F (0.2134)	c (0.3226)
B♭	s	B♭ (0.1895)	F (0.2299)	c (0.2996)
C	s	B♭ (0.2038)	F (0.2128)	c (0.3147)
D	s	B♭ (0.2043)	F (0.2252)	c (0.2985)
E♭	s	B♭ (0.1987)	F (0.2427)	c (0.3112)
D	s	B♭ (0.1997)	F (0.2544)	g (0.2920)
C	s	B♭ (0.2120)	F (0.2376)	c (0.2787)
E♭	s	B♭ (0.2079)	F (0.2551)	c (0.2876)

B.2.22 Bach WTC Book I: Fugue No. 22 in B♭ minor

Fig. B.22 Fugue No. 22 in B♭ minor

Table B.22 CEG key analysis of the fugue subject from the WTC Book I No. 22

Fugue subject		Key choice		
Pitches	Durations	First	Second	Third
B♭	m	B♭ (0.6117)	b♭ (0.6121)	e♭ (0.6140)
E♭	m	b♭ (0.0797)	B♭ (0.1542)	f (0.8120)
G♭	c	b♭ (0.1334)	B♭ (0.3989)	e♭ (0.5819)
F	c	b♭ (0.0772)	B♭ (0.3233)	e♭ (0.7262)
E♭	c	b♭ (0.0456)	B♭ (0.2727)	e♭ (0.3925)
D♭	c	b♭ (0.0870)	B♭ (0.4331)	e♭ (0.4636)
C	c	b♭ (0.0377)	B♭ (0.3173)	e♭ (0.4521)
D♭	c	b♭ (0.0885)	D♭ (0.3866)	B♭ (0.4581)
E♭	c	b♭ (0.1024)	e♭ (0.3613)	D♭ (0.3674)
F	c	b♭ (0.0748)	D♭ (0.3960)	B♭ (0.4046)

B.2.23 Bach WTC Book I: Fugue No. 23 in B major

Fig. B.23 Fugue No. 23 in B major

Table B.23 CEG key analysis of the fugue subject from the WTC Book I No. 23

Fugue subject		Key choice		
Pitches	Durations	First	Second	Third
B	q	B (0.6117)	b (0.6121)	e (0.6140)
A♯	q	B (0.3579)	d♯ (0.4277)	f♯ (0.5384)
B	q	B (0.1462)	b (0.4400)	f♯ (0.7930)
C♯	c	F (0.0843)	f♯ (0.1856)	B (0.1961)
F♯	q	F (0.0714)	f♯ (0.1559)	B (0.1899)
G♯	s	F (0.0777)	f♯ (0.1751)	B (0.1911)
A♯	s	F (0.0515)	f♯ (0.2156)	B (0.2438)
B	q	B (0.1132)	F♯ (0.1423)	f♯ (0.2718)
C♯	s	F (0.1049)	B (0.1509)	f♯ (0.2179)
D♯	s	F (0.1294)	B (0.1445)	f♯ (0.2872)
E	q	B (0.1134)	F♯ (0.2014)	f♯ (0.2656)
D	q	B (0.1130)	F♯ (0.2396)	f♯ (0.3813)
C♯	m	F (0.1036)	f♯ (0.1678)	c♯ (0.2089)
B	dc	B (0.1452)	F♯ (0.2080)	f♯ (0.2427)

B.2.24 Bach WTC Book I: Fugue No. 24 in B minor

Fig. B.24 Fugue No. 24 in B minor

Table B.24 CEG key analysis of the fugue subject from the WTC Book I No. 24

Fugue subject		Key choice		
Pitches	Durations	First	Second	Third
F♯	q	F (0.6117)	f♯ (0.6121)	b (0.6140)
D	q	D (0.5487)	b (0.6953)	d (1.0642)
B	q	b (0.1935)	D (0.5801)	B (0.6363)
G	q	G (0.3579)	b (0.4277)	D (0.5384)
F♯	q	b (0.2516)	D (0.5279)	G (0.6272)
B	q	b (0.1932)	B (0.6110)	G (0.6629)
A♯	q	b (0.0704)	B (0.3027)	f♯ (0.8502)
E	q	b (0.0167)	B (0.2340)	e (0.5266)
D♯	q	b (0.0253)	B (0.1040)	E (0.5514)
C	q	b (0.0477)	B (0.2328)	e (0.3092)
B	q	b (0.0655)	B (0.2338)	e (0.3034)
F♯	q	b (0.0301)	B (0.1968)	e (0.3858)
E♯	q	b (0.0205)	B (0.0757)	E (0.3757)
D	q	b (0.0038)	B (0.1393)	E (0.4310)
C♯	q	b (0.0181)	B (0.1277)	E (0.4034)
B♯	q	B (0.0570)	b (0.0760)	f♯ (0.3260)
C♯	q	B (0.0794)	b (0.1120)	f♯ (0.2502)
A	q	b (0.1201)	B (0.1324)	f♯ (0.2180)
F♯	q	b (0.1103)	B (0.1298)	f♯ (0.2030)
G♯	m	B (0.1480)	f♯ (0.2433)	E (0.2753)
F♯	dq	B (0.1281)	f♯ (0.2083)	b (0.2536)

Appendix C
Glossary

Accidental An *accidental* is the sign indicating momentary departure from the key signature by means of a sharp (♯), flat (♭) or natural (♮).

Bar A *bar* is a metrical unit of time designated in the score by a certain number of beats as indicated in the time signature or perceived naturally by means of the grouping of strong and weak beats.

Chord A *chord* is any simultaneous combination of three or more pitches, often a triad or tetrachord based on intervals of fifths and thirds. Chords are notated using Roman numerals; uppercase Roman numerals denote major chords, and lowercase Roman numerals represent minor chords. The number marks the scale degree corresponding to the root of the chord.

Dyad A *dyad* is a chord with two pitches.

Equal temperament *Equal temperament* refers to the practice of tuning a keyboard "out of tune" so that one can play in any key on the same instrument. To demonstrate that this was possible, Bach wrote the *Well-Tempered Clavier*. In this system of tuning, enharmonic (pitches that differ from each other in name, but not in any other way when keyboards are concerned) pairs of notes such as A♯ and B♭ are treated as one and the same.

Interval Excerpted from the Oxford Dictionary of Music, an *interval* is the distance between any two pitches, expressed by a number. For example, C to G is a 5th, because if we proceed up the major scale of C, the fifth pitch is G. The 4th, 5th, and octave are all called Perfect. The other intervals, measured from the first pitch, in the ascending major scale are all called Major. Any Major interval can be chromatically reduced by a semitone (distance of a half step) to become Minor. If any Perfect or Minor interval is so reduced it becomes Diminished; if any Perfect or Major interval be increased by a semitone it becomes Augmented.

Key Signature A *key signature* is a sign placed at the opening of a composition or of a section of a composition, indicating the key. This sign consists of one or more sharps or flats.

E. Chew, *Mathematical and Computational Modeling of Tonality*,
International Series in Operations Research & Management Science 204,
DOI: 10.1007/978-1-4614-9475-1,

Modulation From the Oxford Dictionary of Music: A *modulation* is the changing from one key to another in the course of a section of a composition by evolutionary musical means and as a part of the work's formal organization.

MIDI *MIDI* refers to Musical Instrument Digital Interface, a convention for representing discrete music information comprising note numbers (representing pitch), and note onset/offset times and velocities, amongst other things.

Octave An *octave* is the distance between one pitch and another that is double its frequency. It is an interval comprising eight scale steps, hence the name.

Parallel major/minor triad For a major (minor) triad, its *parallel* minor (major) is the corresponding minor (major) triad with the same tonic. The two triads share the root and fifth, differing only in the major/minor (minor/major) third, which determines the major/minor (minor/major) modality.

Pitch A *pitch* is a sound of some frequency. High frequency sounds produce a high pitch, and low frequency sounds produce a low pitch. A *note* is a symbol that represents two properties, pitch and duration.

Relative major/minor triad For a major (minor) triad, its *relative* minor (major) triad is defined such that the two scales for which the two triads are the tonic triad share the same pitch set. For example, C major is the relative major of A minor because neither has any sharps or flats; and, vice versa, A minor is the relative minor of C major.

Scale A *scale* is a series of single pitches progressing up or down stepwise. Here, a scale always refers to the diatonic scale, as distinct from the chromatic scale (which uses nothing but semitones), the pentatonic scale (which uses five pitches) or the whole-tone scale (which is free of semitones). When ascending, the diatonic scale degrees are labeled $\hat{1}, \hat{2}, \ldots, \hat{7}$.

The *major scale* has semitone intervals between ($\hat{3}$–$\hat{4}$) and ($\hat{7}$–$\hat{1}$), the two halves thus being alike. The *natural minor scale* has semitone intervals between ($\hat{2}$–$\hat{3}$), ($\hat{5}$–$\hat{6}$). The *harmonic minor scale* has semitone intervals between ($\hat{2}$–$\hat{3}$), ($\hat{5}$–$\hat{6}$) and ($\hat{7}$–$\hat{1}$); and, the *melodic minor scale* has semitones between ($\hat{2}$–$\hat{3}$) and ($\hat{7}$–$\hat{1}$) ascending, and ($\hat{6}$–$\hat{5}$) and ($\hat{3}$–$\hat{2}$) descending.

Each scale degree has a name that reflects its function with respect to the first scale degree. The name or key of the scale is the "tonic", the first scale degree. The fifth is the "dominant", and the fifth below (the fourth) the "subdominant." Halfway between the tonic and dominant is the third scale degree, the "mediant". On the opposite side, halfway between the tonic and subdominant, the sixth scale degree, is the "submediant". The seventh, "leading note", is thus named for its tendency to move towards the tonic. And the second scale degree, the one above the tonic, is the supertonic.

Score A *score*, synonymous with manuscript, is the notated manifestation of a piece of music.

Solfège According to the Merriam-Webster Dictionary, *solfège* is the application of the sol-fa syllables to a musical scale or to a melody. The scale degrees $\hat{1}, \hat{2}, \ldots, \hat{7}$ are assigned the solfege syllables *do*, *re*, *mi*, *fa*, *sol*, *la* and *ti*. *Movable do* solfège refers to the practice of always making the tonic of the melody "do", as opposed to always setting C to "do", in the assignment of solfège syllables to the notes.

Tetrachord A *tetrachord* is a chord with four pitches.

Time Signature The *time signature* is a sign at the beginning of a score indicating the number of beats in a bar, and the note values of these beats.

Tonal center The *tonal center* is the pitch that has attained greatest stability in a musical passage. The tonal center is also the tonic of the key.

Triad A *triad* is a chord with three pitches.

Triad A *triad* is a chord of three notes, basically a "root" and the notes a third and a fifth above it, forming two superimposed thirds, e.g. F-A-C. If lower third is major and the upper minor, the triad is major. If lower third is minor and the upper major, the triad is minor. If both are major the triad is augmented. If both are minor, the triad is diminished.

Trills *Trills* are musical ornaments consisting of rapid alternations between two usually adjacent notes or sets of notes.

Reference

1. Bach, J.S.: The Well-tempered Clavier - Part I, BWV 846–869. Henle Urtext Edition, Munich (1997)

Author Index

E. Chew, *Mathematical and Computational Modeling of Tonality*,
International Series in Operations Research & Management Science 204,
DOI: 10.1007/978-1-4614-9475-1,

Subject Index

E. Chew, *Mathematical and Computational Modeling of Tonality*,
International Series in Operations Research & Management Science 204,
DOI: 10.1007/978-1-4614-9475-1, © Springer Science+Business Media New York 2014

MIX
Papier aus verantwortungsvollen Quellen
Paper from responsible sources
FSC® C105338

If you have any concerns about our products,
you can contact us on
ProductSafety@springernature.com

In case Publisher is established outside the EU,
the EU authorized representative is:
Springer Nature Customer Service Center GmbH
Europaplatz 3, 69115 Heidelberg, Germany

Printed by Libri Plureos GmbH
in Hamburg, Germany